Tissue Engineering
for
The Hand

Research Advances
and Clinical Applications

Tissue Engineering
_{for}
The Hand

Research Advances
and Clinical Applications

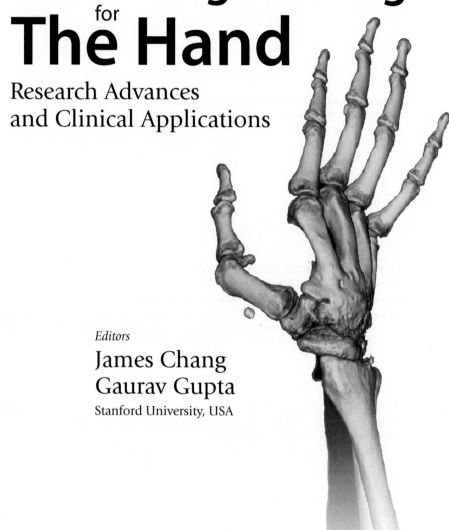

Editors

James Chang
Gaurav Gupta
Stanford University, USA

World Scientific

NEW JERSEY · LONDON · SINGAPORE · BEIJING · SHANGHAI · HONG KONG · TAIPEI · CHENNAI

Published by

World Scientific Publishing Co. Pte. Ltd.

5 Toh Tuck Link, Singapore 596224

USA office: 27 Warren Street, Suite 401-402, Hackensack, NJ 07601

UK office: 57 Shelton Street, Covent Garden, London WC2H 9HE

British Library Cataloguing-in-Publication Data
A catalogue record for this book is available from the British Library.

ISBN-13 978-981-4313-55-1
ISBN-10 981-4313-55-6

Typeset by Stallion Press
Email: enquiries@stallionpress.com

Printed by FuIsland Offset Printing (S) Pte Ltd. Singapore

Acknowledgments

This textbook was conceived during the 2006 Sterling Bunnell Traveling Fellowship in Hand Surgery.

With great thanks to the American Society for Surgery of the Hand and the American Foundation for Surgery of the Hand.

With admiration and appreciation for the authors: students, research fellows, and principal investigators from around the world who strive to translate tissue engineering into clinical use.

With sincere thanks for the love and support of our families, Ambrish, Jyotsna, and Monica Gupta; James Sr., Lily, Anthony, Barbara, and Cecilia Chang; Harriet Roeder, and Julia, Kathleen, and Cecilia Roeder Chang.

Gaurav Gupta MSE
James Chang MD
Stanford, California
2010

Contents

Contributors

Ioannis K. Angelidis, M.D.
Research Fellow
Stanford University School of Medicine
Stanford, California

Gregory H. Borschel, M.D.
Assistant Professor, Division of Plastic Surgery
University of Toronto
Hospital for Sick Children
Toronto, Ontario, Canada

Yilin Cao, M.D., Ph.D.
Professor, Department of Plastic and Reconstructive Surgery
Shanghai Jiao Tong University School of Medicine
9th People's Hospital
Director, Shanghai Tissue Engineering Center
Shanghai, China

James Chang, M.D.
Professor and Chief
Division of Plastic and Reconstructive Surgery
Stanford University School of Medicine
Stanford, California

Harvey Chim, M.D.
Resident, Department of Plastic Surgery
Case Western Reserve University School of Medicine
Cleveland, Ohio

Jennifer Elisseeff, M.D., Ph.D.
Associate Professor, Department of Biomedical Engineering
Johns Hopkins University
Baltimore, Maryland

Laurence A. Galea, M.D.
Research Fellow
Bernard O'Brien Institute of Microsurgery
Melbourne, Australia

Arun K. Gosain, M.D.
De Wayne Richey Professor and Vice Chair
Department of Plastic Surgery
Case Western Reserve University School of Medicine
Cleveland, Ohio

Deepak Gupta, M.D.
Resident, Division of Plastic and Reconstructive Surgery
Stanford University School of Medicine
Stanford, California

Geoffrey C. Gurtner, M.D.
Professor, Division of Plastic and Reconstructive Surgery
Stanford University School of Medicine
Stanford, California

Ryosuke Kakinoki, M.D., Ph.D.
Chief of Hand Surgery & Microsurgery
Department of Orthopedic Surgery & Rehabilitation Medicine
Kyoto University
Kyoto, Japan

James D. Kretlow, B.S.
Research Fellow, Department of Bioengineering
Rice University
Houston, Texas

L. Scott Levin, M.D.
Chairman, Department of Orthopedic Surgery
University of Pennsylvania
Philadelphia, Pennsylvania

Wei Liu, M.D., Ph.D.
Professor, Department of Plastic and Reconstructive Surgery
Shanghai Jiao Tong University School of Medicine
9th People's Hospital
Shanghai, China

Michael Longaker, M.D., M.B.A.
Deane P. and Louise Mitchell Professor
Department of Surgery
Director, Pediatric Surgical Research Laboratories
Stanford University School of Medicine
Stanford, California

Antonios G. Mikos, Ph.D.
Louis Calder Professor of Bioengineering and Chemical and Biomolecular Engineering
Director, Center of Excellence in Tissue Engineering
Rice University
Houston, Texas

Wayne Morrison, M.D.
Head of Plastic and Reconstructive and Hand Surgery, St. Vincent's Hospital
Director
Bernard O'Brien Institute of Microsurgery
Melbourne, Australia

Teruo Okano, Ph.D.
Professor & Director
Institute of Advanced Biomedical Engineering and Science
Tokyo Women's Medical University
Tokyo, Japan

Nicholas J. Panetta, M.D.
Research Fellow, Division of Plastic and Reconstructive Surgery
Stanford University School of Medicine
Stanford, California

Katie L. Pricola, B.S.
Research Fellow, Division of Plastic and Reconstructive Surgery
Stanford University School of Medicine
Stanford, California

Brian Pridgen, B.S.
Research Fellow, Division of Plastic and Reconstructive Surgery
Stanford University School of Medicine
Stanford, California

Mark A. Randolph, M.A.S.
Instructor in Surgery, Harvard Medical School
Laboratory Director, Plastic Surgery Research Laboratory
Massachusetts General Hospital
Boston, Massachusetts

Ashley R. Rothenberg, B.S
Research Fellow, Department of Biomedical Engineering
Johns Hopkins University
Baltimore, Maryland

Johan Thorfinn, M.D., Ph.D.
Department of Plastic Surgery, Hand Surgery and Burns
University Hospital
Linköping, Sweden

Masayuki Yamato, Ph.D.
Professor, Institute of Advanced Biomedical Engineering and Science
Tokyo Women's Medical University
Tokyo, Japan

Xing Zhao, M.D.
Research Fellow, Plastic Surgery Research Laboratory
Massachusetts General Hospital
Boston, Massachusetts

CHAPTER 1

Tissue Engineering: A Historical Perspective

James D. Kretlow[†] *and Antonios G. Mikos*[*,†]

"Begin at the beginning and go on till you get to the end: then stop."

— Lewis Carroll,
Alice's Adventures in Wonderland

Introduction

Tissue engineering and the synonymous or closely related field of regenerative medicine have drawn the widespread attention of clinicians, scientists, policy makers, investors, and the general public for over a decade dating back to the early to mid 1990s. Earlier, pioneering efforts in and related to the field of tissue engineering, including many of the key discoveries that later became the foundation for this field, were made long before this more generalized recognition and subsequent definition of this field. Because the field of tissue engineering, as recognized specifically by term itself, is relatively young but draws upon substantial, sometimes centuries old efforts in many related fields, adequately providing an overview of the history of tissue engineering is a difficult endeavor.

Providing a truly complete history of any field so closely tied to the understanding of biological processes, human physiology, materials science and engineering, and the practice of medicine within the confines of a single book chapter would prove to be impractical due to the magnitude of existing and relevant knowledge within these fields. Conversely, a historical perspective that examines tissue engineering through some rigidly defined criteria such as recent definitions of the term tissue engineering, formulated only once the field has become well demarcated within related areas of science and engineering, would fail to give proper credit to nor engender any appreciation of the substantial efforts related to but not wholly within the present day confines of tissue engineering.

Because of this, we will first aim to provide a working definition of tissue engineering that somewhat limits our scope but allows for deviations into related fields where appropriate. Subsequent sections of this chapter will examine the history of the field within this context, organized broadly in chronological order but with an attempt to also look at certain key areas of scientific exploration within these chronological boundaries.

Defining tissue engineering

In the 1993 *Science* article that is often regarded as having brought tissue engineering into focus within the general scientific community, Robert Langer and Joseph Vacanti loosely defined the field as one that combines engineering and the life sciences towards developing

---

*Corresponding author.
[†]Department of Bioengineering, Rice University, P.O. Box 1892, MS 142, Houston, TX 77251–1892 (U.S.A.). E-mail: mikos@rice.edu.

biological tissue substitutes.[1] They went on to identify cells and cell substitutes, growth factors and delivery vehicles, and cell-scaffold constructs as general strategies for tissue engineering. Today, researchers often refer to the tissue engineering paradigm when defining the field — some combination of bioactive factors, cells, and matrices for the generation or regeneration of living tissues.

Both definitions, while accurate and widely accepted today, prove too limiting within a historical context. Many seminal discoveries related to the field of tissue engineering were not aimed at conception towards generating tissue substitutes. Many of the earliest discoveries or observations also fail to appreciate or involve any of the critical aspects that now form the tissue engineering paradigm. Nonetheless, these efforts deserve mention.

For the sake of this chapter, tissue engineering can be defined as *an attempt, technique, or technology made or at some point applied towards the preternatural generation, regeneration, or restoration of native tissue structure and/or function using biological components.* This definition is purposefully broad, allowing for a wide contextual scope highlighting many predecessors to today's field of tissue engineering. This definition also includes the field of regenerative medicine, which, while often considered identical or overlapping with tissue engineering, is more correctly regarded by those in the field as the products and techniques developed through tissue engineering processes that are directly involved in the support and practice of medicine.[2]

Overview: Why tissue engineering?

What can be gleaned from examining the history of tissue engineering, or rather, why should a historical perspective be of interest to those reading this book? First, even a cursory outline of tissue engineering in the historical sense provides strong evidence confirming the interdisciplinary nature of the field. Contributions from all areas of the natural sciences, engineering, and medicine will be presented both in this chapter and in the rest of this textbook.

Second, the problems that today's tissue engineers seek to tackle have existed for thousands of years, and while etiologies may have changed and existing therapies have improved over time, the need for solutions has not. Slowly, some of these problems are being incrementally yet effectively addressed through tissue engineering approaches.

Finally, while specific and more recent advances will be covered in this chapter and others, historical perspectives are necessary to realize and appreciate that what may often be presented as seemingly science fiction is actually grounded in years, decades, and in some cases, centuries worth of scientific exploration. We do not aim to present tissue engineering as a panacea for the destruction or dysfunction of all tissues — it is not. Rather, we aim to provide some level of historical context such that, when considered along with the other knowledge presented within this book, one can examine recent advances and potential applications with a hopeful yet critical eye.

Conception to Birth

Adopting the definition laid out in section 1.1, one can loosely trace the early history of tissue engineering along a similar path as the history of surgery. Areas such as wound healing and organ transplantation are critical in both fields and played a large role in forming the foundations of tissue engineering. This section will highlight the earliest efforts related to tissue engineering and then examine in greater detail the advances of the 20th century and particularly the 1990s that led to the recognition of tissue engineering as a distinct field.

Ancient foundations

Beginning any discussion of tissue engineering with a section describing ancient history seems a bit exaggerated. Despite the fact that tissue engineering as a formal field is still relatively new, many principles of tissue engineering have been applied dating back to ancient times. Some authors have even noted that the first verse of Genesis in the Christian Bible mentions tissue engineering in the form of extraction and expansion of a man's rib resulting in a complete woman.[3] A brief examination highlighting some ancient predecessors to today's tissue engineers proves that while technologies and our understanding of biology and engineering have greatly expanded, many of the problems from ancient times have persisted to this day.

The Edwin Smith papyrus (Figure 1), an Egyptian text dating between 2600–2200 B.C., describes sutured closure of wounds allowing for subsequent healing by primary intention.[4] Although wound healing is of interest in many fields, sutured closure of wound represents one of the earliest examples of tissue engineering. Because many of the wounds described would otherwise have healed by secondary intention, wound closure and healing represents a preternatural regeneration of native tissue structure/function as it minimizes scarring and thus improves the strength of the healed wound. Additionally, suturing with silk or linen/gum combinations was likely among the first successful uses of biomaterials. Despite a greater than four millennia history, wound healing remains a heavily investigated topic today as researchers and clinicians attempt to both gain improved understanding of its regulation[5,6] and develop better therapies to aid in this process.[7,8]

Figure 1. Plates vi and vii of the Edwin Smith Papyrus. Thought to be written by Imhotep, the document is the first known textbook of surgery and describes wound closure, one of the earliest forms of tissue engineering. Today the Edwin Smith Papyrus can be found at the New York Academy of Medicine.

Working around 600 B.C., the Indian professor and clinician Sushruta provided the first written record of the use of rotational flaps to reconstruct amputated noses — the result of corporal punishment meted out in ancient India.[9] These early flaps, taken from the forehead with the angular artery serving as blood supply, used autologous tissue to restore the aesthetic function of the mutilated nose.[10] This technique also represents one of the earliest uses of autologous tissue for restorative purposes, and similar techniques remain the gold standards of treatment for many cases of tissue loss or dysfunction today.

A host of other important discoveries upon which today's field of tissue engineering lies began during antiquity. While significant efforts towards understanding anatomy, physiology, and various disease processes were made during ancient times and are no doubt critical in modern tissue engineering, the scope of these subjects makes them best left to other, dedicated texts.

Renaissance through 19th century

The Renaissance period was not without contributions in the area of tissue engineering. Modified approaches to rhinoplasty, similar to those of Sushruta, emerged in Italy at the hands of the 15th century Branca and Viano families,[4,11] followed by the 16th century efforts of Giulio Cesare Arantius (also referred to as Aranzio),[12] and finally popularized (and often solely attributed to) by the famous Renaissance surgeon Gaspare Tagliacozzi (Figure 2) in *De Curtorum Chirurgia per insitionem*, a 700-page surgical treatise published in 1597, two years before Tagliacozzi's death.[13] The technique used in these rhinoplasties, now referred to as surgical or vascular delay, at the time involved the elevation of a bipedicled flap from the forearm and later the anterior brachium. Medicated linen or cotton lint was then placed under the flap to prevent reattachment, and after two weeks to one month, the flap was attached to the patient's nose.

Delay represents a major step forward in the history of tissue engineering. Whereas wound edge approximation to allow for closure and healing by primary intention seems an obvious (although mechanistically complex) technique, delay represents one of the first non-obvious techniques resulting in tissue manipulation or truly engineering beyond that visible with the naked eye. Although the mechanism involved in delay was almost certainly not understood by its earliest practitioners, for one of the first times, autologous tissue was being modified through a complex interplay between sympathetic tone, vascular tone, tissue metabolism, and ultimately neovascularization via endothelial progenitor cell recruitment.[14] It has taken nearly 500 years since Tagliacozzi's publication detailing these methods for them to be well understood, and vascular delay remains an accepted practice towards conditioning flaps today.[15,16] For tissue engineering, vascularization of engineered tissues and angiogenic control remain one of the greatest and most unmet challenges facing the field.[17] Significant research efforts continue towards determining effective material and cellular parameters and growth factor regimes to induce neovascularization and angiogenesis such that *ex vivo* or *ex situ* engineered tissues of a clinically relevant size can be constructed.[18,19]

Allograft transplantation was the next major historical step towards tissue engineering. Although evidence of allografts from earlier points in history are apparent, notably including a lower limb allograft transplanted by Saints Cosmas and Damian and captured in a famous painting by Fra Angelico (Figure 3), most early allografts were subject to rejection and disease transmission. Some of the first successful transplantations were performed by the British surgeon John Hunter in the late 18th century. Hunter focused primarily on transplanting teeth, first using a rooster as an animal model but later transplanting human teeth between subjects.[20]

Figure 2. (A) Posthumously painted (1599) portrait of Gaspare Tagliacozzi from the University of Bologna. Gaspare Tagliacozzi, shown in a portrait painted posthumously in 1599 from the University of Bologna (A) advanced reconstructive surgery through techniques such as delay (B), where a flap is lifted and separated from the underlying tissue prior to reconstruction, enabling a complex but beneficial series of changes within the flap. (From *De Curtorum Chirurgia per insitionem*. Venice Italy: Bindoni, 1597.)

This practice was short lived due to the spread of diseases such as syphilis through the practice rather than rejection.

Skin grafts were likely the first allografts to gain more widespread use. Giuseppe Baronio, a contemporary of John Hunter, published extensively on skin autografting in animals, likely paving the way for experiments with allografts.[4] Winston Churchill served as a skin graft donor for a fellow soldier in 1898,[21] although it would be nearly 50 years and two world wars before a better understanding of immune rejection existed.[22]

Corneal transplants were also investigated during the 19th century. Samuel Bigger performed a successful allograft cornea transplantation between gazelles while a prisoner in Egypt in 1837;[23] however, the first successful transplant in humans would not occur until the early 20th century.

Around this time, Nicholas Senn found that decalcified bone induced healing within bone cavities;[24] demineralized bone matrix is now a widely used product for bone healing and regeneration. The mechanism for this effect would not be known until the middle of the next century, and Senn's approach was impressive because, similar to that of Tagliacozzi's delay, he was modifying a human tissue to increase its effectiveness in regenerating an orthotopic defect. The next two sections will further discuss the use of the extracellular matrix and matrix components as powerful tissue engineering tools.

20th century advances

The 20th century saw significant advances in every area of science and technology, including tissue engineering. Progress in the first half of the century was similar to that of the

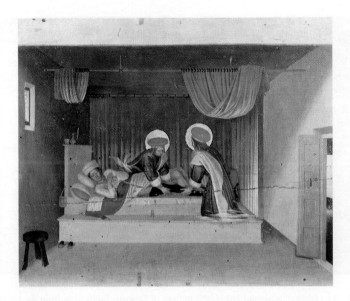

Figure 3. "The healing of Deacon Justinian" by Fra Angelico (*ca.* 1440) depicts the transplantation of a man's leg, taken from an Ethiopian donor, onto the injured deacon. (From *Museo di San Marco*, Florence, Italy.)

decades before in that it was rooted deeply but disparately in clinical medicine, biology, and engineering. During the second half of the century, however, important advances in a variety of fields began to be integrated in what would eventually lead to the birth of tissue engineering as a unique discipline.

Organ transplantation

In 1905, Austrian physician Eduard Zirm performed the first successful, full thickness human corneal allograft transplantation.[25] That same year, Alexis Carrel and Charles Guthrie reported a method to anastomose vessels,[26] a discovery that would lead to a Nobel Prize in 1912 for Carrel and that would later allow for solid organ transplants. In 1935, Carrel and aviator Charles Lindbergh published *The Culture of Whole Organs*, a book detailing *ex vivo* culture techniques to keep entire solid organs alive, including the use of a perfusion pump.[27]

Significant other contributions were made towards organ transplantation in the early and mid 20th century, including the first successful transplantations of many organs and the elucidation of many key aspects of immune tolerance and rejection. While undoubtedly these contributions are relevant and important within the field of medicine and tissue engineering, we will begin to limit our overview at this point to the first attempts towards tissue engineering in the sense that the term is used today. Organ transplants are a life saving therapy; however, persistent problems with organ shortages and immune rejection can be viewed as partially contributing to the development of tissue engineering.

In combination with medical advances, political and military events of the early 20th century also contributed to the rise of tissue engineering. Antibiotics, the use of plasma, and other life-saving medical and surgical techniques decreased mortality in the first two world wars relative to previous wars. As a result, soldiers were surviving previously fatal injuries but often at the cost of disfiguring tissue loss. Other authors have noted the increased importance of reconstructive surgeons following WWI due to the incidence of craniofacial injuries,[28] and this increased morbidity rather than mortality provided a nidus from which tissue engineering would later emerge.

Vascular grafts

Nearly 50 years after Carrel described vessel anastomosis, Arthur Voorhees made the first synthetic vascular graft, replacing aortic segments in dogs with a poly(vinyl chloride) based fabric.[29] Poly(methyl methacrylate) and other polymeric materials had been used previously in dental applications such as dentures, but Voorhees used a synthetic material to replace a tissue whose functionality was dependent on interactions with both adjacent vascular tissue and blood. Over the next 30 years, endothelial cell seeding of vascular grafts, resorbable graft materials, and finally *in vitro* graft engineering would be studied.

Bone and osteoinduction

Also in the 1950s, Marshall Urist began publishing studies on the phenomenon of bone induction using transplanted bone.[30,31] Urist published groundbreaking papers that described in 1965 the differentiation of osteoprogenitor cells in the presence of decalcified bone matrix[32] and later the isolation of osteoinductive bone morphogenetic proteins (BMPs),[33] confirming a preexisting hypothesis regarding the existence of an osteogenic substance within bone.[34] Sampath and Reddi were later able to isolate this osteogenic substance, bone morphogenetic protein 2 (BMP-2),[35] and it is now the basis for both commercially available products for bone regeneration and the focus of a wide variety of experimental applications within bone tissue engineering.[36–38]

Stem cells

Critical advances in cell biology were being made beginning in the 1960s that would later be integrated into tissue engineering approaches. Researchers were beginning to explore in depth the ability of terminally differentiated tissues to renew and regenerate.[39,40] Recognition of a population of pluripotent circulating progenitor cells gave the first signs of the existence of marrow stromal or mesenchymal stem cells, a cell population that is potentially one of the most powerful tools at the disposal of today's tissue engineers.[41,42] Following up the 1964 work of Kleinsmith and Peirce,[43] Gail Martin[44] and the team of Martin Evans and Matthew Kaufman[45] described in separate studies published in 1981 the first isolation of embryonic stem cells.

Other cell-based approaches

Pluripotent cells represent a promising technology for tissue engineering due to the ability to potentially harvest cells from the patient, then expand, differentiate and possibly form a functional tissue *in vitro*, followed by implantation of the cellularized construct back into the patient with little risk of immune reaction. The use of differentiated cells has the disadvantage of possible lack of availability of autologous cells if an organ has been severely damaged or uniformly diseased. Nevertheless, tissues engineered using expanded allogeneic cells can be used to reduce the shortage of available donor tissue for transplantation or reconstruction.

The birth of tissue engineering

The field of tissue engineering as we know it today was eventually borne from efforts to combine differentiated somatic cells with biomaterials to form new living tissues either *in vitro* or *in vivo*. One such approach using endothelial cells to line synthetic vascular grafts was already, albeit briefly, mentioned. A similar approach towards replacing the function of the pancreas was taken by William Chick and his colleagues. Their approach involved seeding pancreatic beta cells on a semipermeable membrane to create a hybrid pancreas that combined living and artificial components.[46] This hybrid pancreas was used to successfully treat diabetes in rats[47] and later dogs[48] and in 1985 resulted in the formation of BioHybrid Technologies, an early tissue engineering company.

Following a similar path, W.T. Green investigated cartilage formation *in vitro* and *in vivo*.[49] His approach using cells implanted on a specifically tailored scaffold would years later be adopted as the fundamental technique in tissue engineering.[3] This same principal of using tailored scaffolds to support tissue growth was the idea behind the first tissue engineering company (although at the time the founders referred to the idea as tissue gardening), Interpore Cross International, Inc., founded in 1975.[50]

In 1981, the cell-scaffold approach yielded reports of the first *in vitro* generated, full thickness skin grafts.[51,52] One approach that resulted from collaboration between John Burke and Ioannis Yannas used a collagen and silicone based material that served as a dermal template, encouraging the ingrowth of native skin and vessels from surrounding skin. The approach taken by Eugene Bell and Howard Green used autologous dermal fibroblasts and epidermal keratinocytes, expanded *in vitro*, and a collagen matrix to generate a full thickness skin graft or living skin equivalent that can be sutured or stapled in place during surgery without the risk of rejection.

With the ability to grow functional, living tissues *in vitro*, the birth of tissue engineering was well underway. The first use of the term *tissue engineering* appears in a 1984 paper describing the macrophage mediated lining of a corneal prosthesis with a membrane.[53] One year later, in 1985, Y.C. Fung proposed that the NSF establish a new research center to be known as the "Center for the Engineering of Living Tissues".[50] His proposal was rejected, but tissue engineering, both the term and the field, was on the way to becoming established within the scientific community. In 1988, Joseph Vacanti and Robert Langer published what would be the first of many tissue engineering papers together — a study of cell transplantation using bioresorbable scaffold carriers.[54]

The 1990s: Research, recognition, and general interest

Research in the early part of the 1990s continued along a similar path as that of the late 1980s. Groups were beginning to focus on biomaterial scaffold fabrication and to characterize the interaction between cells and these scaffolds,[55–58] an area that remains a critical component of tissue engineering research to this day. Publication of the article titled "Tissue Engineering" in the May 14, 1993 issue of *Science* is viewed as the introduction of the field to the broader scientific community.[1] This article helped define tissue engineering, describe critical areas for research within the new field, and also provided a glimpse into the significant potential held within the field.

It was also around this time that tissue engineering societies began emerging across the U.S. and the rest of the world.[59] The Tissue Engineering Society was launched in 1994 and later evolved into the Tissue Engineering Society International (TESI) and then TERMIS, the Tissue Engineering and Regenerative Medicine International Society. Charles A. Vacanti of the Massachusetts General Hospital and Harvard Medical School and Antonios G. Mikos of Rice University founded the journal Tissue Engineering, published by Mary Ann Liebert, Inc. Publishers, in 1994. The first issue was released in early 1995, and the journal remains the flagship journal dedicated solely to tissue engineering research. The journal now publishes in 3 parts, devoted to research, reviews, and methods, with Antonios Mikos now joined by co-editors-in-chief Peter C. Johnson, who in the early 1990s had formed the Pittsburgh Tissue Engineering Initiative, along with John A. Jansen and John P. Fisher.

Research in the 1990s continued to expand the knowledge of biomaterial-cell interactions. Discoveries in other fields that would later become critical to tissue engineering applications, such as adult pluripotent stem cell characterization,[60,61] were also made during this period.

In 1997, public interest in tissue engineering was sparked by major media outlets' coverage of what has become known as the "Vacanti mouse". A 1997 publication in *Plastic and Reconstructive Surgery* from the laboratory of Charles A. Vacanti described the fabrication of an ear shaped cartilage construct via transplantation of chondrocytes onto auricle-shaped poly(glycolic acid)-poly(lactic acid) fiber meshes that were implanted subcutaneously onto the backs of athymic mice (Figure 4).[62] Subsequent coverage of the story in the New York Times and later on the popular television series *Nip/Tuck* only increased public attention to this result and tissue engineering in general.

In reality, tissue engineering had been impacting patient care before the buzz from advances such as the Vacanti mouse. Charles Vacanti reports using a tissue engineering approach to regenerate a patient's sternum as early as 1991,[59] and FDA regulated tissue engineered procedures and products were available by the late 1990s as well in the form of Carticel®, a chondrocyte expansion procedure offered by Genzyme Biosurgery, and Apligraf®, a collagen based full thickness skin equivalent from Organogenesis used to treat lower extremity diabetic and venous stasis ulcers.

Growth: The 21st Century

The beginning of the 21st century has seen tremendous advances in tissue engineering. The field has established a role within clinical care, and recent advances in the laboratory and through *in vivo* testing using animal models point to exciting further developments over the course of the coming years and decades. A plethora of undeniably important advances in tissue engineering have been made thus far this decade, and for many of these advances, the true magnitude of their importance may not be fully realized for many more years. Many recent advances and the state of the art in most areas of tissue engineering will be discussed in the remaining chapters of this book; therefore, the next section will highlight a few select examples of clinically used tissue engineering strategies.

Continued advancement

Earlier this decade, a number of experimental techniques bridged the gap from laboratory to clinic. Reports of tissue engineered pulmonary arteries in the laboratory[63] quickly resulted in major publications reporting clinical use of these arteries.[64] At the same time, tissue engineering techniques were applied clinically to utilize autologous cells to regenerate bone.[65] More recently and based on work performed in the early 1990s,[66] autologous bone flaps have been created *in vivo* using tissue engineering strategies and then used to engineer a functional hemimandible capable of accepting dental implants.[67] This technology eliminates the need to harvest the patient's fibula or rib to reconstruct the mandible

Figure 4. Gross appearance of a chondrocyte seeded-poly(glycolic acid)-poly(lactic acid) scaffold in the shape of a human ear, implanted subcutaneously on the dorsum of an athymic mouse for 12 weeks. Reproduced with permission from Ref. 62.

but still relies on the use of autologous material, thus eliminating concern of disease transmission or rejection.

The use of biological scaffolds is an exciting technology that deserves special note in any discussion of recent tissue engineering trends and advances. While not a new idea, as demonstrated by the aforementioned studies by Nicholas Senn and Marshall Urist, the processing and use of the extracellular matrix has led to a number of exciting advances in tissue engineering this decade. Related to the work of Senn and Urist, a large number of decellularized, demineralized bone matrix products are commercially available and approved by regulatory agencies worldwide. Extracellular matrix based strategies are now expanding to include tissue engineering of other organs.

A group led by Anthony Atala published a landmark study in 2006 describing the use of tissue engineered autologous bladders to replace congenitally defective bladders in 7 patients.[68] The engineered bladders consisted of scaffolds made from decellularized bladder submucosa or a combination of collagen and poly(glycolic acid) seeded with autologous cells expanded *in vitro* from a punch biopsy taken from each patient's bladder (Figure 5). This study marked the first time a whole organ had been engineered for clinical use and garnered well-deserved attention and acclaim.

A related approach was used earlier this year (2008) to replace a patient's left bronchus using a tissue-engineered trachea.[69] The team of researchers, led by Paolo Macchiarini, used a bioreactor to culture autologous bone marrow derived mesenchymal stem cells seeded on a decellularized tracheal matrix taken from a donor. After 96 hours of culture within the bioreactor, the engineered trachea was surgically implanted to replace the patient's stenotic left main bronchus (Figure 6).

As a final example of the promising capabilities of the extracellular matrix, building on reports that transplanted hearts display donor-recipient chimerism indicating autologous seeding of the transplanted organs,[70] researchers have shown that decellularized cardiac matrices can be recellularized *in vitro* and can regain some of the pumping ability of functional hearts.

Commercial products that utilize extracellular matrices for tissue regeneration are also available. Two such products, Alloderm® and Strattice®, both manufactured by LifeCell, Inc., provide decellularized human and porcine matrices, respectively, for skin regeneration *in vivo*.

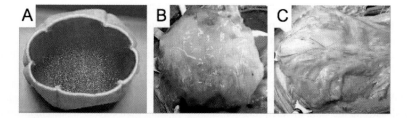

Figure 5. Generation and implantation of a tissue engineered bladder. (A) Scaffold seeded with autologous cells. (B) Tissue engineered bladder being anastomosed to the remaining native bladder for reconstruction. (C) Fibrin glue and omentum covering the reconstructed bladder. Reproduced with permission from Ref. 68.

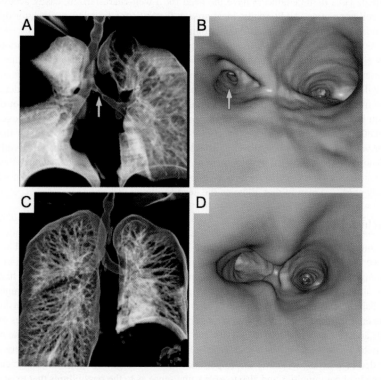

Figure 6. Computed tomography volume rendering (A and C) and bronchoscopic reconstructions (B and D) of a patient with left main bronchus stenosis (arrows) before (A and B) and 1 month after (C and D) resection of the bronchus and reconstruction with a tissue engineered trachea made using bone marrow mesenchymal stem cells seeded onto a decellularized tracheal matrix. Reproduced with permission from Ref. 69.

The Future and the Past

An attempt to predict the future of tissue engineering with any level of specificity or temporal conjecture would be foolish; the field continues to advance at a rapid but ever changing pace. As with any multidisciplinary field, advances will be predicated not only on the efforts of tissue engineers, but also on scientists from other fields, including chemists, materials scientists, molecular and cell biologists, pharmaceutical engineers, and clinicians. Discernible trends, however, do exist within the field and shed possible light on future directions.

One such trend is the use of autologous cells to regenerate patient-specific tissues. While development of embryonic stem cell technology is hampered by ethical and political debates in the U.S., the identification and use of adult stem cells in tissue engineering research is becoming increasingly popular and promising. Stem cell populations in bone marrow, adipose tissue, dental tissues, and the skin represent potential targets upon which tissue engineering strategies will be based. While harvest and expansion of autologous, terminally differentiated cells is possible in many cases, in the future tissue engineering strategies may be used to address clinical problems for which no such tissue can safely be acquired. Thus tissue engineers have the need for a well-characterized, easily accessible stem cell population. Additionally, ideal growth factor or conditioning regimes for differentiation and maintenance of differentiation are necessary.

As previously mentioned, organ and tissue transplantation as long-term solutions to many clinical problems are limited due to the lack of supply of donor organs/tissues and also concern over rejection and long-term immunosuppression. Tissue engineering strategies will ideally overcome this hurdle. The use of autologous cells for tissue regeneration is one method to eliminate the need for immunosuppression. Antigen suppression or elimination is another potential solution to this problem. For strategies that rely on allogeneic extracellular matrices, the supply of available materials is certainly greater than that of donor organs and tissues but still somewhat limited relative to solutions that use synthetic biomaterials. The continued development of methods to create functional synthetic extracellular matrices and to generate *de novo* allogeneic extracellular matrices, many of which already exist,[71–74] may further alleviate the need for donor tissues.

Conclusions

The common opinion of tissue engineering holds that it is a relatively new field within science. While this may be true of the name and more precise definition of the field, as with any multidisciplinary field, tissue engineering has roots that extend far into history. Early attempts at and advances in tissue engineering-related sciences were examined to establish the existence of this long history and also to exemplify the close relationship of tissue engineering to other fields within science and medicine. The defining studies of the 1980s and 1990s were discussed, as this can be viewed as the time period when the modern field of tissue engineering emerged. Highlights of more recent work were also provided as a means to justify the importance of tissue engineering within the scientific and medical communities and also to give some sense as to the possibilities that exist with the continued pursuit of research within the field.

References

1. Langer R, Vacanti JP. Tissue engineering. *Science* **260** (1993) 920–926.
2. Mikos AG, Johnson PC. Redefining tissue engineering... and our new rapid publication policy. *Tissue Eng* **12** (2006) 1379–1380.
3. Vacanti CA. History of tissue engineering and a glimpse into its future. *Tissue Eng* **12** (2006) 1137–1142.
4. Santoni-Rugiu P, Sykes PJ. *A History of Plastic Surgery*. Berlin: Springer-Verlag (2007).

5. Aarabi S, Bhatt KA, Shi Y, Paterno J, Chang EI, Loh SA, *et al.* Mechanical load initiates hyper-trophic scar formation through decreased cellular apoptosis. *FASEB J* **21** (2007) 3250–3261.

6. Derderian CA, Bastidas N, Lerman OZ, Bhatt KA, Lin SE, Voss J, *et al.* Mechanical strain alters gene expression in an *in vitro* model of hypertrophic scarring. *Ann Plast Surg* **55** (2005) 69–75; discussion.

7. Aarabi S, Longaker MT, Gurtner GC. Hypertrophic scar formation following burns and trauma: new approaches to treatment. *PLoS Med* **4** (2007) e234.

8. Fu X, Li H. Mesenchymal stem cells and skin wound repair and regeneration: possibilities and questions. *Cell Tissue Res* 2008.

9. Sushruta. The Sushruta Samhita. In: Bhishagratna KK, ed. *An English Translation of the Sushruta Samhita.* Calcutta: Kaviraj Kunjalal Bhishagratna (1911).

10. Eisenberg I. A history of rhinoplasty. *S Afr Med J* **62** (1982) 286–292.

11. Myers MB, Cherry G. Augmentation of tissue survival by delay: an experimental study in rabbits. *Plast Reconstr Surg* **39** (1967) 397–401.

12. Gurunluoglu R, Gurunluoglu A. Giulio Cesare Arantius (1530–1589): a surgeon and anatomist: his role in nasal reconstruction and influence on Gaspare Tagliacozzi. *Ann Plast Surg* **60** (2008) 717–722.

13. Zimbler MS. Gaspare Tagliacozzi (1545–1599): renaissance surgeon. *Arch Facial Plast Surg* **3** (2001) 283–284.

14. Ghali S, Butler PE, Tepper OM, Gurtner GC. Vascular delay revisited. *Plast Reconstr Surg* **119** (2007) 1735–1744.

15. Ribuffo D, Atzeni M, Corrias F, Guerra M, Saba L, Sias A, *et al.* Preoperative Angio-CT pre-liminary study of the TRAM flap after selective vascular delay. *Ann Plast Surg* **59** (2007) 611–616.

16. Kajikawa A, Ueda K, Tateshita T, Katsuragi Y. Breast reconstruction using tissue expander and TRAM flap with vascular enhancement procedures. *J Plast Reconstr Aesthet Surg* (2008).

17. Johnson PC, Mikos AG, Fisher JP, Jansen JA. Strategic Directions in Tissue Engineering. *Tissue Eng* **13** (2007) 2827–2837.

18. Patel ZS, Mikos AG. Angiogenesis with biomaterial-based drug- and cell-delivery systems. *J Biomater Sci Polym Ed* **15** (2004) 701–726.

19. Patel ZS, Young S, Tabata Y, Jansen JA, Wong M, Mikos AG. Dual delivery of an angiogenic and an osteogenic growth factor for bone regeneration in a critical size defect model. *Bone* **43** (2008) 931–940.

20. Dobson J. *John Hunter.* Edinburgh: E. & S. Livingstone Ltd. (1969).

21. Churchhill W. *My Early Life.* London: Butterworth (1930).

22. Gibson T, Medawar PB. The fate of skin homografts in man. *J Anat* **77**(4) (1943) 299–310.

23. George AJ, Larkin DF. Corneal transplantation: the forgotten graft. *Am J Transplant* **4** (2004) 678–685.

24. Senn N. On the healing of aseptic bone cavities by implantation of antiseptic decalcified bone. *Am J Med Sci* **98** (1889) 219–240.

25. Zirm EK. Eine erfolgreiche totale Keratoplastik (A successful total keratoplasty) (1906). *Refract Corneal Surg* **5** (1989) 258–261.

26. Carrel A, Guthrie CC. Functions of a transplanted kidney. *Science* **22** (1905) 473.

27. Carrel A, Lindbergh CA. The Culture of Whole Organs. *Science* **81** (1935) 621–623.

28. Saltzman WM. Tissue Engineering. New York: Oxford University Press (2004).

29. Voorhees AB, Jr., Jaretzki A, 3rd, Blakemore AH. The use of tubes constructed from vinyon "N" cloth in bridging arterial defects. *Ann Surg* **135** (1952) 332–336.

30. Urist MR, Mc LF. Osteogenetic potency and new-bone formation by induction in transplants to the anterior chamber of the eye. *J Bone Joint Surg Am* **34-A** (1952) 443–476.

31. Urist MR, McLean FC. The local physiology of bone repair with particular reference to the process of new bone formation by induction. *Am J Surg* **85** (1953) 444–449.

32. Urist MR. Bone: Formation by autoinduction. *Science* **150** (1965) 893–899.

33. Urist MR, Mikulski A, Lietze A. Solubilized and Insolubilized Bone Morphogenetic Protein. *Proc Natl Acad Sci USA* **76** (1979) 1828–1832.

34. Lacroix P. Recent investigations of the growth of bone. *Nature* **156** (1945) 576.

35. Sampath TK, Reddi AH. Dissociative extraction and reconstitution of extracellular matrix components involved in local bone differentiation. *Proc Natl Acad Sci USA* **78** (1981) 7599–7603.

36. Govender S, Csimma C, Genant HK, Valentin-Opran A, Amit Y, Arbel R, *et al*. Recombinant human bone morphogenetic protein-2 for treatment of open tibial fractures: a prospective, controlled, randomized study of four hundred and fifty patients. *J Bone Joint Surg Am* **84** (2002) 2123–2134.

37. Lutolf MP, Weber FE, Schmoekel HG, Schense JC, Kohler T, Muller R, *et al*. Repair of bone defects using synthetic mimetics of collagenous extracellular matrices. *Nat Biotechnol* **21** (2003) 513–518.

38. Ruhe PQ, Hedberg EL, Padron NT, Spauwen PH, Jansen JA, Mikos AG. rhBMP-2 release from injectable poly(DL-lactic-co-glycolic acid)/calcium-phosphate cement composites. *J Bone Joint Surg Am* **85-A**(Suppl. 3) (2003) 75–81.

39. Altman J. Are new neurons formed in the brains of adult mammals? *Science* **135** (1962) 1127–1128.

40. Altman J, Das GD. Post-natal origin of microneurones in the rat brain. *Nature* **207** (1965) 953–956.

41. Till JE, Mc CE. Early repair processes in marrow cells irradiated and proliferating *in vivo*. *Radiat Res* **18** (1963) 96–105.

42. Friedenstein AJ, Deriglasova UF, Kulagina NN, Panasuk AF, Rudakowa SF, Luria EA, *et al*. Precursors for fibroblasts in different populations of hematopoietic cells as detected by the *in vitro* colony assay method. *Exp Hematol* **2** (1974) 83–92.

43. Kleinsmith LJ, Pierce GB, Jr. Multipotentiality of Single Embryonal Carcinoma Cells. *Cancer Res* **24** (1964) 1544–1551.

44. Martin GR. Isolation of a pluripotent cell line from early mouse embryos cultured in medium conditioned by teratocarcinoma stem cells. *Proc Natl Acad Sci USA* **78** (1981) 7634–7638.

45. Evans MJ, Kaufman MH. Establishment in culture of pluripotential cells from mouse embryos. *Nature* **292** (1981) 154–156.

46. Chick WL, Like AA, Lauris V, Galletti PM, Richardson PD, Panol G, *et al*. A hybrid artifical pancreas. *Trans Am Soc Artif Intern Organs* **21** (1975) 8–15.

47. Whittemore AD, Chick WL, Galletti PM, Mannick JA. Function of hybrid artificial pancreas in diabetic rats. *Surg Forum* **28** (1977) 93–97.

48. Maki T, Ubhi CS, Sanchez-Farpon H, Sullivan SJ, Borland K, Muller TE, *et al*. Successful treatment of diabetes with the biohybrid artificial pancreas in dogs. *Transplantation* **51** (1991) 43–51.

49. Green WT, Jr., Ferguson RJ. Histochemical and electron microscopic comparison of tissue produced by rabbit articular chondrocytes *in vivo* and *in vitro*. *Arthritis Rheum* **18** (1975) 273–280.

50. Viola J, Lal B, Grad O. The emergence of tissue engineering as a research field. arlington, VA: The National Science Foundation (2003).

51. Bell E, Ehrlich HP, Buttle DJ, Nakatsuji T. Living tissue formed *in vitro* and accepted as skin-equivalent tissue of full thickness. *Science* **211** (1981) 1052–1054.

52. Yannas IV, Burke JF, Warpehoski M, Stasikelis P, Skrabut EM, Orgill D, *et al*. Prompt, long-term functional replacement of skin. *Trans Am Soc Artif Intern Organs* **27** (1981) 19–23.

53. Wolter JR, Meyer RF. Sessile macrophages forming clear endothelium-like membrane on inside of successful keratoprosthesis. *Trans Am Ophthalmol Soc* **82** (1984) 187–202.

54. Vacanti JP, Morse MA, Saltzman WM, Domb AJ, Perez-Atayde A, Langer R. Selective cell transplantation using bioabsorbable artificial polymers as matrices. *J Pediatr Surg* **23** (1988) 3–9.

55. Cima LG, Ingber DE, Vacanti JP, Langer R. Hepatocyte culture on biodegradable polymeric substrates. *Biotechnol Bioeng* **38** (1991) 145–158.

56. Vacanti CA, Langer R, Schloo B, Vacanti JP. Synthetic polymers seeded with chondrocytes provide a template for new cartilage formation. *Plast Reconstr Surg* **88** (1991) 753–759.

57. Cohen S, Bano MC, Cima LG, Allcock HR, Vacanti JP, Vacanti CA, *et al*. Design of synthetic polymeric structures for cell transplantation and tissue engineering. *Clin Mater* **13** (1993) 3–10.

58. Mikos AG, Bao Y, Cima LG, Ingber DE, Vacanti JP, Langer R. Preparation of poly(glycolic acid) bonded fiber structures for cell attachment and transplantation. *J Biomed Mater Res* **27** (1993) 183–189.

59. Vacanti CA. The history of tissue engineering. *J Cell Mol Med* **10** (2006) 569–576.

60. Gronthos S, Graves SE, Ohta S, Simmons PJ. The STRO-1+ fraction of adult human bone marrow contains the osteogenic precursors. *Blood* **84** (1994) 4164–4173.

61. Pittenger MF, Mackay AM, Beck SC, Jaiswal RK, Douglas R, Mosca JD, *et al.* Multilineage potential of adult human mesenchymal stem cells. *Science* **284** (1999) 143–147.

62. Cao Y, Vacanti JP, Paige KT, Upton J, Vacanti CA. Transplantation of chondrocytes utilizing a polymer-cell construct to produce tissue-engineered cartilage in the shape of a human ear. *Plast Reconstr Surg* **100** (1997) 297–302; discussion 3–4.

63. Shinoka T, Shum-Tim D, Ma PX, Tanel RE, Isogai N, Langer R, *et al.* Creation of viable pulmonary artery autografts through tissue engineering. *J Thorac Cardiovasc Surg* **115** (1998) 536–545; discussion 45–46.

64. Shin'oka T, Imai Y, Ikada Y. Transplantation of a tissue-engineered pulmonary artery. *N Engl J Med* **344** (2001) 532–533.

65. Vacanti CA, Bonassar LJ, Vacanti MP, Shufflebarger J. Replacement of an avulsed phalanx with tissue-engineered bone. *N Engl J Med* **344** (2001) 1511–1514.

66. Miller MJ, Goldberg DP, Yasko AW, Lemon JC, Satterfield WC, Wake MC, *et al.* Guided bone growth in sheep: a model for tissue-engineered bone flaps. *Tissue Eng* **2** (1996) 51–59.

67. Cheng MH, Brey EM, Ulusal BG, Wei FC. Mandible augmentation for osseointegrated implants using tissue engineering strategies. *Plast Reconstr Surg* **118** (2006) 1e–4e.

68. Atala A, Bauer SB, Soker S, Yoo JJ, Retik AB. Tissue-engineered autologous bladders for patients needing cystoplasty. *Lancet* **367** (2006) 1241(6).

69. Macchiarini P, Jungebluth P, Go T, Asnaghi MA, Rees LE, Cogan TA, *et al.* Clinical transplantation of a tissue-engineered airway. *Lancet* (2008).

70. Quaini F, Urbanek K, Beltrami AP, Finato N, Beltrami CA, Nadal-Ginard B, *et al.* Chimerism of the transplanted heart. *N Engl J Med* **346** (2002) 5–15.

71. Meng Y, Qin YX, Dimasi E, Ba X, Rafailovich M, Pernodet N. Biomineralization of a Self-Assembled Extracellular Matrix for Bone Tissue Engineering. *Tissue Eng Part A* (2008).

72. Woodrow KA, Wood MJ, Saucier-Sawyer JK, Solbrig C, Saltzman WM. Biodegradable meshes printed with extracellular matrix proteins support micropatterned hepatocyte cultures. *Tissue Eng Part A* (2008).

73. Datta N, Holtorf HL, Sikavitsas VI, Jansen JA, Mikos AG. Effect of bone extracellular matrix synthesized *in vitro* on the osteoblastic differentiation of marrow stromal cells. *Biomaterials* **26** (2005) 971–977.

74. Pham QP, Kasper FK, Scott Baggett L, Raphael RM, Jansen JA, Mikos AG. The influence of an *in vitro* generated bone-like extracellular matrix on osteoblastic gene expression of marrow stromal cells. *Biomaterials* **29** (2008) 2729–2739.

CHAPTER 2

Current Clinical Needs in Hand Surgery

L. Scott Levin*

By definition, hand surgery is an integration of concepts that include principles and practices of multiple surgical disciplines, centering on the form and function of the human hand. Current clinical needs in hand surgery can be divided into medical, social, and economic needs.

Dr. Sterling Bunnell developed the principles of hand surgery, based on the battlefield experience of World War II. Much was done to define the practice of hand care by Bunnell during World War II and beyond. Bunnell stated, "Hand surgery is a combination of neurosurgery, vascular surgery, orthopedic surgery, and plastic surgery." Over the last several decades, microsurgical tools and techniques have revolutionized care in all aspects of hand surgery. Now, history is repeating itself. We have had a resurgence of devastating upper extremity injuries as a result of Operation Enduring Freedom in Afghanistan and Operation Iraqi Freedom. Soldiers, (both men and women), have returned home with devastating problems related to missile injuries resulting from rocket propelled grenades, and blast injuries from improvised explosive devices (IED's).

We are now on the brink of a new era in hand and upper extremity reconstruction. Current needs are to create a generation of hand surgeons that are familiar with immuno-logic aspects of composite tissue allotransplantation in order to advance the specialty. Hand transplantation has already been performed in multiple countries around the world over the last decade.[1,7] Still the problems of immunology, indications and ethics have slowed the progress in this field.[5,11,12] Despite advances and early surgical success, much work needs to be done. Clearly, tissue engineering and tissue allotransplantation appear to be pathways to the future. The highways of information and science run parallel at times, but hopefully will intersect in the future. Rather than being competitive, they should be complementary.

The era of tissue engineering has already begun in hand surgery. Skin substitutes such as Integra have revolutionized the care of the burned hand and have provided tissue engi-neered substrates for autogenous skin grafting.[2] Neural conduits for digital nerve repair have already been in use for some time.[9] While we have not been able to add commercially available nerve growth products to these conduits, this concept remains a possibility in the future. Neural regeneration remains an unsolved problem in hand surgery, and repair tech-niques have altered very much in the last 50 years following the advent of the operating microscope by Jacobsen and Suarez in 1960. Attempts have also been made to create small vascular conduits. Dr. Niklason and colleagues have attempted to fabricate tissue engi-neered vascular prostheses that would have wide implications for traumatic injuries to the hand, specifically in replantation.[6,10]

Progress has been made on tissue engineered tendons by Chang and others.[3,13] However, the problems of tendon adhesions and fibrosis still exist. In part, these have been

---

* Corresponding author.

addressed by materials that wrap tendon repairs and try to improve gliding, but are a far cry from solving the "one wound, one scar" problem described by Peacock decades ago. Although an implantable tendon prosthesis was introduced by Hunter thirty years ago, we await the development of a reliable tendon replacement. In the face of our aging population, hand arthritis has become a limiting factor in the older patient's independence. Further development in joint replacements and even tissue engineered joints that have cartilage and improved ability to integrate or osteointegrate into bone would be desirable.

Current clinical needs in hand surgery also include the development of more sophisticated prostheses, and prostheses that can be biologically integrated with nerves and muscles to restore function and communication with the cerebral cortex. Dr. Dumanian and colleagues at Northwestern have been instrumental in providing improved neural control of terminal devices of prosthetic arms using nerve transfers.[4,8] This holds promise to allow interfaces with mechanical prostheses, or even tissue engineered interfaces that allow cortical input for control of a mechanical hand, with neural feedback.

Industrial injuries have decreased over the last thirty years with Occupational Safety and Health Administration guidelines. However, traumatic injuries still occur in the hand on the farm and in the industrial workplace. Prevention of hand injuries in the workplace is a clinical need that would be critical to help reduce costly Worker's Compensation claims.

Another clinical need in hand surgery is improvement in current training programs. Hand and microvascular fellowships that exist after orthopedic or plastic surgery residencies, more often than not, fail to provide adequate microvascular training. These skills are necessary to perform complex procedures such as toe-to-hand transfer, brachial plexus reconstruction, vascularized nerve grafts, functioning free muscle transfers, and perforator flaps. Because fewer surgeons are adequately trained in reconstructive microsurgery, the workforce in hand surgery is diminishing. While this is perhaps the most pressing clinical need in hand surgery, the hope is that a new generation will enter the specialty of hand surgery with a profound interest in all the areas of tissue engineering, transplantation, and microvascular surgery. They will contribute to the specialty both in the research laboratory and in the clinical arena.

References

1. Brandacher G, Ninkovic M, Piza-Katzer H, Gabl M, Hussl H, Rieger M, Schocke M, Egger K, Loescher W, Zelger B, Ninkovic M, Bonatti H, Boesmueller C, Mark W, Margreiter R, Schneeberger S. The Innsbruck hand transplant program: update at 8 years after the first transplant. *Trans Proc* **41** (2009) 491–494.
2. Choi M, Armstrong MB, Panthaki ZJ. Pediatric hand burns: thermal, electrical, chemical. *Cranio Sur* **20** (2009) 1045–1048.
3. Chong AK, Riboh J, Smith RL, Lindsey DP, Pham HM, Chang J. Flexor tendon tissue engineering: acellularized and reseeded tendon constructs. *Plas Recons Surg* **123** (2009) 1759–1766.
4. Dumanian GA, Ko JH, O'Shaughnessy KD, Kim PS, Wilson CJ, Kuiken TA. Targeted reinnervation for transhumeral amputees: current surgical technique and update on results. *Plastic and Reconstructive Surgery* **124** (2009) 863–869.
5. Gabl M, Rennekampff O. [Handtransplantation and Composite Tissue Allotransplantation (CTA)]. *Handchir Mikrochir Plast Chir* **41** (2009) 203–204.
6. Gong Z, Nikklason LE. Small-diameter human vessel wall engineered from bone marrow-derived mesenchymal stem cells (hMSCs). *FASEB J* **22** (2008) 1635–1648.
7. Jones JW, Gruber SA, Barker JH, Breidenbach WC. Successful hand transplantation. One-year follow-up. Louisville Hand Transplant Team. *The New Eng J Medi* **343** (2000) 468–473.

8. Kuiken T, Miller L, Lipschutz R, Stubblefield K, Dumanian G. Prosthetic command signals following targeted hyper-reinnervation nerve transfer surgery. *Conf Proc IEEE Eng Med Biol Soc* **7** (2005) 7652–7655.

9. Meek MF, Nicolai JP, Robinson PH. Secondary digital nerve repair in the foot with resorbable p(DLLA-epsilon-CL) nerve conduits. *J Reconst Microsurg* **22** (2006) 149–151.

10. Norotte C, Marga FS, Niklason LE, Forgacs G. Scaffold-free vascular tissue engineering using bioprinting. *Biomaterials* **30** (2009) 5910–5917.

11. Piza-Katzer H, Wechselberger G, Estermann D, Gabl M, Arora R, Hussl H. [Ten years of hand transplantation experiment or routine?]. *Handchir Mikrochir Plast Chir* **41** (2009) 210–216.

12. Sucher R, Hautz T, Brandacher G, Lee WP, Margreiter R, Schneeberger S. [Immunosuppression in hand transplantation: state of the art and future perspectives]. *Handchir Mikrochir Plast Chir* **41** (2009) 217–223.

13. Zhang AY, Bates SJ, Morrow E, Pham H, Pham B, Chang J. Tissue-engineered intrasynovial tendons: optimization of acellularization and seeding. *J Rehab Res Dev* **46** (2009) 489–498.

CHAPTER 3

Principles of Tissue Engineering for Reconstruction of the Hand

Ryosuke Kakinoki*

Preface

Hand surgeons often encounter tissue losses or shortages, most of which are caused by trauma, tumor or congenital abnormalities. Such tissue losses are usually reconstructed using autogenous tissue transplantation, and bone grafts taken from the distal radius or iliac crest are still the gold standard for reconstructing bone shortages. Sensory cutaneous nerves such as the sural or medial forearm antebrachial nerves are popular donor sources for autogenous nerve grafts. Palmaris longus and plantaris tendon grafts are used to bridge tendon defects. Skin defects are covered with split- or full-thickness skin grafts. After the invention of operative microscopes and the subsequent development of microsurgical techniques in the 1960s, the treatment of tissue shortages had the benefit of a new concept: reconstruction using vascularized tissue transplantation. In hand surgery, finger defects caused by traumatic amputation or congenital anomalies began to be reconstructed by the transplantation of toe digits. Using free or pedicled skin grafts containing subcutaneous nerve branches, sensitized skin with subcutaneous adipose tissue could also be reconstructed. Bone defects adjacent to joints in children can now be reconstructed using a vascularized bone graft including an epiphysis such as the proximal fibular head, metatarsal or phalangeal bones of the foot, and normal bone growth of the transplanted bone can be expected. Recently, with the development of immunosuppression, allogeneic tissue transplantations have been used in the reconstruction of tissue loss. In 1998, allogeneic hand transplantation was performed in France on a patient with bilateral arm amputation. Since then, allogeneic upper limb transplantation has been performed on more than 27 patients in several countries.[1] However, this technique is still associated with several immunological, social, ethical and psychological problems. For example, immunological problems include graft rejection, side effects of the immunosuppressants used, and graft-versus-host disease.[2] It is known that the transfer of musculoskeletal and cutaneous tissues is less likely to achieve immunological tolerance or tissue chimerism than the transplantation of internal organs such as the liver, kidney or heart. Thus, reduction of immunosuppressant dosage depending on the postoperative time course is more difficult in the transplantation of bone, muscle, peripheral nerves and skin compared with internal organ transplantation.[1]

Tissue engineering is defined as a technology to improve or replace biological functions by using cells, tissue, biochemical and physiochemical factors or artificial materials. The term regenerative medicine can be used synonymously with tissue engineering, but the former places more emphasis on the use of stem cells that have the capacity to differentiate into various cell types leading to the production of diverse tissues or organs.[3] Tissue

* Corresponding author.

engineering is a medical field where the principles of tissue growth are applied to produce functional replacement of tissue clinically. Because tissue engineering is multidisciplinary and readily applicable clinically, academic and commercial interests are currently focusing on this field.[4]

Tissue Engineering

Because the purpose of tissue engineering is to produce new body architecture that can be replaced by original tissue, it requires three major factors: cells, a scaffold providing the place for cell growth or differentiation, and an appropriate environment including some chemical or growth factors that can trigger cell differentiation, or external mechanical forces such as compressive force for cartilage production, and tensile forces for tendinous tissue production.

Cell Transplantation

Cell transplantation techniques are categorized according to the cell source and are divided into autologous, allogeneic, xenogeneic and syngeneic types. In autologous transplantation, cells are harvested from the same individual who will receive the implant. This is not associated with immune rejection. However, tissue that might be affected by systemic or genetic diseases cannot be used as a source of cells, and very senile or seriously ill people cannot be donors. Harvesting autologous cells is always associated with donor site morbidity including pain, infection and loss of some function. Moreover, autologous cells must be cultured from samples and proliferated *in vitro* before they can be transplanted. This process usually takes time, which can be a disadvantage. In allogeneic transplantation, cells are harvested from members of the same species, but this is associated with immunological and ethical problems. In xenogeneic transplantation, cells are taken from an individual of another species. Problems with interspecies infections and hyperacute immune rejection must be solved before any clinical use of xenogeneic cells. Although xenogeneic cell transplantation has been performed for cardiovascular tissue implants, it has minimal potential for clinical use at the present. Syngeneic or isogeneic cells are isolated from individuals with the same genotypes, such as in the transplantation of embryonic stem cells between monozygotic (identical) twins.

In tissue engineering, cells that induce a low immunological response and that have the potential for great proliferation and differentiation into a variety of cells are required. Autologous cells are widely used at present, but further research is needed for the practical use of allogeneic or xenogeneic cells in tissue engineering.

Totipotent and pluripotent stem cells

The human body comprises 60 trillion cells, which originate from a single fertilized oocyte (zygote) and proliferate by repeated mitotic divisions. The zygote and early morula-stage embryo are the only cells to have the potential to produce all sorts of cells constituting a human being, including the placenta. This capacity is called totipotency. When the embryo develops into the blastocyst stage, it divides into two cell types: one proliferates to form an inner cell mass that will differentiate into all the cells constituting the body except for the placenta, and the other proliferates into trophoblastic cells that will form the placenta. This feature of cells of the inner cell mass is called pluripotency, the potential to differentiate into all sorts of cells except the placenta. Embryonic stem (ES) cells extracted from the inner cell mass, endogenous neural or hematopoietic stem cells, and mesenchymal stem cells harvested from the bone marrow or fat tissue are called pluripotent cells.[5] Note that some authors have adopted a different classification of totipotent cells to include ES cells, because these have the capacity to differentiate into all cell types except for the placenta, but the other pluripotent stem cells (endogenous stem cells or mesenchymal stem cells) have less capacity for differentiation.

ES cells

ES cells are harvested from cell lines derived from the inner cell mass of cultured blastocyst stage embryos and demonstrate the capacity to differentiate into a variety of tissue types (pluripotency).[5] The successful production of an entire mouse from an ES cell[6] and the successful isolation of human ES cells[7] attracted great attention to the concept that one could use ES cells medically for repairing tissue shortage or damage. However, the use of human ES cells is associated with critical ethical problems. First, ES cells can be harvested only by destroying or seriously compromising a living embryo with the potential to become a human being. In addition, generating ES cells runs the risk of injuring healthy young donor women.[8] Therefore, recent studies have focused on using stem cells or creating cells with a level of pluripotency similar to ES cells by transferring several specific genes to somatic cells; these are termed induced pluripotent stem (iPS) cells.

iPS cells

Unlike somatic cells, iPS cells and ES cells can proliferate and maintain their pluripotency, although there is no difference in the genomic DNA sequence between ES and somatic cells. ES cells express specific genes including those for the transcription factors Oct3/4 and Nanog to maintain pluripotent differentiation, but somatic cells do not express these genes. All somatic cells possess the genes for Oct3/4 and Nanog, but the expression of these genes is inhibited by the action of transcription factors and by epigenetic mechanisms that regulate the methylation of specific DNA bases. In 2006, Yamanaka *et al.* created a pluripotent cell line resembling true ES cells by transferring four genes (*Oct3/4*, *Sox2*, *Klf4* and *c-Myc*) into the genomes of mouse fibroblasts.[9] In 2007, they successfully established a human iPS cell line from fibroblasts harvested from the skin of the forehead of a 36-year-old woman.[10] This development of iPS cells has apparently solved the ethical problems associated with the use of human ES cells; however, there are still major problems to be overcome before clinical use of the cells, namely the oncogenicity of such iPS cells. It is reported that 20% of chimeric mice born from embryos following the transfusion of iPS cells generated carcinomas. The rate of cancer development using iPS cells seemed to be greater than when using ES cells. It was suspected that *c-Myc* — itself an oncogene — might be responsible, or that the retroviruses used during gene transfer might have induced gene transformation leading to carcinogenesis. Therefore, current research focuses on the development of iPS cell lines with less oncogenicity. Recently, iPS cells have been produced successfully without transferring the *c-Myc* gene that was suspected of playing a role in oncogenicity, but the rate of iPS cell production using this new method decreased to ~1% of that using *c-Myc*.[11,12] Although there are several hurdles to overcome, there are great expectations for the clinical application of iPS cells in tissue engineering.

Adult stem cells

Several undifferentiated cell types are found among the differentiated cells in mature tissues and organs. Adult stem cells can be isolated from bone marrow, adipose tissue, articular synovial tissue, neural tissue, umbilical cord blood and even peripheral blood. These cells are pluripotent and can differentiate into a variety of tissue types, including bone, cartilage, fat, tendon, vessels and nerves. However, such adult stem cells have less capacity for differentiation than ES or iPS cells. Thus, neural stem cells can be found in the central nervous system but are committed to a neural fate. They can differentiate into astrocytes and oligodendrocytes but not into bone, fat, muscle or cartilage.[13] Hematopoietic stem cells taken from blood in the umbilical cord can differentiate into various types of blood cells including erythrocytes, lymphocytes, neutrophils, macrophages and platelets. Mesenchymal stem cells (bone marrow derived or fat tissue derived) are known to differentiate into osteocytes, chondrocytes, adipocytes and other connective tissue cells but not into hematopoietic cells.[14-16]

Mesenchymal stromal cells

Mesenchymal stromal cells (MSCs) are multipotent stem cells that can differentiate into a variety of cell types, such as osteoblasts, chondrocytes and adipocytes. It was suggested that they transdifferentiated into neuronal cells; however, it remains controversial whether this was from spontaneous cell fusion between damaged neurons and the transplanted mesenchymal stromal cells or evidence of stem cell plasticity.[17-19] The term "mesenchyme" refers to embryonic connective tissue derived from the mesoderm. Stromal cells are connective tissue cells that form a structure supporting functional cells or tissues. The term MSC is sometimes used for multipotent cells harvested from the muscle, umbilical cord or dental pulp. The Mesenchymal and Tissue Stem Cell Committee of the International Society for Cellular Therapy has listed the minimum criteria to define a human MSC as follows.[20]

1. MSCs must adhere to plastic when maintained in standard culture conditions.
2. MSCs must express the surface molecules CD105, CD73 and CD90 and lack expression of CD45, CD34, CD14, CD11b, CD79α, CD19 or HLA-Dr.
3. MSCs must differentiate into osteoblasts, adipocytes and chondroblasts *in vitro*.

A recent study has revealed that MSCs originate from the pericytes around blood vessels.[21] Conventionally, MSCs are isolated from tissue by relying on their ability to adhere to plastic. This method requires prolonged culture *in vitro*, rendering the entire procedure costly and time consuming. Recent studies indicated that MSCs could be isolated from bone marrow samples by using specific antibodies that interact with CD106, CD146 and Stro-1 cell surface antigens.[22] This method is known an immunoisolation. Immunoisolated MSCs exhibit more colony-forming units than MSCs isolated using the plastic adherence method.[23] Recent reports indicate that the capacity of MSCs to proliferate and differentiate decreases with the age of the donor as well as the time in culture. They may lose some of their stem cell properties after prolonged culture and may become tumorigenic.[24-27] Conventional methods of adult stem cell isolation include an *in vivo* culture period that can increase the number of cells available for implantation. Other novel strategies include isolation techniques that allow the immediate use of adult stem cells for *in vivo* implantation.[30]

The immunological properties of MSCs have been revealed recently. Several studies have indicated that human MSCs avoid allorecognition, interfere with dendritic cell and T-cell function, and generate a local immunosuppressive microenvironment by secreting cytokines.[28] A recent report has also indicated that the immunomodulatory function of human MSCs is enhanced when the cells are exposed to an inflammatory environment characterized by the presence of elevated local interferon-gamma levels.[29] There are also experimental trials to reduce immune responses in allogeneic organ transplantation using MSCs.[28]

Tissue Scaffolds

Extracellular matrix formation is important in tissue reconstruction. Pluripotent cells can create extracellular matrix during their proliferation and differentiation. However, materials to provide transplanted cells with a place for the proliferation and differentiation and to hold the cells in the target area are needed until the extracellular matrix produced by the cells can obtain sufficient structural integrity to support newly formed tissue. Thus, scaffolds are essential in tissue reconstruction.

Generally, there are two types of scaffolds used in tissue engineering: artificial and biological scaffolds. Artificial scaffolds need a high porosity with an adequate pore size that will allow transplanted cells to extend through the whole structure and enable nutrients and oxygen to reach the cells. Biodegradability of the scaffolds is often essential, because preferably scaffolds should be absorbed by the surrounding tissues, to avoid the need for surgical removal and to allow newly formed tissue to integrate with the surrounding tissue. Ideally, the rate of degradation should coincide with the rate of tissue formation.[30–32] The materials polylactic acid (PLA), polyglycolic acid (PGA) and polycaprolactone (PCL) are commonly used. They degrade within the human body to form materials that are normally absorbed by the body. In practice, mixtures of those materials are often used as scaffolds in tissue engineering. The rate of degradation can be determined by the ratio of the components.

The other type of scaffold is made of natural materials. Basement membranes fabricated from isogeneic or allogeneic muscles or nerves in which the original cells are devitalized and the allogeneicity is reduced using some chemical agents or the repetitive freeze–thaw technique can support cell growth.[34–38] Allogeneic protein materials (collagen or fibrin) and polysaccharidic materials (chitosan or glycosaminoglycans) demonstrate good cell compatibility and are preferably used as scaffolds in tissue engineering after treatment to reduce allogeneicity.

Cell Environments

Growth factors, hormones, specific metabolites or nutrients, chemical and physical stimuli are sometimes required to determine the fate of pluripotent cells, which can then proliferate and differentiate into various cells with different properties. When MSCs are transplanted to a site with adequate growth factors or hormones related to bone development, they will differentiate into osteogenic cells. When they are transplanted in the environment with factors for nerve growth, they will differentiate into neuronal cells. Physical stimuli such as pressure, vibration and rotation also play an important role in determining the types of cells that pluripotent cells can differentiate into.[31,32] Thus, hydrostatic pressure can drive pluripotent stem cells to differentiate into cartilaginous cells. Some cells differentiate into endothelial cells, responding to shear stress from the fluid flow encountered in blood vessels. Such physical stimuli can be generated by bioreactor systems that mimic the *in vivo* conditions with hydrostatic pressure, vibration or cell rotation.

Appropriate nutrition and oxygen delivery to the transplanted cells is essential. When pluripotent cells are transplanted to hypoxic regions, the proliferation and differentiation potential of the cells can be impaired leading to cell death. It is known that peripheral nerve regeneration occurs in a tube-like material: a phenomenon called tubulation. To promote nerve regeneration by tubulation using tissue-engineering technology, several studies have implanted pluripotent cells within tubular chambers. When the tube is short, the seeded cells can obtain nutrients and oxygen from blood capillaries extending from transected nerve stumps connected to either end of the tube. However, when the interstump gap is longer than the distance over which capillaries can extend from the nerve stumps, the tube needs some devices to deliver oxygen and nutrients to the transplanted cells. Tubes with micropores may facilitate the delivery of nutrients and oxygen to the transplanted cells inside.[39,40] However, if the tube is placed in a hypoxic area such as scar tissue, the micropores may not be able to deliver sufficient nutrients and oxygen for the transplanted cells to survive. In a clinical setting, bone necrosis is often a target for tissue engineering repair efforts. There are several experimental and clinical studies of treating bone necrosis using pluripotent cell transplantation. Because the pluripotent cells are to be transplanted to a region of bone necrosis, good vascularization of the area is also necessary to allow the cells to survive. The combination of vascularized tissue and cell transplantation should be an option for the treatment of bone necrosis.[41]

Peripheral Nerve Engineering

Autologous nerve transplantation using cutaneous sensory nerve grafts such as sural or forearm cutaneous nerves is still the gold standard to repair a nerve deficit in the hand. Using the recent development of immunosuppressants, allogeneic nerve grafts are now an option. However, there are several problems to be solved before allogeneic nerve transplantation can be accepted as a popular technique; for example, acute rejection of the grafts, side effects of immunosuppressants, and graft-versus-host disease.[42] It is still controversial to use allogeneic nerve grafts that have life-threatening risks to treat peripheral nerve injuries, most of which do not cause mortality in themselves. For peripheral nerve tissue engineering, as with other tissues, cells, scaffolds, and chemical and growth factors are key components. Any scaffold must support axon and Schwann cell extension. Moreover, it must also play the role of holding transplanted cells in the place where nerve regeneration occurs. Chemical and growth factors such as glial cell-derived nerve growth factor (GDNF), brain-derived neurotrophic factor, acidic and basic fibroblast growth factors, neurotrophin-3 (NT-3), insulin-like growth factor, platelet-derived growth factor, ciliary neurotrophic factor, interleukin-1 and transforming growth factor are necessary to vitalize Schwann cells and to extend axons.[43] There is a spectrum of axonal sensitivity to various growth factors in the injured peripheral nerves. GDNF is preferable to support the growth of motor axons, and NT-3 supports the extension of sensory axons. Not only growth factors but also some elements of the extracellular matrix such as collagen, laminin and fibronectin are known to accelerate peripheral nerve regeneration; these are cell-adherent factors that guide axons in peripheral nerve regeneration.[44-47] While many factors promote axon regeneration, some materials are found to inhibit peripheral nerve regeneration. For instance, chondroitin sulfate proteoglycan is known to be an inhibitor of axon extension and Schwann cell migration in peripheral nerves.[48]

In tissue-engineered peripheral nerves, scaffolds with transplanted *in vitro* cultured Schwann cells have been employed in many experimental models and have demonstrated improved nerve regeneration.[49–51] Because the culture of Schwann cells is lengthy and technically demanding, recently, mesenchymal stem cells including bone-marrow-derived (BMC) or fatty-tissue-derived cells have been preferably used in nerve tissue engineering. Several studies demonstrated better nerve regeneration through nerve conduits containing BMCs than those without.[52–54] Generally, scaffolds used to support tissue-engineered nerves take two forms. One is a scaffold made of a tube-like material, and the other is a scaffold made of devitalized basal lamina of allogeneic or isogeneic muscles and nerves with reduced or diminished immunogenicity. It is known that fewer axons, and axons with smaller diameters, undergo regeneration through unmodified tube-like materials compared with autogenous nerve grafts. Moreover, as discussed above, there is a limit to the length through which nerve regeneration can occur through tube-like scaffold materials. For the rat sciatic nerve, the maximum distance across which nerve fibers can regenerate through an unmodified silicone tube is known to be 10 mm.[55] Several bioengineering modifications have been applied to improve nerve regeneration through tubular scaffolds. Biodegradable tubes are beneficial in nerve regeneration because they produce a minimal late foreign body response.[39,56] Micropores that allow the delivery of nutrients and oxygen into the chamber space can be created in the wall of nerve conduits to keep cells alive within the conduits.[39] Tubes containing some neurochemical factors or some cell adhesion factors help to promote nerve regeneration in animal models.[44–47,57,58] Better nerve regeneration was demonstrated through tubes with cultured Schwann cells or with MSC transplantation.[49–54] Clinically, polyglycolic acid bioabsorbable conduits are available commercially (Neurotube; Neuroregen LLC, Bel Air, MD, USA). A clinical study demonstrated that sensory recovery through tubes used to repair digital nerve deficits less than 3 mm long was better than that through autogenous nerve grafts.[56] At present, however, it is not clear whether these findings can be extrapolated to longer nerve defects and larger mixed nerves.

Basement membrane is a natural scaffold of peripheral nerves. Several researchers have demonstrated successful nerve regeneration through a devitalized basal laminar scaffold that was created from autologous or allogeneic nerves, muscles and tendons using repetitive freeze–thawing or some chemical treatments.[35–38] Allogeneic peripheral nerve basal laminar scaffolds with devitalized chondroitin sulfate proteoglycan and chemically reduced immunogenicity are now available clinically (Avance Nerve Grafts, Axogen Inc., FL, USA).

For either the nerve conduit or basal laminar scaffold-type nerve tissue engineering, nutrients and oxygen must be delivered to the transplanted cells. Thus, the tissue-engineered nerves must be transplanted on a well-vascularized bed. In clinical application, when long nerve grafts are used in scar tissue, reconstruction of the vascularity of the recipient site or vascularization of the grafts (vascularized nerve grafts) is needed. One experimental study[54] revealed better nerve regeneration through a silastic conduit containing a vascular pedicle and cultured BMCs. Molecular biology studies demonstrated that the transplanted BMCs differentiated into cells with Schwann cell-like phenotypes and that about 30% of cells in the regenerated nerve originated from the seeded BMCs 12 weeks after the tubulation. Vascular bundle implantation and BMC transplantation are rational choices for intrachamber nutritional and oxygen delivery. The chamber space in the study was filled with a variety of neurochemical factors secreted from the transected sciatic nerve stumps, and these might have helped drive the transplanted BMCs to differentiate into Schwann cell-like cells that facilitated nerve regeneration. The implanted vascular pedicle delivered nutrition and oxygen to the intrachamber BMCs and might have also acted as a scaffold to hold the seeded cells within the tube.

Bone Tissue Engineering

The carpal, metacarpal and phalangeal bones in the hand are relatively small, and a bone source supplied by the radius, iliac crest or fibula is usually sufficient to repair bone defects in the hand. Most hand surgeons have experienced difficulties in treating scaphoid nonunions and osteonecrosis of the lunate (Kienboeck disease). Tissue engineering in hand surgery is expected to be feasible for treating these pathologic conditions.

Bone morphogenetic proteins (BMPs) were discovered by Urist *et al.* in 1965 as proteins that induce strong osteoinductivity.[59,60] They are produced by osteoblasts and are classified into several subtypes. Among them, BMPs-2, -4, -6 and -7 are known to have strong osteoconductive capabilities.[61] Recently, bone repair using gene- and cell-based therapies has been attempted in animal models. Generally, there are two strategies: one is direct gene delivery to the site of a bone repair with the aid of viral and nonviral vectors; the other is a cell-based gene therapy in which cells having been engineered *in vitro* to express certain transgenes to facilitate bone formation are implanted to the site of bone repairs. Direct gene delivery to bone defects or ectopic sites has been performed in numerous animal models using various viral[62–73] and non-viral vectors[74–85] that encode for different BMPs and that were shown to be effective and straightforward for bone formation. Cell based gene-delivery exploits the ability of adult stem cells to express a transgene that can both affect the cells themselves and recruit host stem cells to achieve bone formation. MSCs genetically engineered to over express an osteogenic gene such as BMP-2 were hypothesized to secrete the protein, which would lead to the differentiation of the engineered MSCs into bone-forming cells and also recruit host MSCs to the site of implantation.[84]

Some studies used other genes to induce osteogenic differentiation. One example is the gene for LIM mineralization protein-1 (LMP-1), which is considered a critical regulator of osteoblast differentiation.[85] BMCs into which LMP-1 cDNA was delivered using an adenovector which overexpressed LMP-1. When these were implanted ectopically, they increased BMP-4 and -7 expression and finally yielded ectopic bone. Runx-2, a transcription factor produced downstream of BMP signaling pathways, also demonstrated osteogenic effects upon overexpression in adipose-derived stem cells, both *in vivo* and *in vitro*.[86]

Scaffolds in tissue engineering are generally biodegradable and should in theory be replaced by newly formed tissue. Biodegradable materials used for bone formation are expected to provide local structural support and to degrade as new bone is deposited. The most important role of bone is to provide structural support to resist physical forces applied to the skeleton. However, scaffolds used in bone fabrication do not necessarily degrade when new bone is formed. The scaffold may remain in newly formed bone and continue to provide structural support by cooperating with the bone. Natural biodegradable polymers such as collagen and chitosan and synthetic biodegradable polymers such as polylactic or polyglycolic acid[61,87,88] have been used as scaffolds in fabricating bone tissue; however, nonbiodegradable scaffolds that have osteoconductive or osteoinductive properties including calcium phosphate ceramics[89–91] or some metals such as titanium alloy can be options for the scaffolds used in bone formation.[92,93]

Biomaterials used for artificial osteogenesis need an interconnective porous structure and specific surface properties.[92,93] The porous structure allows osseous cells to go into the scaffold and to grow and form bone tissue. The optimal pore diameter for *in vivo* osteoconduction is thought to be in the range of 150–500 μm.[94,95] New bone with a lamellar structure is formed on the inner surface of the micropores. The interconnectivity of the porous structure allows the cells and bone tissue to invade deep into the pores, thereby promoting osteoconduction.[96] The three-dimensionally porous structure of the scaffolds is closely related to apatite formation. The chemical and thermal treatment on the surface of the scaffolds forms an apatite to promote osteogenesis. Fujibayashi *et al.*[92,93] reported that a porous titanium block with a chemical treatment (immersion in 5M aqueous NaOH solution at 60°C for 24 h) followed by a hot water treatment (immersion in distilled water at 40°C for 48 h) and a thermal treatment (heating to 600°C at a rate of 5°C/min, maintained at 600°C for 1 h and then allowed to cool at the natural rate of the furnace) demonstrated better osteoinductivity than an untreated porous titanium block. Thus, such treated porous titanium implants might be good alternatives to direct bone grafting.

Bone tissue engineering has been promoted combined with the use of mesenchymal stem cells. Bone-marrow-derived or adipose-tissue-derived cells successfully differentiated into bone tissue in appropriate scaffolds such as hydroxyapatite, ceramics or beta-tricalcium phosphate (β-TCP). In hand surgery, successful bone formation is needed in the treatment of scaphoid nonunions or osteonecrosis of the lunate (Kienboeck disease) or scaphoid (Priser disease). A condition mimicking osteonecrosis was also treated using a tissue engineering approach. Transplantation of vascularized bone and an MSC-seeded β-TCP scaffold demonstrated successful bone formation and remodeling in an osteonecrosis model using canine carpal bones, whereas transplantation of a vascularized bone graft and fibroblast-seeded β-TCP did not produce osseous tissue.[41] When cells are transplanted to avascular or hypovascular areas, vascularization is essential to allow the transplanted cells to survive. The BMCs in this study might have differentiated into osseous tissue after being influenced by BMPs or other osteogenesis-related factors that were produced by the transplanted vascularized bone.

Tendon Tissue Engineering

There are three major problems in tendon surgery for hand surgeons to overcome. One is how to reduce the adhesion of a tendon repair site to the surrounding tendon sheath. Avoiding tendon adhesion is very important in tendon surgery, especially in zone 2 flexor tendon injuries. In this region, there are two flexor tendons (deep and superficial flexor tendons) that are surrounded by the fibro-osseous pulleys. Zone 2 injuries often involve multiple pulleys of the flexor tendons. In repair of these, either the A1 or the A2 pulley and either the A3 or the A4 pulley should be preserved to prevent bowstring deformation of the flexor tendons. If these pulleys are not preserved, a new pulley has to be reconstructed using tendon transplantation. The two injured tendons and their pulleys adhere easily to each other in patients with zone 2 injuries. The site of tenorrhaphy is repaired by neovascularization from the surrounding tissues (extrinsic healing) and by intratendinous neovascularization (intrinsic healing). Therefore, to reduce tendon adhesion, extrinsic healing should be suppressed and intrinsic healing promoted. Several methods have been reported to suppress extrinsic healing. Thus, hyaluronate injection into the space between repaired tendons and the surrounding tendon sheath was shown to reduce adhesions.[97,98] One report showed that transplantation of plasmids encoding the production of platelet-derived growth factor to the sites of the tenorrhaphy could reduce tendon adhesion.[99] However, at present, there is no other way that is more reliable to avoid tendon adhesion following tenorrhaphy than early controlled active exercise of the joint.

The second problem is to how to increase tensile strength at the site of tenorrhaphy after tendon injuries. Tensile stimuli to injured tendons can promote the maturation and longitudinal extension of collagen fibers and help prevent extrinsic collagenous linkages between the tendon sheath and the repaired tendons. Recent biomechanical studies on tenorrhaphy revealed that increasing the number of intertendinous sutures provides greater initial tensile strength. Controlled early active movement of tendons that have been repaired using four or six strands of intratendinous sutures and additional several epitendon sutures is the best method to repair tendons.[100–102] However, no successful studies using tissue-engineering techniques have reported the provision of tendons with an initial tensile strength that is great enough to perform early active exercises after tenorrhaphy.

The final major problem for the hand surgeon is how to create tendon tissue that has been lost following trauma or in patients with congenital anomalies. The ordinary source for tendon transplantation is the palmaris longus, plantaris or extensor tendons of the toes. These are not intrasynovial but extrasynovial tendons, which are more prone to adhesions than intrasynovial tendons in canine models.[103,104] In clinical settings, harvesting intrasynovial tendons is almost impossible, considering the likelihood of morbidity at donor sites.

There have been several reports on the production of tendons using tissue-engineering techniques. Fibroblasts taken from tendons or skin, or tenocytes have been used commonly. Because of the low mitotic activity of fibroblasts and tenocytes and the limited donor sites for tenocyte harvest, alternatives such as MSCs have been used.[105] Cross-linked type-1 collagen, human or bovine dura mater, small intestinal submucosa and various biosynthetic polymers including polyglycolic acid, polylactic acid and polypropylene have been used extensively as scaffolds for tendon tissue engineering.[106] Cao *et al.*[107] grew autologous tenocytes in culture and mixed them with unwoven polyglycolic acid fibers *in vitro* to form cell-seeded constructions with the shape of a tendon. The construction was reinforced with intestinal submucosa and implanted to flexor tendon defects in chicken models. They reported that newly formed tendon consisting of longitudinally oriented collagen fibers had a ratio of cells to collagen similar to that of normal flexor tendons and demonstrated the tensile strength of 83% of normal tendons 14 weeks after the implantation. Several authors reported that delivering MSCs using collagen or biosynthetic biodegradable implants to reconstruct large tendon defects markedly improved the structure and biomechanical properties and function of the reconstruction.[105,108,109] Young *et al.* demonstrated that collagen gel containing MSCs implanted in a 1 cm-long defect of the rabbit patellar tendon turned into tendon-like tissue with appropriate collagen fiber alignment and biomechanical properties.[105] Chong *et al.* transplanted fibrin gel matrices containing bone marrow-derived MSCs (treatment group) or fibrin gel alone (control group) to the site of Achilles tendon repair in rabbits. Cell tracing showed that labeled MSCs were retained in the intratendinous region of the treatment group for at least six weeks, becoming more diffuse at later times. At three weeks, the treatment group revealed more organized collagen fibers and showed significantly better nuclear morphometric parameters than the control group. This significance was not observed at later times. Biomechanical testing showed a better modulus in the treatment group than the control group at three weeks but not at subsequent times.[109] Recently, Murphy *et al.* repaired a 1 cm-long flexor tendon defect using bioengineered porcine small-intestinal submucosa (SIS) without cell seeding in a rat model. The engineered tendon demonstrated compatible tensile strength and concentrations of hydroxyproline, transforming growth factor beta (TGF-β) and TGF-β receptor as compared to autogenous tendon graft eight weeks after transplantation, although the postoperative values of the SIS grafts at one week were inferior to those of the autologous tendon grafts.[110]

In tendon construction using tissue engineering, newly formed tendons should be exposed to stimuli of tensile forces, which facilitate the maturation of intratendinous collagen fibers and promote longitudinal alignment of the fibers and increasing tensile strength. It might be a solution in tendon construction to fabricate scaffolds that can provide an initial strength that is sufficient to resist the tensile force of the tendon and to transmit the force to the transplanted cells.

Cartilaginous Tissue Engineering

The minimal intrinsic healing capacity of cartilage and continuous physical forces on the articular surface make the repair of cartilage injuries difficult. Hand surgeons often confront traumatic cartilage defects in the interphalangeal joints of fingers. Arthrodesis is an option for treatment, but it leaves great morbidities in patients, especially arthrodesis in the ulnar fingers. Artificial joints can be another option for senile patients or for patients with chronic inflammatory diseases such as rheumatoid arthritis, but they never become an option for repairing traumatic cartilage defects in young patients. Transplantation of composite cartilage–bone matrices harvested from carpometacarpal joints, rib cartilage or cartilage in non-weight-bearing regions has been used for the treatment of traumatic cartilage defects in the finger joints or osteochondromal diseases of elbow joints.[111–115] Several authors have reported short-term satisfactory results from osteochondral transplantation, but no long-term results have been reported.

There are several clinical reports of cell-based therapy for the treatment of cartilage defects in knee joints. Chondrocytes harvested from non-weight-bearing portions of joints were cultured, multiplied *in vitro* and implanted to sites of cartilage defects that were then covered by periosteal flaps.[116] This procedure was effective for focal cartilaginous defects but ineffective for massive defects of articular cartilage.

Cartilage repair using pluripotent stem cells is an attractive prospect. However, one of the biggest problems is the fabrication of scaffolds on which mesenchymal stem cells could differentiate into cartilaginous cells. It is known that MSCs subjected to compression forces can differentiate into cartilaginous cells and form a cartilage pellet. MSCs can also differentiate into chondrocytes in optimal scaffolds when they are administered in chondrogenic differentiation media in the presence of TGF-β.[117] There are several materials that can be suitable scaffolds for MSCs and that permit chondrogenic differentiation, such as hyaluronic acid, chitosan, type 2 collagen, chondroitin sulfate (CS) and alginate. A recent study demonstrated that CS was the best extracellular matrix for BMCs to differentiate into cartilaginous tissue. BMCs cultured in medium with both CS and type 2 collagen produced glycosaminoglycan, but BMCs cultured with type 2 collagen expressed more type 1 collagen mRNA than those with CS. Because this mRNA is a marker of chondrocyte hypertrophy associated with endochondral ossification, chondrocytes produced by BMCs cultured with CS would be a better candidate for cartilage formation than those with type 2 collagen. Moreover, the BMCs cultured with CS are more efficient in suppressing type 1 collagen mRNA expression than BMCs cultured with type 2 collagen, indicating that cartilaginous tissue produced by BMCs with CS would be more hyaline-like than that produced by BMCs with type 2 collagen.[117] Several authors have demonstrated that stem cells harvested from the articular synovium differentiate into chondrocytes and produce successful cartilage *in vivo*.[118,119]

Currently, tissue-engineered cartilage cannot be a fully functional substitute for articular cartilage. Future attention should focus on creating cartilaginous tissue with a zonal distribution in fabricating tissue-engineered articular cartilage.

Skin Tissue Engineering

The skin of the hand is often subjected to burns or trauma, and hand surgeons often encounter situations when they need to deal with such skin defects. Skin defects can be obliterated by split-thickness or full-thickness skin grafts, when the base of the skin defects is well vascularized. The skin grafts are usually harvested from the forehand, abdomen or buttocks. Skin defects in the exposed area should be covered by skin grafts harvested from the medial malleolus of the foot to get a good color match; however, the source is limited. Split- or full-thickness skin grafts often fail to survive or to adhere to bone, nerve or tendons. Vascularized cutaneous grafts or split-thickness skin grafts on vascularized fascia grafts can be options for treatment.

Artificial skin substitutes can be expected to bring plenty of benefits, including an unlimited supply and the elimination of donor site morbidities such as pain and scar formation. A method of clonal keratinocyte culture has been developed[120] and collagen hydrogel sheets on which autologous *in vitro* cultured and expanded keratinocytes had been seeded were applied to burns.[121] However, these demonstrated some drawbacks, such as the time-consuming procedure of keratinocyte expansion, limited survival, liability to infections, poor resistance to contraction, fragility of the substances, faulty anchorage of the epidermis to the collagen matrix, and lack of dermal support.[122] Subsequent attention to skin substitutes focused on the importance of the basement membrane and dermis, followed by the invention of dermal substitutes such as cadaveric skin, collagen-based dermal substitutes, hyaluronic acid-based membranes and synthetic polymer substitutes. Acellular dermal substitutes retain the dermal elements and basement membranes that act as a base for subsequent thin skin grafting or application of autologous keratinocyte cell culture to facilitate wound healing. Acellularized human cadaveric dermis (Alloderm: Life Cell Corp., Branchiburg, NJ, USA) has been used clinically and has demonstrated good results for the treatment of burns to the face, hand and foot.[123,124] Biological or synthetic dermal substitutes supporting epithelial cultures have overcome the drawbacks of autologous keratinocyte cell culture, thus enabling applications in the treatment of extensive loss or removal of skin. Application of cultured autologous keratinocytes constructed onto collagen–glycosaminoglycan composite substrates demonstrated an outcome similar to those treated with mesh or with split-thickness skin autografts, providing the benefits of minimizing the requirement for harvesting a donor skin.[125] Hyaluronic acid, a naturally occurring molecule present in all soft tissue, plays an essential role in maintenance of the normal extracellular matrix structure.[126,127] Recently, a three-dimensional artificial skin based on cocultured keratinocytes and melanocytes seeded on a surface of a biodegradable fabric scaffold made of a nonwoven mesh of hyaluronic acid containing *in vitro* resuspended fibroblasts has been used clinically and has demonstrated good clinical results.[128] However, bioengineered skin grafts lack the important adnexa and accessory epithelial structures, such as sweat glands and hairs, with negative repercussions on skin lubrication and heat exchange. Experimental *in vitro* models seem to suggest that the passive transport routes through skin from bioengineered epidermal cultures might be different from those through natural skin. In fact, while natural skin routes are limited to the extracellular space, epidermal cultures seem to follow both extracellular and intracellular routes.[129]

Skin equivalents could be improved further by the generation of a dermal–epidermal architecture including a microcapillary network in a three-dimensional biomaterial, by inserting autologous endothelial cells.

Small Blood Vessel Engineering

Hand surgeons often encounter situations where they need to repair blood vessels to retrieve peripheral blood circulation. Autogenous vein implantation is the gold standard for repairing defects of blood vessels in the hand. Acellular synthetic vascular vessels are used clinically in the reconstruction of large vessels, such as lower limb vessels, aorta, venae cavae and their direct branches. The inner walls of synthetic acellular vessels become coated by endothelial cells that migrate from the adjacent vessel stumps. The lack of toxicity, antigenicity and oncogenicity, absence of thrombogenesis and absence of degradation in the human body, minimum blood leakage from the vessel wall, and good handling are necessary to fabricate synthetic vessels. Several types of synthetic vessels are available clinically, although each has several advantageous and disadvantageous material properties. Grafts made of expanded polytetrafluoroethylene (e-PTEF) can be used as a substitute for an artery greater than 8 mm in diameter; however, in general, acellular synthetic vessels less than 6 mm in diameter are known to have more chance of occlusion by thrombosis formation.[130–132] One study on the reconstruction of lower limb vessels revealed that the patency rate of saphenous vein grafts was better than that of synthetic vessels.[133]

Tissue-engineered vessels have been used in limited clinical settings. The early generation of tissue-engineered cellular vessels had a simple structure. *In vitro* cultured endothelial cells harvested from peripheral veins were seeded onto a synthetic scaffold. Today, several modifications have been applied to this simple vessel substitute. However, the concept that the endothelium is a key structure in synthetic vessels remains unchanged, because it has antithrombotic and anticoagulant functions and also resists intimal hyperplasia.[134] Efforts to increase the mechanical strength of synthetic vessels have been made by creating the medial and adventitial layers using smooth muscle cells and fibroblasts[131] by using collagen-based scaffolds reinforced by polyester fibers[134,135] smooth muscle cells reoriented with bioreactors.[136,137] or synthetic biodegradable based scaffolds such as polyglycolic acid, polytetrafluoroethylene or polycaprolactone–polylactic acid copolymer.[138] Bioreactors are useful for applying stress to cells to force cell orientation and to produce an extracellular matrix, resulting in improvement of the material properties of the vessels.[130,139] Biochemical modifications using ascorbic acid, proline and ribose have also been applied to reinforce collagen fiber production and crosslinking.[139–142] Reduction of thrombogenesis and reinforcement of mechanical stress is still needed for the successful production of tissue-engineered vascular vessels.

The Future

Tissue-engineering techniques have just started to be applied to hand surgery based on animal models. Although they need continuous improvement of materials, techniques and concepts, these approaches are advancing steadily. We hope and believe that sural nerve, palmaris longus or plantaris tendons, radial or iliac bone graft and full- or split-thickness skin grafts will soon be replaced by tissue-engineered nerves, tendons, bone and skin.

References

1. Swearingen B, Ravindra K, Xu H, Wu S, Breidenbach WC, Ildstad ST. Science of composite tissue allotransplantation. *Transplant* **86** (2008) 627–635.
2. Steven HP, Hovius SE, Heeney JL. Immunological aspects and complications of composite tissue allografting for upper extremity reconstruction. A study in the rhesus monkey. *Trans Proc* **23** (1991) 623–625.

3. Langer R, Vacanti JP. Tissue engineering. *Science* **260** (1993) 920–926.
4. MacArthur, Ofero BD, ROC. Bridging the gap. *Nature* **433** (2005) 19.
5. Evans M, Kaufman M. Establishment in culture of pluripotential cells from mouse embryos. *Nature* **292** 154–156.
6. Nagy A, Rossant J, Nagy R, Abramow-Newerly W, Roder JC. Derivation of completely cell culture-derived mice from early-passage embryonic stem cells. *Proc Natl Acad Sci USA* **90** (1993) 385–396.
7. Thompson JA, Itskovitz-Eldor J, Shapiro SS, Waknitz MA, Swiergiel JJ, Marshall VS, Jones JM. Embryonic stem cell lines derived from human blastocysts. *Science* **282** (1998) 1145–1147.
8. Zeng X, Rao MS. Human embryonic stem cells: long term stability, absence of senescence and a potential cell source for neural replacement. *Neuroscience* **145** (2007) 1348–1358.
9. Takahashi K, Yamanaka S. Introduction of pluripotent sstem cells from mouse embryonic and adult fibroblast cultures by defined factors. *Cell* **126** (2006) 663–676.
10. Takahashi K, Tanabe K, Ohnuki M, Narita M, Ichisaka T, Tomoda K, Yamanaka S. Induction of pluripotent stem cells from adult human fibroblasts by defined factors. *Cell* **131** (2007) 861–872.
11. Nakagawa M, Koyanagi M, Tanabe K, Takahashi K, Ichisaka T, Aoi T, Okita K, Mochiduki Y, Takizawa N, Yamanaka S. Generation of induced pluripotent stem cells without Myc from mouse and human fibroblasts. *Nat Biotech* **26** (2008) 101–106.
12. Wering M, Meissner A, Cassady JP, Laenisch R. c-Myc is dispensable for direct reprogramming of mouse fibroblasts. *Cell Stem Cell* **2** (2008) 10–12.
13. Enzmann GU, Benton RL, Talbott JF, Cao Q, Whittemore SR. Tunctional considerations of stem cell transplantation therapy for spinal cord repair. *J Neurotrauma* **23** (2006) 479–495.
14. Gosain AK. What's new in plastic surgery, *J Am Coll Surg* **192** (2001) 356–365.
15. Weissman L. Translating stem and progenitor cell biology to the clinic barriers and opportunities, *Science.* **287** (2000) 1442–1446.
16. E.L. Herzog, L. Chai, D.S. Krause, Plasticity of marrow-derived stem cells, *Blood* **102** (2003) 3483–3493.
17. Mezey E, Key S, Vogelsang G, Szalayova I, Lange GD, Crain B. Transplanted bone marrow generates new neurons in human brains. *Proc Natl Acad Sci USA* **100** (2003) 1364–1369.
18. Weinmann JM, Charlton CA, Brazelton TR, Hackman RC, Blau HM. Contribution of transplanted bone marrow cells to Purkinje neurons in human adult brains. *Proc Natl Acad Sci USA* **100** (2003) 2088–2093.
19. Cogle CR, Yachinis AT, Laywell ED, Zander DS, Wingard JR, Steindler DA, Scott EW. Bone marrow transdifferentiation in brain after transplantation: a retrospective study. *Lancet* **363** (2004) 1432–1437.
20. MsNiece I. Subsets of mesenchymal stromal cells. *Cytotherapy* **9** (2007) 301–302.
21. Crisan M. A perivascular origin for mesenchymal stem cells in multiple human organs. *Cell Stem Cell* **3** (2008) 301–313.
22. Gronthos S, Grave SE, Ohta S, Simmons PJ. The STRO-1+ fraction of adult human bone marrow contains the osteogenic precursors. *Blood* **84** (1995) 4164–4173.
23. Aslan H, Zilberman Y, Kendel A, Liebergal M, Oskouian R, Gazit D, Gazit Z. Osteogenic differentiation of noncultured immunoisolated bone marrow-derived CD 105+ cells. *Stem Cells* **24** (2006) 17281–1737.
24. Rombouts WJ, Ploemacher RE. Primary murine MSC show high efficient homing to the bone marrow but lose homing ability following culture. *Leukemia* **17** (2003) 160–170.
25. Rubio D, Garcia-Castro J, Martin MC, de la Fuente R, Cigudosa JC, Lloyd AC, Bernad A. Spontaneous human adult stem cell transformation. *Cancer Res* **65** (2005) 3035–3039.
26. Serakinci N, Guldberg P, Burns JS, Abdallah B, Schrodder H, Jensen T, Kassem M. Adult human mesenchymal stem cell as a target for neoplastic transformation. *Oncogene* **23** (2004) 5095–5098.
27. Wang Y, Huso DL, Harrington J, Kellner J, Jeong DK, Turner J, McNiece IK. Outgrowth of a transformed cell population derived from normal human BM mesenchymal stem cell culture. *Cytotherapy* **7** (2005) 509–519.

28. Ryan JM, Barry FP, Murphy JM, Mahon BP. Mesenchymal stem cells avoid allogenic rejection. *J Inflamm* **2** (2005) 8.

29. Ryan JM, Barry FP, Murphy JM, Mahon BP. Interferon-gamma does not break, but promotes the immunosuppressive capacity of adult human mesenchymal stem cells. *Clin Exp Immunol* **149** (2007) 353–363.

30. Ito K. Development of a new device using the nonwoven fabrics for isolation of mesenchymal stem cells from bone marrow. International Society for Stem Cell Research annual meeting (2008) June 11–14 Philadelphia.

31. Gomes ME, Reis RL. Tissue engineering: key elements and some trends. *Macromol Biosci* **4** (2004) 737–742.

32. Sikavitsas VI, Bancroft GN, Mikos AG. Formation of three-dimensional cell/polymer constructs for bone tissue engineering in a spinner flask and a rotating wall vessel bioreactor. *J Biomed Mater Res* **62** (2002) 136–148.

33. Mooney DJ, Organ G, Vacanti JP, Langer R. Design and fabrication of biodegradable polymer devices to engineer tubular tissues. *Cell Transplant* **3** (1994) 203–210.

34. Glasby MA, Gschmeissner SE, Huang CL, De Souza BA. Degenerated muscle grafts used for peripheral nerve repair in primates. *J Hand Surg* **11B** (1986) 347–351.

35. Zhong H, Chen B, Lu S, Zhao M, Guo Y, Hou S. Nerve regeneration and functional recovery after a sciatic nerve gap is repaired by an acellular nerve allograft made through chemical extraction in canines. *J Reconstr Microsurg* **23** (2007) 479–87.

36. Sondell M, Lundborg G, Kanje M. Regeneration of the rat sciatic nerve into allografts made acellular through chemical extraction. *Brain Res* **795**(1–2) (1998) 44–54.

37. Neto HS, Sabha MJ Jr, Marques MJ. Cryopreserved muscle basal lamina grafts retain their grafting potential for nerve repair. *Cryobiology* **50** (2005) 112–115.

38. Rao P, Kotwal PP, Farooque M, Dinda AK. Muscle autografts in nerve gaps. Pattern of regeneration and myelination in various lengths of graft: an experimental study in guinea pigs. *J Orthop Sci* **6** (2001) 527–534.

39. Chang CJ, Hsu SH. The effect of high outflow permeability in asymmetric poly(dl-lactic acid-co-glycolic acid) conduits for peripheral nerve regeneration. *Biomater* **27** (2006) 1035–1042.

40. Patist CM, Mulder MB, Gautier SE, Maquet V, Jérôme R, Oudega M. Freeze-dried poly(D,L-lactic acid) macroporous guidance scaffolds impregnated with brain-derived neurotrophic factor in the transected adult rat thoracic spinal cord. *Biomater* **25** (2004) 1569–1582.

41. Ikeguchi R, Kakinoki R, Aoyama T, Shibata KR, Otsuka S, Fukiage K, Ishibe T, Shima Y, Ohtsuki B, Azuma T, Tsutsumi S, Nakayama T, Nakamura T, Toguchida J. Regeneration of osteonecrosis model of canine scapho-lunate by bone marrow stromal cells: possible therapeutic approach for Kienböck disease. *Cell Transplant* **15**(5) (2006) 411–422.

42. Siemionow M, Sonmez E. Nerve allograft transplantation: a review. *J Reconstr Microsurg* **23** (2007) 511–520.

43. Raivich G, Kreutzberg GW. Peripheral nerve regeneration role of growth factors and their receptors. *Int J Dev Neurosci* **11** (1993) 311–324.

44. Trigg DJ, O'Grady KM, Bhattacharyya T, Reinke M, Toriumi DM. Peripheral nerve regeneration: comparison of laminin and acidic fibroblast growth factor. *Am J Otolaryngol* **19** (1998) 29–32.

45. Itoh S, Matsuda A, Kobayashi H, Ichinose S, Shinomiya K, Tanaka J. Effects of a laminin peptide (YIGSR) immobilized on crab-tendon chitosan tubes on nerve regeneration. *J Biomed Mater Res B Appl Biomater* **73** (2005) 375–382.

46. Donzelli R, Maiuri F, Piscopo GA, de Notaris M, Colella A, Divitiis E. Role of extracellular matrix components in facial nerve regeneration: an experimental study. *Neurol Res* **28** (2006) 794–801.

47. Vogelezang M, Forster UB, Han J, Ginsberg MH, ffrench-Constant C. Neurite outgrowth on a fibronectin isoform expressed during peripheral nerve regeneration is mediated by the interaction of paxillin with alpha4beta1 integrins. *BMC Neurosci* **8** (2007) 44.

48. Neubauer D, Graham JB, Muir D. Chondroitinase treatment increases the effective length of acellular nerve grafts. *Exp Neurol* **207** (2007) 163–170.

49. Sun XH, Che YQ, Tong XJ, Zhang LX, Feng Y, Xu AH, Tong L, Jia H, Zhang X. Improving Nerve Regeneration of Acellular Nerve Allografts Seeded with SCs Bridging the Sciatic Nerve Defects of Rat. *Cell Mol Neurobiol* **29** (2009) 347–353.

50. Hedayatpour A, Sobhani A, Bayati V, Abdolvahhabi MA, Shokrgozar MA, Barbarestani M. A method for isolation and cultivation of adult Schwann cells for nerve conduit. *Arch Iran Med* **10**(4) (2007) 474–480.

51. Kim SM, Lee SK, Lee JH. Peripheral nerve regeneration using a three dimensionally cultured schwann cell conduit. *J Craniofac Surg* **18**(3) (2007) 475–488.

52. Dezawa M, Takahashi I, Esaki M, Takano M, Sawada H. Sciatic nerve regeneration in rats induced by transplantation of *in vitro* differentiated bone-marrow stromal cells. Eur. *J. Neurosci* **14** (2001) 1771–1776.

53. Tohill M, Mantovani C, Wiberg M, Terenghi G. Rat bone marrow mesenchymal stem cells express glial markers and stimulate nerve regeneration *Neurosci Lett* **362** (2004) 200–203.

54. Yamakawa T, Kakinoki R, Ikeguchi R, Nakayama K, Morimoto Y, Nakamura T. Nerve regeneration was promoted by bone-marrow-derived cells transplanted in a tube containing pedicled vascular vessels. *Cell Transplant* **16** (2007) 811–822.

55. Lundborg G, Dahlin LB, Danielson N, Gelberman H, Longo FM, Powell HC, Varon S. Nerve regeneration in silicone chambers; Influence of gap length and of distal stump component. *Exp Neurol* **76** (1982) 361–365.

56. Weber RA, Breidenbach WC, Brown RE, Jabaley ME, Mass DP. A randomized prospective study of polyglycolic acid conduits for digital nerve reconstruction in humans. *Plast Reconstr Surg* **106** (2000) 1036–1046.

57. Fujimoto E, Mizoguchi A, Hanada K, Yajima M, Ide C. Basic fibroblast growth factor promotes extension of regenerating axons of peripheral nerve. *In vivo* experiments using a Schwann cell basal lamina tube model. *J Neurocytol* **26** (1997) 511–528.

58. Toba T, Nakamura T, Lynn AK, Matsumoto K, Fukuda S, Yoshitani M, Hori Y, Shimizu Y. Evaluation of peripheral nerve regeneration across an 80-mm gap using a polyglycolic acid (PGA) — collagen nerve conduit filled with laminin-soaked collagen sponge in dogs. *Int J Artif Organs.* **25** (2002) 230–237.

59. Urist MR. Bone formation by autoinduction, *Science* **150** (1965) 893–899.

60. Urist MR, DeLange RJ, Finerman GA, Bone cell differentiation and growth factors, *Science* **220** (1983) 680–686.

61. Salgado AJ, Coutinho OP, Reis RL. Bone tissue engineering state of the art and future trends. *Macromol Biosci* **4** (2004) 743–765.

62. Bertone AL, Pittman, DD, Bouxsein ML, Li J, Clancy B, Seeherman HJ. Adenoviral-mediated transfer of human BMP-6 gene accelerates healing in a rabbit ulnar osteotomy model. *J Orthop Res* **22** (2004) 1261–1270.

63. Betz OB, Betz VM, Nazarian A, Pilapil CG, Vrahas MS, Bouxsein ML, Gerstenfeld LC, Einhorn TA, Evans CH. Direct percutaneous gene delivery to enhance healing of segmental bone defects. *J Bone Joint Surg* **88A** (2006) 355–365.

64. Egermann M, Baltzer AW, Adamaszek S, Evans C, Robbins P, Schneider E, Lill CA. Direct adenoviral transfer of bone morphogenetic protein-2 cDNA enhances fracture healing in osteoporotic sheep. *Hum Gene Ther* **17** (2006) 507–517.

65. Ito H, Koefoed M, Tiyapatanaputi P, Gromov K, Goater JJ, Carmouche J, Zhang X, Rubery PT, Rabinowitz J, Samulski RJ, Nakamura T, Soballe K, O'Keefe RJ, Boyce BF, Schwarz EM. Remodeling of cortical bone allografts mediated by adherent rAAVRANKL and VEGF gene therapy. *Nat Med* **11** (2005) 291–297.

66. Kang Q, Sun MH, Cheng H, Peng Y., Montag AG, Deyrup AT, Jiang W, Luu HH, Luo J, Szatkowski JP, Vanichakarn P, Park JY, Li Y, Haydon RC, He TC. Characterization of the distinct orthotopic boneforming activity of 14 BMPs using recombinant adenovirusmediated gene delivery. *Gene Ther* **11** (2004) 1312–1320.

67. Koefoed M, Ito H, Gromov K, Reynolds DG, Awad HA, Rubery PT, Ulrich-Vinther M, Soballe K, Guldberg RE, Lin AS, O'Keefe RJ, Zhang X, Schwarz EM. Biological effects of rAAV-caAlk2 coating on structural allograft healing. *Mol Ther* **12** (2005) 212–218.

68. Laurent JJ, Webb KM, Beres EJ, McGee K, Li J, van Rietbergen B, Helm GA. The use of bone morphogenetic protein-6 gene therapy for percutaneous spinal fusion in rabbits. *J Neurosurg Spine* **1** (2004) 90–94.

69. Li JZ, Hankins GR, Kao C, Li H, Kammauff J, Helm GA. Osteogenesis in rats induced by a novel recombinant helper-dependent bone morphogenetic protein-9 (BMP-9) adenovirus. *J Gene Med* **5** (2003) 748–756.

70. Li JZ, Li H, Hankins GR, Dunford B, Helm GA. Local immunomodulation with CD4 and CD8 antibodies but not cyclosporine A, improves osteogenesis induced by ADhBMP9 gene therapy. *Gene Ther* **12** (2005) 1235–1241.

71. Li JZ. LI H, Sasaki T, Holman D, Beres B, Dumont RJ, Pittman DD, Hankins GR, Helm GA. Osteogenic potential of five different recombinant human bone morphogenetic protein adenoviral vectors in the rat. *Gene Ther* **10** (2003) 1735–1743.

72. Schreiber RE, Blease K, Ambrosio A, Amburn E, Sosnowski B, Sampath TK. Bone induction by AdBMP-2/collagen implants. *J Bone Joint Surg* **87A** (2005) 1059–1068.

73. Zhu W, Rawlins BA, Boachie-Adjei O, Myers ER, Arimizu J, Choi E, Lieberman JR, Crystal RG, Hidaka C. Combined bone morphogenetic protein-2 and — 7 gene transfer enhances osteoblastic differentiation and spine fusion in a rodent model. *J Bone Miner Res* **19** (2004) 2021–2032.

74. Aslan H, Zilberman Y, Arbeli V, Sheyn D, Matan Y, Liebergal M, Li J, Helm GA, Gazit D, Gazit Z. Nucleofection-based *ex vivo* nonviral gene delivery to human stem cells as a platform for tissue regeneration. *Tissue Eng* **12** (2006) 877–889.

75. Endo M, Kuroda S, Kondo H, Maruoka Y, Ohya K, Kasugai S. Bone regeneration by modified geneactivated matrix: effectiveness in segmental tibial defects in rats. *Tissue Eng* **12** (2006) 489–497.

76. Fang J, Zhu YY, Smiley E, Bonadio J, Rouleau JP, Goldstein SA, McCauley LK, Davidson BL, Roessler BJ. Stimulation of new bone formation by direct transfer of osteogenic plasmid genes. *Proc Natl Acad Sci USA* **93** (1996) 5753–5758.

77. Hosseinkhani H, Yamamoto M, Inatsugu Y, Hiraoka Y, Inoue S, Shimokawa H, Tabata Y. Enhanced ectopic bone formation using a combination of plasmid DNA impregnation into 3-D scaffold and bioreactor perfusion culture. *Biomaterials* **27** (2006) 1387–1398.

78. Huang YC, Simmons C, Kaigler D, Rice KG, Mooney DJ. Bone regeneration in a rat cranial defect with delivery of PEI-condensed plasmid DNA encoding for bone morphogenetic protein-4 (BMP-4). *Gene Ther* **12** (2005) 418–426.

79. Kawai M, Bessho K, Kaihara S, Sonobe J, Oda K, Iizuka T, Maruyama H. Ectopic bone formation by human bone morphogenetic protein-2 gene transfer to skeletal muscle using transcutaneous electroporation. *Hum Gene Ther* **14** (2003) 1547–1556.

80. Kawai M, Bessho K, Maruyama H, Miyazaki J, Yamamoto T. Human BMP-2 gene transfer using transcutaneous *in vivo* electroporation induced both intramembranous and endochondral ossification. *Anat Rec A Discov Mol Cell Evol Biol* **287** (2005) 1264–1271.

81. Kishimoto KN, Watanabe Y, Nakamura H, Kokubun S. Ectopic bone formation by electroporatic transfer of bone morphogenetic protein-4 gene. *Bone* **31** (2002) 340–347.

82. Nakashima M, Iohara K, Ishikawa M, Ito M, Tomokiyo A, Tanaka T, Akamine, A. Stimulation of reparative dentin formation by *ex vivo* gene therapy using dental pulp stem cells electrotransfected with growth/differentiation factor 11 (Gdf11). *Hum Gene Ther* **15** (2004) 1045–1053.

83. Nakashima M, Tachibana K, Iohara K, Ito M, Ishikawa M, Akamine, A. Induction of reparative dentin formation by ultrasound-mediated gene delivery of growth/differentiation factor 11. *Hum Gene Ther* **14** (2003) 591–597.

84. Gazit D, Turgeman G, Kelley P, Wang E, Jalenak,M, Zilberman Y, Moutsatsos I. Engineered pluripotent mesenchymal cells integrate and differentiate in regenerating bone: a novel cell-mediated gene therapy. *J Gene Med* **1** (1999) 121–133.

85. Minamide A, Boden SD, Viggeswarapu M, Hair GA, Oliver C, Titus L. Mechanism of bone formation with gene transfer of the cDNA encoding for the intracellular protein LMP-1. *J Bone Joint Surg* **85-A** (2003) 1030–1039.

86. Zhang X, Yang M, Lin L, Chen P, Ma KT, Zhou CY, Ao YF. Runx2 overexpression enhances osteoblastic differentiation and mineralization in adipose — derived stem cells *in vitro* and *in vivo*. *Calcif Tissue Int* **79** (2006) 169–178.

87. Liu X, Ma PX. Polymeric scaffolds for bone tissue engineering. *Ann Biomed Eng* **32** (2004) 477–486.

88. Yang S, Leong KF, Du Z, Chua CK. The design of scaffolds for use in tissue engineering. Part I. Traditional factors. *Tissue Eng* **7** (2001) 679–689.

89. Kokubo T. Bioactive plass ceramics; properties and applications. *Biomater* **12** (1991) 155–163.

90. Hench LL. Bioactive materials; the potential for tissue regeneration. *J Biomed Mater Res* **41** (1998) 511–518.

91. Yuan H, de Bruijin JD, Zhang X, van Blitterswijk CA, de Groot K. Bone induction by porous glass ceramic made from Bioglass (45S5). *J Biomed Master Res (Appl Biomater)* **58** (2001) 270–276.

92. Fujibayashi S, neo M, Kim H-M, Kokubo T, Nakamura T. Osteoinduction of porous bioactive titanium metal. *Biomaterials* **25** (2004) 443–450.

93. Takemoto M, Fujibayashi S, Neo M, So K, Akiyama N, Matsushita T, Kokubo T, Nakamura T. A porous bioactive titanium implant for spinal interbody fusion: an experimental study using a canine model. *J Neurosurg Spine* **4** (2007) 435–443.

94. Hulbert SF, Morrison SJ, Klawitter JJ. Tissue reaction to three ceramics of porous and nonporous structures. *J Biomed Mater Res* **6** (1972) 347–374.

95. Flatley TJ, Lynch KL, Benson M. Tissue response to implants of calcium phosphate ceramic in the rabbit spine. *Clin Orthop* **179** (1983) 246–274.

96. Tamai N. Novel hydroxyapatite ceramics with an interconnective porous structure exhibit superior osteoconduction *in vivo*. *J Biomed Mater Res* **59** (2002) 110–117.

97. Liu Y, Skardal A, Shu XZ, Prestwich GD. Prevention of peritendinous adhesions using a hyaluronan-derived hydrogel film following partial-thickness flexor tendon injury. *J Orthop Res* **26**(4) (2008) 562–569.

98. Akasaka T, Nishida J, Imaeda T, Shimamura T, Amadio PC, An KN. Effect of hyaluronic acid on the excursion resistance of tendon graft: a biomechanical *in vitro* study in a modified human model. *Clin Biomech (Bristol, Avon)* **21**(8) (2006) 810–815.

99. Wang XT, Liu PY, Tang JB. Tendon healing *in vitro*: modification of tenocytes with exogenous vascular endothelial growth factor gene increases expression of transforming growth factor beta but minimally affects expression of collagen genes. *J Hand Surg* **30A** (2005) 222–229.

100. Viinikainen A, Göransson H, Huovinen K, Kellomäki M, Törmälä P, Rokkanen P. The strength of the 6-strand modified Kessler repair performed with triple-stranded or triple-stranded bound suture in a porcine extensor tendon model: an *ex vivo* study. *J Hand Surg* **32A** (2007) 510–517.

101. Dovan TT, Ditsios KT, Boyer MI. Eight-strand core suture technique for repair of intrasynovial flexor tendon lacerations. *Tech Hand Up Extrem Surg* **7** (2003) 70–74.

102. Xie RG, Xue HG, Gu JH, Tan J, Tang JB. Effects of locking area on strength of 2- and 4-strand locking tendon repairs. *J Hand Surg* **30A** (2005) 455–460.

103. Seiler III JR, Chu CR, Amiel D, Woo SL RH. Gelberman RH. Autogenous flexor tendon grafts. Biologic mechanisms for incorporation, *Clin Orthop* **345** (1997) 239–247.

104. Peacock Jr. EE, Biological principles in the healing of long tendons, *Surg Clin North Am* **45** (1965) 461–476.

105. Young RG, Butler DL, Weber W, Caplan AI, Gordon SL, Fink DJ. Use of mesenchymal stem cells in a collagen matrix for Achilles tendon repair, *J Orthop Res* **16** (1998) 406–413.

106. Tendon tissue engineering and gene transfer: the future of surgical treatment. *J Hand Surg* **31B** (2006) 693–704.

107. Cao Y, Liu Y, Liu W, Shan Q, Buonocore SD, Cui L. Bridging tendon defects using autologous tenocyte engineered tendon in a hen model, *Plast Reconstr Surg* **110** (2002) 1280–1289.

108. Awad HA, Butler DL, Boivin FN, Smith P, Malaviya B, Huibregtse B, Caplan AI. Autologous mesenchymal stem cell-mediated repair of tendon. *Tissue Eng* **5** (1999) 267–277.

109. Chong AK, Ang AD, Goh JC, Hui JH, Lim AY, Lee EH, Lim BH. Bone marrow-derived mesenchymal stem cells influence early tendon-healing in a rabbit achilles tendon model. *J Bone Joint Surg* **89A** (2007) 74–81.

110. Murphy KD, Mushkudiani IA, Kao D, Levesque AY, Hawkins HK, Gould LJ. Successful incorporation of tissue-engineered porcine small-intestinal submucosa as substitute flexor tendon graft is mediated by elevated TGF-b1 expression in the rabbit. *J Hand Surg* **33A** (2008) 1168–1178.

111. Hasegawa T, Yamano Y. Arthroplasty of the proximal interphalangeal joint using costal cartilage grafts. *J Hand Surg* **17B** (1992) 583–585.

112. Sato K, Sasaki T, Nakamura T, Toyama Y, Ikegami H. Clinical outcome and histologic findings of costal osteochondral grafts for cartilage defects in finger joints. *J Hand Surg* **33A** (2008) 511–515.

113. Katsaros J, Milner R, Marshall NJ. Perichondrial arthroplasty incorporating costal cartilage. *J Hand Surg* **20B** (1995) 137–142.

114. Wahegaonkar AL, Doi K, Hattori Y, Addosooki A. Technique of osteochondral autograft transplantation mosaicplasty for capitellar osteochondritis dissecans. *J Hand Surg* **32A** (2007) 1454–1461.

115. Ansah P, Vogt S, Ueblacker P, Martinek V, Woertler K, Imhoff AB. Osteochondral transplantation to treat osteochondral lesions in the elbow. *J Bone Joint Surg* **89A** (2007) 2188–2194.

116. Brittberg M, Peterson L, Sjogren-Jansson E, Tallheden T, Lindahl A. Articular cartilage engineering with autologous chondrocyte transplantation. A review of recent developments, *J Bone Joint Surg* **85A** (2003) 109–115.

117. Wu Y-N, Yang Z, Hui JHP, Ouyang H-W, Lee EH. Cartilagenous ECM component-modification of the micro-bead culture system for chondrogenic differentiation of mesenchymal stem cells. *Biomater* **28** (2007) 4056–4067.

118. Ando W, Tateishi K, Katakai D, Hart DA, Higuchi C, Nakata K, Hashimoto J, Fujie H, Shino K, Yoshikawa H, Nakamura N. *In vitro* generation of a scaffold-free tissue-engineered construct (TEC) derived from human synovial mesenchymal stem cells: biological and mechanical properties and further chondrogenic potential. *Tissue Eng* Part A **14** (2008) 2041–2049.

119. Sakao K, Takahashi KA, Arai Y, Inoue A, Tonomura H, Saito M, Yamamoto T, Kanamura N, Imanishi J, Mazda O, Kubo T. Induction of chondrogenic phenotype in synovium-derived progenitor cells by intermittent hydrostatic pressure. *Osteoarthr Carti* **16** (2008) 805–814.

120. Rheinwald JG, Green H. Serial cultivation of strains of human epidermal keratinocytes the formation of keratinizing colonies from single cells. *Cell* **6** (1975) 331–343.

121. Gallico 3rd GG, O'Connor NE, Compton CC, Kehinde O, Green H. Permanent coverage of large burn wounds with autologous cultured human epithelium. N Engl *J Med* **311** (1984) 448–451.

122. Bannasch H, Fohn M, Unterberg T, Bach AD, Weyand B, Stark GB. Skin tissue engineering *Clin Plast Surg* **30** (2003) 573–579.

123. Lattari V, Jones LM, Varcelotti JR, Latenser BA, Sherman HF, Barrette RR. The use of a permanent dermal allograft in full-thickness burns of the hand and foot a report of three cases. *J Burn Care Rehabil* **18** (1997) 147–155.

124. Achauer BM, VanderKam VM, Celikoz B, Jacobson DG. Augmentation of facial soft-tissue defects with Alloderm dermal graft. *Ann Plast Surg* **41** (1998) 503–507.

125. Boyce ST, Kagan RJ, Yakuboff KP, Meyer NA, Rieman MT, Greenhalgh DG, et al. Cultured skin substitutes reduce donor skin harvesting for closure of excised, full-thickness burns. *Ann Surg* **235** (2002) 269–279.

126. Campoccia D, Doherty P, Radice M, Brun P, Abatangelo G, Williams DF. Semisynthetic resorbable materials from hyaluronan esterification. *Biomater* **19** (1998) 2101–2127.

127. Zacchi V, Soranzo C, Cortivo R, Radice M, Brun P, Abatangelo G. *In vitro* engineering of human skin-like tissue. *J Biomed Mater Res* **40** (1998) 187–194.

128. Scuderi N, Onesti MG, Bistoni G, ceccarelli S, Rotolo S, Angeloni A, Marchese C. The clinical application of autologous bioengineered skin based on a hyaluronic acid scaffold. *Biomater* **29** (2008) 1620–1629.

129. Visualization of diffusion pathways across the stratum corneum of native and *in vitro* reconstructed epidermis by confocal laser scanning microscopy. *Arch Dematol Res* **287** (1995) 465–473.

130. Nerem RM, Seliktar D. Vascular tissue engineering. *Annu Rev Biomed Eng* **3** (2001) 225–243.

131. Kakisis JD, Liapis CD, Breuer C, Sumpio BE. Artificial blood vessel the Holy Grail of peripheral vascular surgery. *J Vasc Surg* **41** (2005) 349–354.

132. Heyligers JM, Arts CH, Verhagen HJ, Groot PG, Moll FL. Improving small-diameter vascular grafts from the application of an endothelial cell lining to the construction of a tissue-engineered blood vessel. *Ann Vasc Surg* **19** (2005) 448–456.

133. Klinkert P, Post PN, Breslau PJ, van Bockel JH. Saphenous vein versus PTFE for above-knee femoropopliteal bypass. A review of the literature. *Eur J Vasc Endovasc Surg* **27** (2004) 357–362.

134. Hirai J, Matsuda T. Venous reconstruction using hybrid vascular tissue composed of vascular cells and collagen tissue regeneration process. *Cell Transplant* **5** (1996) 93–105.

135. Hirai J, Kanda K, Oka T, Matsuda T. Highly oriented, tubular hybrid vascular tissue for a low pressure circulatory system. *ASAIO J* **40** (1994) M383–M388.

136. Kanda K, Matsuda T, Oka T. *In vitro* reconstruction of hybrid vascular tissue. Hierarchic and oriented cell layers. *ASAIO J* **39** (1993) M561–M565.

137. Tranquillo RT, Girton TS, Bromberek BA, Triebes TG, Mooradian DL. Magnetically orientated tissue-equivalent tubes application to a circumferentially orientated media-equivalent. *Biomater* **17** (1996) 349–357.

138. Shin'oka T, Imai Y, Ikada Y. Transplantation of a tissue-engineered pulmonary artery. *N Engl J Med* **344** (2001) 532–533.

139. Niklason LE, Gao J, Abbott, WM, Hirschi KK, Houser S, Marini R, *et al*. Functional arteries grown *in vitro*. *Science* **284** (1999) 489–493.

140. L'Heureux N, Paquet S, Labbe R, Germain L, Auger FA. A completely biological tissue-engineered human blood vessel. *FASEB J* **12** (1998) 47–56.

141. Girton TS, Oegema TR, Grassl ED, Isenberg BC, Tranquillo RT. Mechanisms of stiffening and strengthening in media-equivalents fabricated using glycation. *J Biomech Eng* **122** (2000) 216–223.

142. Girton TS, Oegema TR, R.T. Tranquillo RT. Exploiting glycation to stiffen and strengthen tissue equivalents for tissue engineering. *J Biomed Mater Res* **46** (1999) 87–92.

CHAPTER 4

Primary Cell Lines and Stem Cells

Katie L. Pricola* and Hermann Peter Lorenz*

Introduction

As new frontiers are discovered and explored in the field of tissue engineering and stem cell biology, the therapeutic potential of these advancements are only beginning to be realized. A number of issues must still be solved for cell-based tissue engineering strategies to become clinically useful. These problems include: selecting the optimal cell type, maximizing cell numbers while minimizing donor site morbidity, maintaining and manipulating cells *in vitro* without altering their *in vivo* characteristics, and reintroducing cells and cell-based constructs to the *in vivo* environment without loss of their *in vitro* capabilities. Current research efforts focus on three broad directions: 1) isolation of primary cells with regenerative potential, 2) development of biocompatible scaffolds, and 3) control of the environmental cues and stimuli that drive the generation of functional tissue.

In this chapter, we will explore the stem cell populations that provide the basis for cell-based tissue engineering strategies. Stem cells are broadly defined as cells that both self-renew and have multi-lineage differentiation potential.[1] Three categories of stem cells are defined: 1) totipotent stem cells, which can form both embryonic and extra-embryonic tissue such as the placenta; 2) pluripotent stem cells, which can form all three embryonic tissues and all tissues of the developing fetus, but cannot form the placenta; and 3) multipotent stem cells, which can develop into more than one tissue type but not all. The three major sources of stem cells for cell-based tissue engineering are pluripotent human embryonic stem cells (ESCs), induced pluripotent cells (iPS), and multipotent human adult tissue-derived stem cells. We will begin by discussing each of these sources separately, with particular attention to how they are defined and maintained, what their *in vivo* and *in vitro* capabilities are, and finally, how they can be utilized for tissue engineering strategies and other therapeutic applications.

Human Embryonic Stem Cells

Much of our present day insight regarding the existence of pluripotent stem cells derives from the 1954 study of murine teratocarcinoma-derived embryonal carcinoma (EC) cells.[3] It was found that EC cells are capable of unlimited self-renewal and multilineage differentiation,[2-4] and express surface antigens similar to those present on cells of the inner cell mass (ICM).[5,6]

In the earliest stages of development, fertilization results in the formation of a zygote, which undergoes 7–8 cell divisions to then become a bilayered blastocyst. The outer trophoblast layer of the blastocyst forms the placenta, while all tissues and organs of the developing fetus are derived from the ICM. Embryonic stem cells (ESCs) were first

---

* Corresponding authors.

derived from the ICM of pre-implantation mouse blastocysts in 1981.[7,8] However, successful isolation and maintenance of human ESCs was not achieved until almost two decades later, when Thomson *et al.* optimized the necessary *in vitro* culture conditions.[9] Human ESCs are defined as pluripotent cells that are isolated from the ICM of the blastocyst, proliferate indefinitely without karyotypic change,[10] and differentiate along all three germ layer lineages (as evidenced by teratoma formation following injection of ESCs into immunosuppressed mice).[11]

Maintenance of ESCs

In vitro derivation of ESCs starts with cells from the ICM, which form embryoid bodies. — i.e. clusters of clonogenic cells from which the ESCs develop. The first reports on the successful maintenance of human ESCs described optimal *in vitro* culture conditions requiring both a mitotically-inactive fibroblast feeder layer and serum-containing media.[9] Since then murine (MEFs),[12] human,[13] and synthetic scaffolds such as Matrigel[14] have been described. A feeder layer is a basal layer of cells (fibroblasts typically), which provides secreted factors, extracellular matrix, and cellular contacts which maintain stem cells in their undifferentiated state. Animal serum contributes similarly by recreating the *in vivo* milieu. A better understanding of the signaling pathways and stimuli that maintain human ESCs would allow us to limit the use of serum and feeder layers, and instead directly stimulate ESCs with discrete, purified chemicals and growth factors. This would eliminate cross-species contamination that occurs when animal serum[12] and feeder cell layers[15–18] are used to maintain ESCs in culture.

Multiple growth factors have been reported to play an integral role in human ESC self-renewal, which include members of the FGF and TGFβ/Activin[19] signaling pathways. These inhibit BMP signaling and up-regulate TGFβ expression.[18–23] Additional growth factors that have been studied and found to contribute to ESC self-renewal include Wnts, IGF1, pleiotrophin, sphingosine-1-phosphate and PDGF.[24–28] Interestingly, some growth factors that support renewal in mouse ESCs have the opposite effect in human ESCs, underscoring the complexity of ESCs. For example, the LIF pathway does not have a significant role in the maintenance of human ESCs but does for mouse ESCs. BMPs, which support self-renewal in mouse ESCs, actually initiate differentiation of human ESCs.[29]

The clinical use of ESCs remains hindered by their predilection for teratoma formation *in vivo*, risk of malignant transformation, and issues involving host immune recognition.[30] However, the major obstacle preventing the therapeutic use of ESCs are the political and ethical controversies surrounding them.[31,32] As a result, much research investment- both intellectual and monetary- has instead been funneled toward reprogramming adult cells to become more embryonic-like, i.e. induced pluripotent stem (iPS) cells, as well as identifying and optimizing *adult* tissue-derived stem cell lines.

Induced Pluripotent Stem Cells (iPS Cells)

Reprogramming adult cells to become more embryonic-like began once it was discovered that somatic cell nuclear transfer (SCNT)[33,34] and somatic/ESC fusions[35–38] could induce a differentiated adult cell to become embryonic-like. What is now commonly called "reprogramming" actually refers to a technique that, instead of transferring the ESC nuclei entirely, transfers only the specific transcription factors (TFs) that preferentially regulate embryonic cells. This technique has garnered more recent attention than SCNT and ESC fusion due to its reproducibility and comparatively-decreased cost. Working independently,

Yamanaka and Thomson performed the initial screens to identify TFs capable of reprogramming differentiated somatic cells into pluripotent stem cells. In 2006, Yamanaka's group showed that the addition of select transcription factors-*Oct4, Klf4, c-Myc*, and *Sox-2-* to mouse fibroblasts was sufficient to create embryonic-like cells.[39] In December of 2007, both Yamanaka's and Thomson's groups demonstrated that virally-induced expression of these select transcription factors allowed adult human fibroblasts and other somatic cells to behave like human embryonic stem cells.[39–41] Other groups have since repeated this work with similar success.[42–46] This exciting advancement in the field of stem cell biology has the potential to accelerate progress toward cell-based tissue engineering, with particular applications in the field of plastic and reconstructive surgery that include new treatments for burn injury, improvements in soft tissue coverage and reconstruction, and novel regenerative therapies in hand surgery.

Before translation of iPS cells into the clinical sector can be realized, safety issues must be addressed. The first major safety concern is the tumorigenic potential of iPS cells. Tumors were observed to develop in 20% of germline chimeric mice generated by injecting mouse iPS cells into C57BL/6-derived blastocysts.[47] This phenomenon is most likely due to the reactivation of the *c-Myc* oncogene. The second major safety concern is the use of retroviruses as a vector for inducing gene expression. Retroviral integration of TFs may activate or inactivate host genes or cause insertional mutagenesis, which may then result in tumorigenicity. Previous failures with retroviral treatment strategies, such as the development of leukemia in three children, who received retroviral gene transfer further defines the risks associated with viral-mediated therapies.[48–50]

Current research in the field of reprogramming focuses on ways to refine this technology to address these safety and efficacy concerns. One approach is to optimize reprogramming in the absence of oncogenic TFs in order to decrease the tumorigenicity of the resulting iPS cells.[47] The second major effort is to move away from viral to nonviral methods of reprogramming, such as high efficiency delivery methods like electroporation of DNA into the nucleus. Recently this approach successfully derived mouse and human iPS cells in the absence of the *c-Myc* retrovirus.[51] The concept of reprogramming is also being applied to *adult* stem cell lines.[52] Currently, it is not possible to harvest sufficient numbers of adult stem cells from primary sources and simultaneously limit donor site morbidity without relying on *in vitro* expansion. This is particularly challenging given that adult stem cells eventually reach senescence during *in vitro* passaging. Therefore, researchers are attempting to "reprogram" adult stem cells, with the aim of maintaining their undifferentiated state at the time of harvest, through culture expansion without senescence. This would achieve better yields of high quality, multipotent cells.

Adult Tissue-Derived Stem Cells

Both ESCs and iPS cells portend an exciting future where complex tissue and organ regeneration may be an everyday reality. However, adult stem cells are the only current cells with immediate clinical application due to the aforementioned safety, ethical and political controversies of both ESCs and iPS cells. Adult stem cells function in normal cell turnover and possess differentiation potentials that reflect their *in vivo* microenvironment; they have been identified in most adult tissues including the bone marrow, heart, central nervous system, liver, lungs, kidney, adipose, skin, and muscle.[53–61] Once activated, these resident stem cells can stimulate tissue regeneration and repair *in vivo*, possibly by proliferating, releasing soluble cytokines or growth factors to stimulate other nearby cells, and/or fusing with differentiated adult cells.[62,63] The best studied adult stem cells are two types of bone mar-row stem cells: hematopoietic stem cells (HSCs) and mesenchymal stem cells (MSCs). In this chapter, we will focus on MSCs, which are found mainly in the bone marrow and can migrate to distant tissues for repair.[64] MSCs are also resident in musculoskeletal tissues and are particularly relevant to cell-based tissue engineering of the hand.

Mesenchymal stem cells (MSCs)

MSCs were first postulated to exist more than thirty years ago when Friedenstein estab-lished a population of non-hematopoietic, adherent, fibroblastic bone marrow (BM)-derived cells that gave rise to bone and cartilage-like colonies.[65] Although initially called "colony-forming unit-fibroblasts" (CFU-F), these cells have come to be referred to interchangeably as bone marrow stromal cells, mesenchymal progenitor cells, and mes-enchymal stem cells. Due to their ease of isolation and expansion, multilineage differentiation capabilities, and unique immunomodulatory effects, MSCs are a promising cell source for tissue engineering and *in vivo* stimulation of other tissue resident stem cells.

Defining MSCs: Isolation, identification, and selection

Classically, MSCs are isolated from BM[66] but can also be harvested from muscle,[67,68] umbilical cord,[69] trabecular bone,[70] periosteum,[71] synovial membrane,[72,73] periodontal lig-ament,[74] dental pulp,[75] palatine tonsil,[76] the perivascular regions of most tissues,[77,78] and adipose tissue.[79,80] Identification and selection of MSCs is complex. They must be plastic-adherent, form colonies by clonogenic assay *in vitro*, differentiate into osteo-, adipo-, and chondrogenic cells, and positively and negatively express certain cell markers. During harvesting procedures, MSCs are released from the tissue source by a process of homog-enization and/or centrifugation, then grown in tissue culture conditions that foster proliferation, and isolated based on clonogenicity and plastic adherence. The relatively non-specific isolation and culture methods are thought to reflect the heterogeneity of MSC populations.

Controversy exists over whether or not MSCs are a homogenous stem cell population because a reliable phenotype has not been reproducibly demonstrated. MSCs variably express CD105, CD73, CD90, Stro-1, and CD44.[66,79,81–94] MSCs consistently lack expres-sion of the hematopoietic stem cell markers CD34 and CD45 as well as the immune cell markers CD14, CD11b, and HLA-DR surface molecules.[66,79,81–95] This is notable when compared to the more well-characterized HSC populations, which consistently express CD34$^+$, Thy1$^+$, CD38$^{lo/-}$, C-kit$^{-/lo}$, CD105$^+$, Lin$^-$.[96,97]

The absence of definitive surface markers has made identifying the MSC *in vivo* and purifying MSC populations *in vitro* a controversial and important area of research. In 2006, the International Society for Cytotherapy suggested the following guidelines to stan-dardize identification of MSCs- (i) plastic adherent under standard culture conditions, (ii) CD105$^+$/CD73$^+$/CD90$^+$ and CD45$^-$/CD34$^-$/CD14$^-$/CD11b$^-$/CD79$^-$/CD19$^-$, and lacking HLA-DR surface molecules, and (iii) differentiate along osteo-, chondro-, and adipogenic lineages.[98] However, controversy still exists because this identification system does not preclude the possibility of a heterogeneous population of cells. For this reason, clonogenic assays are critical for strict identification of MSCs.

An additional controversy surrounding MSCs is whether or not MSCs isolated from different sources are in fact the same cells. Many studies have shown that the cell integrity and differentiation capabilities are comparable across the varying sources and isolation techniques,[99] while other studies have shown variability.[100] Much of the characterization of these cells and their niche, as well as our understanding of the biological properties of MSCs, however, draw from study of the bone marrow-derived MSC population.

MSC Niche: The Bone Marrow (BM) microenvironment

The BM microenvironment is a hypoxic, heterogeneous milieu of HSCs, MSCs, adipocytes, osteoblasts, endothelial cells, smooth muscle cells macrophages, and fibroblasts.[70,101] As a result, MSC maintenance depends on a complex combination of low oxygen tension, ECM and cell-cell interactions, as well as soluble/cell-secreted factors. Hypoxia increases MSC proliferation and plasticity *in vitro* through a proposed HIF-2α induced upregulation of *Oct-4* and *Rex-1*.[102,103] Interactions with neighboring cells also play an important role; MSCs are integral to HSC maintenance,[104–106] and presumably MSCs are similarly influenced by HSCs. The mechanism by which these cell-cell interactions affect MSC maintenance is unknown although the role of cadherins is being explored. Cadherins are Wnt-interacting proteins known to regulate cell-cell adhesion, migration, differentiation, and polarity of MSCs,[107] and they play a role in the maintenance of other stem cell niches.[108] The specific ECM components and soluble, secreted factors[109,110] that maintain naïve MSCs still have not been completely identified, although we will discuss this further with regard to MSC self-renewal. Our modest understanding of the MSC niche, and how it is maintained, is matched by a similarly modest understanding of what triggers MSC homing to distant sites of injury.[111] MSC migration *in vitro* is regulated by *Stro-1* expression,[112] stromal derived factor 1/CXCR4 signaling, hepatocyte growth factor/*c-Met* signaling, and matrix metalloproteinases.[113] Continued study of the *in vivo* MSC niche will further elucidate the mechanisms by which MSC self-renewal and multipotency are maintained as well as that which triggers their regenerative response to injury.

In vivo and in vitro capabilities

MSCs are a product of their microenvironment and a great cellular example of the balance between nature and nurture. MSCs retain their self-renewal and multipotency when cultured in basal conditions, differentiate in the presence of growth factors, trans-differentiate across cell lineages when culture conditions are appropriately altered, and de-differentiate back to a stem-like state when stimuli are removed.[114] MSCs also immunoregulate their microenvironment. In order to harness the potential clinical utility, safety, and efficacy of MSCs as a primary cell source for tissue engineering, a better understanding of the biological basis of their capabilities is required.

MSC self-renewal

Multiple signaling pathways contribute to MSC self-renewal. We possess a limited understanding of the complex interplay of intrinsic and extrinsic signals that induce or inhibit differentiation and promote MSC self-renewal. In our earlier discussion of the MSC niche, we discussed the importance of the ECM and cell-cell interactions, but did not go into detail regarding the role of soluble and secreted factors. This is in fact a very important area of research as it may enable us to develop methods of enhancing tissue-resident MSCs to repair injury *in vivo*. Simple stimulation with the appropriate growth factors or direct activation of downstream regulators of key involved pathways may be all that is needed to stimulate MSC self-renewal. This would be an exciting adjuvant to surgery by speeding tissue repair and limiting scar formation, which would have obvious benefits for tendon repair. Both hypothesis-driven and hypothesis-generating research has thus far shown a role for a number of candidate genes in maintaining MSC self-renewal. The long-term goal of these candidate gene studies is to identify both 1) soluble factors that, when added to *in vitro* culture, maintain later passage MSC proliferation and self-renewal, and 2) endogenously expressed genes that, when over expressed in MSCs by gene transduction, enhance the intrinsic MSC self-renewal mechanism.

Hypothesis-driven candidate gene studies have shown a functional role for extrinsic growth factors including the Wnt, Leukemia Inhibitory Factor (LIF), Epidermal Growth Factor (EGF),[115] Platelet Derived Growth Factor (PDGF),[116] Fibroblast Growth Factor2 (FGF2),[95,117–121] and Transforming Growth Factor (TGF)-beta cytokine families.[122]

Wnts are a highly conserved family of proteins that play a key role in embryogenesis and normal limb development by maintaining cells in the apical epidermal ridge (AER) in a hyperproliferative, undifferentiated state.[123–127] Wnts also expand gastrointestinal,[128] skin,[129] and hematopoietic stem cell populations.[130,131] MSCs express mediators of the Wnt signaling pathway including Wnt2, Wnt4, Wnt 5a, Wnt11, Wnt16, Fz2, Fz3, Fz4, Fz5, Fz6, and Dkk1.[89] For example, Wnt3 enhances MSC proliferation and reversibly suppresses osteogenic differentiation *in vitro* through canonical Wnt signaling.[132] Although Wnts play a role in MSC self-renewal, the exact mechanism by which this occurs remains to be clarified, particularly given the multiple functions engaged by Wnt signaling.[89,133]

Leukemia Inhibitory Factor is a cytokine and member of the Interleukin-6 (IL-6) family of cytokines, which includes IL-6, IL-11, Oncostatin-M, ciliary neurotrophic factor, and cardiotrophin-1. LIF typically acts through the gp130 pathway in concert with the Erk1/2 /MAPK pathway.[134] LIF stimulation of murine MSCs upregulates MSC cytokine expression and increases MSC proliferation potential.[135] LIF also plays both an inhibitory and stimulatory role in osteoblast and osteoclast function.[136] Bipotency of LIF is dependent on the interplay of other signaling pathways and the activation status of the downstream regulators of the gp130 pathway such as shp2. The relationship between gp130 and Erk/MAPK activation is particularly interesting in MSCs. *In vitro* inhibition of the Erk/MAPK pathway via Erk1/2 inhibition blocks the mitogenic effects of PDGF and EGF.[137] Interestingly, the related cytokine IL-6 has also been shown to enhance MSC self-renewal, inhibit adipogenic differentiation, and decrease chondrogenesis in BM-derived MSCs. IL-6 does so by activating Erk/MAPK in the absence of gp130 activation. We have found that inhibition of Erk1/2 in the presence of IL-6 stimulation abrogates the mitogenic effect of IL-6 (Pricola, *et al.* unpublished).

Hypothesis-generating gene arrays have further expanded the search for candidate "stemness" genes, which may play an intrinsic role in MSC regulation. In 2006, Song *et al.* identified candidate MSC "stemness" genes and "differentiation" genes using gene array technology.[138] This important work has led to the ongoing characterization of many more genes that may play a role in both the positive and negative regulation of MSCs. Further investigation into the regulation of MSC self-renewal will direct future clinical applications.

Immune function

Although self-renewal is a classic stem cell attribute, MSCs are also very unique in that they possess a clinically useful immunomodulatory capability. MSCs have demonstrated immune-evasive and immunosuppressive effects both *in vitro* and *in vivo*. These unique immunologic properties make MSCs ideal for both autologous and allogeneic cell-based tissue engineering. MSCs evade detection by the immune system due to their limited alloantigen expression; they have intermediate levels of MHC class I antigen, but lack MHC class II antigen.[139] MSCs also lack the co-stimulatory molecules CD40, CD80 and CD86, which prevents activation of alloreactive T cells against MSCs.[140] MSCs also inhibit T-cell proliferation and promote T cell anergy. Mixed lymphocyte reactions show a dose-dependent suppression of memory T cells, CD4+, and CD8+ T cells as well as suppression of CD8+ T cell-mediated lysis.[139–145] MSCs inhibit B cell proliferation and antibody production,[146] inhibit dendritic cell proliferation and the maturation and function of all antigen presenting cells.[143,147–149] MSCs suppress NK cell proliferation and pro-inflammatory cytokine secretion, and promote induction of suppressor T regulatory cells.[143,150,151]

The mechanism by which MSCs exert their inflammatory down-regulation is not entirely understood, however this may occur through MSC-mediated soluble factor secretion and cell-cell interaction. Both result in a more immune tolerant environment with decreased production of pro-inflammatory cytokines like TNF-α and IFN-γ and increased production of soluble immunomodulatory factors, including IL-4, IL-6, IL-10, transforming growth factor-β, prostaglandin E2, hepatocyte growth factor, nitric oxide, and indoleamine 2,3-dioxygenase.[141–143] While investigators continue to study the mechanisms of MSC immunoregulation, already a number of clinical applications that hinge on MSC immunomodulation have been done. They are further discussed in the section: "Immune Regulation, Differentiation Potential and Gene Therapy."

Multilineage differentiation potential

By definition, multipotent MSCs are capable of osteogenic, adipogenic, and chondrogenic differentiation. Animal models using MSCs have also shown that these cells can differentiate and/or induce nearby cells to differentiate, and repair other mesodermal tissues including tendon and non-mesodermal tissues including neurons and skin, to list a few.[152] Here we describe MSC multipotency as it pertains to tissue engineering of the hand-bone, cartilage, fat, tendon, and muscle.

Osteogenesis

MSCs generate osteoprecursor cells, which then form osteoprogenitor cells, pre-osteoblasts, osteoblasts, and osteocytes. This progression relies on the temporal activation and inactivation of transcription factors including Cbfa1/Runx2, Msx2, Dlx5, Osx, as well as expression of bone markers including osteopontin, collagen type I, alkaline phosphatase(ALP), bone sialoprotein, and osteocalcin.[153,154] BMPs (BMP-2 and BMP-6) also play an important role in MSC osteogenesis.[155,156] BMP2 activation leads to the acetylation and enhanced activation of Runx2, a major transcriptional activator of osteogenesis.[157] Deacetylated Runx2 is degraded by Smurf1, Smurf2, and E3 ubiquitin ligase, similar to the mechanism by which TNFα down regulates Runx2 expression.[157,158] Wnt family proteins also have a modulatory role in MSC osteogenesis. Canonical Wnt signaling through Wnt3a inhibits early osteogenesis and promotes proliferation,[159] while non-canonical signaling through Wnt5 increases *in vitro* bone formation. Gain of function mutation of LRP5, the Wnt co-receptor, results in increased bone formation; loss of function mutations lead to decreased bone formation.[132] The role of the Wnt pathway in osteogenesis is complicated by dose-dependent Wnt –BetaCatenin signaling, which has been elucidated through the study of transgenic mice for Wnt signaling. Low levels of Wnt inhibit, while high levels of Wnt promote, osteogenesis.[160]

This information drives both *in vitro* differentiation and identification of MSC-derived osteoblasts. *In vitro*, osteogenesis is induced by extended culture in high glucose media supplemented with dexamethasone, ß-glycerolphosphate, ascorbate phosphate, and 1,25 dihydroxyvitamin D_3. Osteoblastic differentiation is assessed by immunohistochemical stain for alkaline phosphatase (ALP) and alizarin red, which stains calcified matrix. Quantification of osteogenesis is also assessed at the mRNA level by measuring levels of Runx2, ALP, osteocalcin, and a number of other surrogate markers for osteogenesis.[114] The *in vivo* applications of MSC-derived osteoblasts have thus far revolved around gene therapy, which we detail further in the section: "Immune Regulation, Differentiation Potential and Gene Therapy."

Chondrogenesis

MSCs are also capable of chondrogenesis and *in vitro* chondrocyte differentiation. Based on our knowledge of limb bud development and embryologic chondrogenesis, we know that cartilage formation requires a suitable three-dimensional environment and various mechanical and biochemical stimuli.[161] MSC-derived chondrocytes have been well characterized and are known to express transcription factors including Sox-9 and scleraxis as well as extracellular matrix (ECM) genes including collagen types II and IX, aggrecan, biglycan, decorin, and cartilage oligometric matrix protein.[162] Some of the signaling pathways and growth factors that regulate MSC chondrogenesis are fibroblast growth factors (FGFs), members of the transforming growth factor beta (TGF-β) superfamily, BMPs, GDFs, and Wnts.[163] FGF2 has been shown *in vitro* to enhance the proliferation and multipotent differentiation potential of human MSCs and improve the proliferative potential of chondrocytes;[164] aberrant FGF2 signaling leads to developmental disorders due to dysregulated chondrogenesis such as craniosynostoses.[165,166] MSCs over-expressing TGF-β1, TGF-β3,[166] BMP2, BMP-4, BMP-6,[167,168] BMP-12,[169] BMP-13,[170] and GDF-5[155] have enhanced chondrogenic potential. The mechanism by which these molecules promote chondrocyte formation in MSCs is relatively unknown; however there is ongoing research into TGF β and BMP signaling through Smad proteins and the MAPK pathway in MSCs.[166,171–173] Wnt signaling plays a bipotential role in chondrogenesis, similar to Wnt modulation of osteogenesis. Pulsatile Wnt7a expression induces chondrogenesis through a TGF-β 1-MAPK dependent mechanism,[107] while Wnt1 inhibits chondrogenesis through upregulation of Twist1.[156] The contradictory effects of cytokines within the same family exemplify the complexity of chondrogenesis regulation.

In vitro, MSCs undergo chondrogenic differentiation in high-density pellet culture (2.5×10^5 cells/pellet) in serum-free high glucose medium supplemented with dexamethasone, ascorbate phosphate, L-proline, sodium pyruvate, insulin, and TGF-ß3. These pellets can also be grown in bioreactors. The undifferentiated cells can also be seeded onto 3D-constructs such as nanofibrous scaffolds prior to chondrogenic induction. Chondrogenic differentiation can be assessed by staining for sulfated matrix proteoglycan with alcian blue, and by immunohistochemical staining for collagen type II. Chondrogenesis by MSCs can be further assessed at the mRNA level by measuring levels of cartilage oligomeric matrix protein (COMP), metalloproteinase 13 (MMP13), aggrecan, and a number of other markers of chondrogenesis.[114] Both autologous and allogeneic chondrocytes that were differentiated from MSCs have had success in models of regeneration *in vivo*. A recent study showed that nanofibrous scaffolds seeded with human MSCs fully repair 7 mm full-thickness porcine articular cartilage defects.[174] A recent study in a rabbit model similarly showed that not only were allogeneic human MSC-based repairs not rejected, but they actually surpassed autologous chondrocyte-based repair.[175]

Adipogenesis

At the molecular level, adipogenic differentiation of MSCs is positively regulated by the nuclear hormone receptor peroxisome proliferator-activated receptor gamma (PPARγ).[176,177] Mechano-induction of differentiation also plays a significant role in MSC adipogenesis; non-stretch conditions promote adipogenesis by inducing round cell morphology and activating specific tension-induced/inhibited proteins (TIPS).[178] *In vitro* adipogenesis of MSCs is induced by growth in high density monolayer culture, which promotes cell-cell interactions, and by stimulation with glucocorticoids and insulin.[179] Adipogenesis can be assessed by the presence of neutral lipids in the cytoplasm stained with Oil Red O. Adipogenesis of MSCs can be further assessed at the mRNA level by measuring levels of lipoprotein lipase (LPL), fatty acid binding protein 4 (FABP4), PPARγ, and various other markers of adipogenesis.[114]

Tenogenesis

MSC tenogenesis is affected by culture conditions and stimulation with growth factors, as well as physical stimuli, including mechanical loading. Little is known about the signaling pathways involved in tenogenesis, even less so than for osteo- and chondrogenesis. Members of the TGF-β superfamily, specifically growth and differentiation factors (GDFs), as well as BMPs have been implicated in MSC tenogenesis. GDFs induce ectopic tendon formation *in vivo*,[180] and treatment with GDFs improves tendon repair *in vivo*.[181,182] Blockade of BMP-related members of the TGF-β family, including BMP4 and activin, suppress tendon formation *in vivo*.[183] Smad8, which modulates BMP signaling,[184] promotes MSC differentiation into tenocytes.[185] Mechanical stimulation is also requisite for *in vitro* tendon formation.[186,187] Tensile strength and stretch loading are vital for biocompatible tendon formation and alignment.[188] *In vivo* and *in vitro* study of environmental cues and stimuli that regulate tenocyte behavior has increased our understanding of the biological basis of tendon repair and adhesion formation, including new evidence of differential lactate and TGF-β receptor expression in tendon sheath fibroblasts, epitenon tenocytes, and endotenon tenocytes.[189,190]

The clinical efficacy of MSCs for *in vivo* tendon repair has been investigated by several animal studies and clinical trials in humans.[191,192,193] Local injection of MSCs improves tendon repair, although there have been reports of ectopic bone formation.[191,192,193] Successive studies have shown that decreasing the number of cells transferred in MSC-collagen constructs minimizes ectopic bone formation.[194] Implantation of MSC constructs (collagen,[195] PLGA scaffolds,[196] and gel-sponge composites[187]) into animal models of tendon defects also improves tendon repair as measured by biomechanical properties and tissue architecture of the repaired tendons.[195,197] In 2007, Kryger *et al* studied a rabbit model of flexor tendon repair, which used decellularized allografted tendon as the biocompatible scaffold. They analyzed the repair capacity of four different primary cell lines-epitenon tenocytes, tendon sheath fibroblasts, bone marrow derived MSCS and adipo-derived mesenchymal stem cells (ASCs). They showed, for the first time, that not only are MSCs capable of tenogenesis and repair, but that they are as effective as tenocytes for repair and engraftment into tendon defects.[198]

The use of MSCs for tendon repair is particularly exciting in the setting of hand injury where tendon repair is challenging for a number of reasons, particularly if it involves Zone II. Modest defects may be repaired primarily with reanastamosis while larger defects require tendon graft or transfer. Complete return of function is restricted by adhesion formation and potential for ruptured repair. The use of grafts, which has been studied since 1912,[199-207] further complicates repair. Autografts are limited by donor site morbidity, allografts pose a risk of rejection and infection,[208] and synthetic grafts introduce foreign body risks such as inflammation and incomplete biocompatibility. Further, the study of intrasynovial vs. extrasynovial tendon grafts, distinguished by the lack of epitenon layer in extrasynovial tendons, has shown that intrasynovial grafts have a greater capacity for intrinsic healing while extrasynovial tendon grafts, such as Palmaris longus, and plantaris, act as more of a scaffold for repair.[209] Clinically this knowledge has not been translated because there is a lack of donor intrasynovial tendons. Moreover, recapitulation of this tendon type *in vitro* through tissue engineering methods remains unsolved.[210] Cell-based tissue engineering can help to overcome some of the aforementioned limitations, however it is constrained by our limited knowledge of embryologic tenogenesis. Thus, further study into the molecular pathways integral to tendon formation is necessary.

Myogenesis

Adult muscle stem cells with myogenic potential include satellite cells,[211] muscle derived stem cells from the interstitial space,[212] muscle side population cells,[213] pericyte-like mesangioblasts,[214] and bone marrow derived MSCs.[215,216,217] Bone marrow-derived muscle precursor cells capable of homing to and repair of muscle were first described by Ferrari *et al* in 1998.[217] BM, adipo, and synovium-derived MSCs have successfully regenerated muscle progenitor cells in dystrophic mouse models.[218,219] Human MSCs differentiate into skeletal muscle cell lineages following Notch1 activation.[220] Study of MSC myogenesis has also focused on the generation of cardiomyocytes or the co-stimulatory effects of MSCs on cardiomyocytes in models of cardiovascular infarct.[221] Further studies are needed to clarify the role of MSCs in promoting skeletal myogenic repair as well as to better understand MSC myogenic differentiation.

Therapeutic potential: Immune regulation, tissue repair, and gene therapy

As we have just reviewed, MSCs are very capable cells with enormous potential for clinical application. To date, MSCs have been used for immune regulation, tissue repair and regeneration, and as a vehicle for gene therapy. Their immunomodulatory effects make MSCs useful adjuvant therapy for immune-mediated pathology, and as the vehicle for gene therapy. *In vivo* animal models have thus far shown that systemic administration of MSCs prolongs skin grafts,[144] quells autoimmune encephalomyelitis,[145] and inhibits bleomycin-induced pulmonary fibrosis.[222] Thus far in human clinical trials, MSCs have been reported to successfully treat therapy- resistant GvHD,[223,224] hemorrhagic cystitis, pneumomediastinum, and perforated colon,[64] and to improve hematopoietic recovery after co-infusion with autologous-blood stem cells in post-chemotherapy cancer patients.[225-227] The immunoevasive ability of MSCs also makes them ideal cells for tissue engineering and gene therapy.

MSC multipotent differentiation capability uniquely qualifies them for tissue engineering of bone, cartilage, fat, tendon, and muscle. *In vivo* models of MSC-based repair have shown not only a role for MSCs during soft tissue and musculoskeletal defect repair, but also a role during cardiac and neurologic defect repair. A number of studies have shown that MSCs improve repair following cardiac injury.[228] One clinical study looked at 20 patients who suffered from transmural infarct and were managed with and without MSC treatment. Approximately one week post-infarct, ten patients had autologous MSCs delivered to the infarcted tissue by balloon catheter. MSC-treated patients had significantly decreased perfusion defects and decreased infarct sizes, compared to the non-MSC treated controls.[229] Other studies have also shown significant improvements in ischemia-induced cardiac injury following MSC infusion.[230–238] Interestingly, the MSCs were shown to differentiate into non-mesodermal derived cell types including neurons and astrocytes.[239] Accordingly, studies have looked at the role of MSCs in improving neurodegenerative disease and neural ischemia. In a study of patients with Amyotrophic Lateral Sclerosis (ALS), a disease for which there is no cure, intrathecal injection of autologous MSCs resulted in a mild delaying of proximal muscle weakness of the lower extremities.[240] Thus, disease progression was slowed, which is an important first step towards an effective treatment.

The potential for human MSC-based gene therapy seems similarly limitless, especially given MSC immunosuppressive and multipotent differentiation capabilities. Transgene delivery using MSCs has been studied for the past ten years,[241] and induction methods for stable transgene expression continues to be optimized. Viral vectors generally allow for more efficient transfection than non-viral methods, but in addition to the clinical limitations of introducing viral vehicles *in vivo*, even viral vectors cannot maintain indefinite transgene expression.[242,243] Some promising non-viral methods that have had preliminary success in MSCs include electric field vibration and electroporation,[244,245] as well as mammalian artificial chromosomes (ACEs).[246] Despite the need for further optimization, MSCs have achieved some gene therapy successes. Animal models have used MSC-delivered gene therapy to treat Parkinson's,[243,247] mucopolysaccharidosis,[248] lysosomal storage disorders including Tay Sachs[249] and Niemann Picks disease,[250,251] and stroke.[252] MSCs also have exciting applications in the treatment of musculoskeletal disease.[253] Human MSCs transfected with adenoviral BMP2 induce heterotopic bone formation following intramuscular injection in an *in vivo* mouse model.[254,255] In critical-size rat cranial defects and swine maxillary defects, BMP2 over-expressing MSCs enhance autologous bone formation. Similarly, injecting BMP4 over-expressing MSCs into critical-size rat cranial defects causes faster, more complete bone formation compared to untransfected controls at four weeks.[256] And the most exciting application to date, MSCs transduced with adenoviral vector encoding mutant dominant negative collagen Type1 have been successful in repairing bones in children with osteogenesis imperfecta.[257,258] In addition to the clinical applications already mentioned, there are currently sixty different human clinical trials listed involving the use of MSCs (http://clinicaltrials.gov).[224]

Limitations of MSCs

The clinical potential for MSCs, both future and realized, is substantial. In order to maximize therapeutic efficacy, however, we must continue to acknowledge their limitations so that we further improve upon them. Primary sources of MSCs are limited by the quantity of cells that can be efficiently harvested. Currently, the yield from primary tissue sources is insufficient to meet the need for clinical applications. Additionally, as with many other stem cells and primary cell lines, there is a lot of variability in proliferation and differentiation potential depending on isolation methods, culture medium, seeding density, donor age and patient variability.[259–261] The need remains for the standardization of isolation techniques, culture conditions, and differentiation protocols in order to establish reproducible results and clinical utility. Additionally, a better understanding of the signaling pathways responsible for MSC self-renewal and differentiation will enhance both our understanding of MSC biology and our ability to promote MSC self-renewal and readiness to differentiate in the future.

Conclusion

Cell-based tissue engineering of the hand is a very achievable reality in this era of stem cell biology. Exciting advances in the field have opened the door to new potential cell sources with the advent of iPS cells. Increased understanding of the molecular biology and signals that govern stem cell fate and behavior has improved our ability to control and manipulate adult stem cells, particularly MSCs. And the unique ability of MSCs to immunoregulate their environment may eliminate or at least decrease concerns of immune-mediated rejection of cell-based tissue engineered constructs. Continued basic science and clinical research portend a bright future with even more exciting advances as we seek to achieve regeneration, not just of cells or tissues, but of human form and function.

References

1. Weissman IL. Stem cells: Units of development, units of regeneration, and units in evolution. *Cell* **100** (2000) 157–168.
2. Stevens LC, Little CC. Spontaneous Testicular Teratomas in an Inbred Strain of Mice. *Proc Natl Acad Sci USA* **40** (1954) 1080–1087.
3. Kleinsmith LJ, Pierce GB, Jr. Multipotentiality of Single Embryonal Carcinoma Cells. *Cancer Res* **24** (1964) 1544–1551.
4. Kahan BW, Ephrussi B. Developmental potentialities of clonal in vitro cultures of mouse testicular teratoma. *J Natl Cancer Inst* **44** (1970) 1015–1036.
5. Gachelin G, Kemler R, Kelly F, Jacob F. PCC4, a new cell surface antigen common to multipotential embryonal carcinoma cells, spermatozoa, and mouse early embryos. *Dev Biol* **57** (1977) 199–209.
6. Solter D, Knowles BB. Monoclonal antibody defining a stage-specific mouse embryonic antigen (SSEA-1). *Proc Natl Acad Sci USA* **75** (1978) 5565–5569.
7. Martin GR. Isolation of a pluripotent cell line from early mouse embryos cultured in medium conditioned by teratocarcinoma stem cells. *Proc Natl Acad Sci USA* **78** (1981) 7634–7638.
8. Evans MJ, Kaufman MH. Establishment in culture of pluripotential cells from mouse embryos. *Nature* **292** (1981) 154–156.
9. Thomson JA, *et al.* Embryonic stem cell lines derived from human blastocysts. *Science* **282** (1998) 1145–1147.
10. Buzzard JJ, Gough NM, Crook, JM, Colman A. Karyotype of human ES cells during extended culture. *Nat Biotechnol* **22** (2004) 381–382 author reply 382.
11. Przyborski SA. Differentiation of human embryonic stem cells after transplantation in immune-deficient mice. *Stem Cells* **23** (2005) 1242–1250.
12. Amit M, *et al.* Clonally derived human embryonic stem cell lines maintain pluripotency and proliferative potential for prolonged periods of culture. *Dev Biol* **227** (2000) 271–278.
13. Richards M, Fong CY, Chan WK, Wong, PC, Bongso A. Human feeders support prolonged undifferentiated growth of human inner cell masses and embryonic stem cells. *Nat Biotechnol* **20** (2002) 933–936.
14. Rosler ES, *et al.* Long-term culture of human embryonic stem cells in feeder-free conditions. *Dev Dyn* **229** (2004) 259–274.

15. Wang G, *et al.* Noggin and bFGF cooperate to maintain the pluripotency of human embryonic stem cells in the absence of feeder layers. *Biochem Biophys Res Commun* **330** (2005) 934–942.

16. Xu C, *et al.* Basic fibroblast growth factor supports undifferentiated human embryonic stem cell growth without conditioned medium. *Stem Cells* **23** (2005) 315–323.

17. Xu RH, *et al.* Basic FGF and suppression of BMP signaling sustain undifferentiated proliferation of human ES cells. *Nat Methods* **2** (2005) 185–190.

18. Levenstein ME, *et al.* Basic fibroblast growth factor support of human embryonic stem cell self-renewal. *Stem Cells* **24** (2006) 568–574.

19. Beattie GM, *et al.* Activin A maintains pluripotency of human embryonic stem cells in the absence of feeder layers. *Stem Cells* **23** (2005) 489–495.

20. Amit M, Shariki C, Margulets V, Itskovitz-Eldor J. Feeder layer- and serum-free culture of human embryonic stem cells. *Biol Reprod* **70** (2004) 837–845.

21. Besser D. Expression of nodal, lefty-a, and lefty-B in undifferentiated human embryonic stem cells requires activation of Smad2/3. *J Biol Chem* **279** (2004) 45076–45084.

22. Greber B, Lehrach H, Adjaye J. Fibroblast growth factor 2 modulates transforming growth factor beta signaling in mouse embryonic fibroblasts and human ESCs (hESCs) to support hESC self-renewal. *Stem Cells* **25** (2007) 455–464.

23. James D, Levine AJ, Besser D, Hemmati-Brivanlou A. TGFbeta/activin/nodal signaling is necessary for the maintenance of pluripotency in human embryonic stem cells. *Development* **132** (2005) 1273–1282.

24. Sato N, Meijer L, Skaltsounis L, Greengard P, Brivanlou AH. Maintenance of pluripotency in human and mouse embryonic stem cells through activation of Wnt signaling by a pharmacological GSK-3-specific inhibitor. *Nat Med* **10** (2004) 55–63.

25. Xiao L, Yuan X, Sharkis SJ. Activin A maintains self-renewal and regulates fibroblast growth factor, Wnt, and bone morphogenic protein pathways in human embryonic stem cells. *Stem Cells* **24** (2006) 1476–1486.

26. Wang L, *et al.* Self-renewal of human embryonic stem cells requires insulin-like growth factor-1 receptor and ERBB2 receptor signaling. *Blood* **110** (2007) 4111–4119.

27. Soh BS, *et al.* Pleiotrophin enhances clonal growth and long-term expansion of human embryonic stem cells. *Stem Cells* **25** (2007) 3029–3037.

28. Pebay A, *et al.* Essential roles of sphingosine-1-phosphate and platelet-derived growth factor in the maintenance of human embryonic stem cells. *Stem Cells* **23** (2005) 1541–1548.

29. Xu RH, *et al.* BMP4 initiates human embryonic stem cell differentiation to trophoblast. *Nat Biotechnol* **20** (2002) 1261–1264.

30. Lees JG, *et al.* Transplantation of 3D scaffolds seeded with human embryonic stem cells: biological features of surrogate tissue and teratoma-forming potential. *Regen Med* **2** (2007) 289–300.

31. Weissman IL. Medicine: politic stem cells. *Nature* **439** (2006) 145–147.

32. Weissman IL. Stem cells — scientific, medical, and political issues. *N Engl J Med* **346** (2002) 1576–1579.

33. Wilmut I, Schnieke AE, McWhir J, Kind AJ, Campbell KH. Viable offspring derived from fetal and adult mammalian cells. *Nature* **385** (1997) 810–813.

34. Vajta G. Somatic cell nuclear transfer in its first and second decades: successes, setbacks, paradoxes and perspectives. *Reprod Biomed Online* **15** (2007) 582–590.

35. Cowan CA, Atienza J, Melton DA, Eggan K. Nuclear reprogramming of somatic cells after fusion with human embryonic stem cells. *Science* **309** (2005) 1369–1373.

36. Silva J, Chambers I, Pollard S, Smith A. Nanog promotes transfer of pluripotency after cell fusion. *Nature* **441** (2006) 997–1001.

37. Tada M, Tada T, Lefebvre L, Barton SC, Surani MA. Embryonic germ cells induce epigenetic reprogramming of somatic nucleus in hybrid cells. *EMBO J* **16** (1997) 6510–6520.

38. Tada M, Takahama Y, Abe K, Nakatsuji N, Tada T. Nuclear reprogramming of somatic cells by *in vitro* hybridization with ES cells. *Curr Biol* **11** (2001)1553–1558.

39. Takahashi K, Yamanaka S. Induction of pluripotent stem cells from mouse embryonic and adult fibroblast cultures by defined factors. *Cell* **126** (2006) 663–676.

40. Takahashi K, *et al.* Induction of pluripotent stem cells from adult human fibroblasts by defined factors. *Cell* **131** (2007) 861–872.
41. Yu J, *et al.* Induced pluripotent stem cell lines derived from human somatic cells. *Science* **318** (2007) 1917–1920.
42. Park IH, *et al.* Reprogramming of human somatic cells to pluripotency with defined factors. *Nature* **451** (2008) 141–146.
43. Lowry WE, *et al.* Generation of human induced pluripotent stem cells from dermal fibroblasts. *Proc Natl Acad Sci USA* **105** (2008) 2883–2888.
44. Maherali N, *et al.* Directly reprogrammed fibroblasts show global epigenetic remodeling and widespread tissue contribution. *Cell Stem Cell* **1** (2007) 55–70.
45. Hanna J, *et al.* Direct reprogramming of terminally differentiated mature B lymphocytes to pluripotency. *Cell* **133** (2008) 250–264.
46. Brambrink T, *et al.* Sequential expression of pluripotency markers during direct reprogramming of mouse somatic cells. *Cell Stem Cell* **2** (2008) 151–159.
47. Okita K, Ichisaka T, Yamanaka S. Generation of germline-competent induced pluripotent stem cells. *Nature* **448** (2007) 313–317.
48. Hacein-Bey-Abina S, *et al.* A serious adverse event after successful gene therapy for X-linked severe combined immunodeficiency. *N Engl J Med* **348** (2003) 255–256.
49. Hacein-Bey-Abina S, *et al.* LMO2-associated clonal T cell proliferation in two patients after gene therapy for SCID-X1. *Science* **302** (2003) 415–419.
50. Check E. Gene therapy put on hold as third child develops cancer. *Nature* **433** (2005) 561.
51. Nakagawa M, *et al.* Generation of induced pluripotent stem cells without Myc from mouse and human fibroblasts. *Nat Biotechnol* **26** (2008) 101–106.
52. Kim JB, *et al.* Pluripotent stem cells induced from adult neural stem cells by reprogramming with two factors. *Nature* **454** (2008) 646–650.
53. van Vliet P, Sluijter JP, Doevendans PA, Goumans MJ. Isolation and expansion of resident cardiac progenitor cells. *Expert Rev Cardiovasc Ther* **5** (2007) 33–43.
54. Schaffler A, Buchler C. Concise review: adipose tissue-derived stromal cells — basic and clinical implications for novel cell-based therapies. *Stem Cells* **25** (2007) 818–827.
55. Fiegel HC, *et al.* Fetal and adult liver stem cells for liver regeneration and tissue engineering. *J Cell Mol Med* **10** (2006) 577–587.
56. Peault B, *et al.* Stem and progenitor cells in skeletal muscle development, maintenance, and therapy. *Mol Ther* **15** (2007) 867–877.
57. Brittan M, Wright NA. Gastrointestinal stem cells. *J Pathol* **197** (2002) 492–509.
58. Kim CF, *et al.* Identification of bronchioalveolar stem cells in normal lung and lung cancer. *Cell* **121** (2005) 823–835.
59. Griffiths MJ, Bonnet D, Janes SM. Stem cells of the alveolar epithelium. *Lancet* **366** (2005) 249–260.
60. Bussolati B, *et al.* Isolation of renal progenitor cells from adult human kidney. *Am J Pathol* **166** (2005) 545–555.
61. Herrera MB, *et al.* Isolation and characterization of a stem cell population from adult human liver. *Stem Cells* **24** (2006) 2840–2850.
62. Moore, KA, Lemischka IR. Stem cells and their niches. *Science* **311** (2006) 1880–1885.
63. Ishikawa F, *et al.* Purified human hematopoietic stem cells contribute to the generation of cardiomyocytes through cell fusion. *FASEB J* **20** (2006) 950–952.
64. Ringden O, *et al.* Tissue repair using allogeneic mesenchymal stem cells for hemorrhagic cystitis, pneumomediastinum and perforated colon. *Leukemia* **21** (2007) 2271–2276.
65. Friedenstein AJ, Petrakova KV, Kurolesova, AI, Frolova, GP Heterotopic of bone marrow. Analysis of precursor cells for osteogenic and hematopoietic tissues. *Transplantation* **6** (1968) 230–247.
66. Pittenger MF, *et al.* Multilineage potential of adult human mesenchymal stem cells. *Science* **284** (1999) 143–147.
67. Young HE, *et al.* Human reserve pluripotent mesenchymal stem cells are present in the connective tissues of skeletal muscle and dermis derived from fetal, adult, and geriatric donors. *Anat Rec* **264** (2001) 51–62.

68. Nesti LJ, *et al.* Differentiation potential of multipotent progenitor cells derived from war-traumatized muscle tissue. *J Bone Joint Surg Am* **90** (2008) 2390–2398.
69. Sarugaser R, Lickorish D, Baksh D, Hosseini MM, Davies JE. Human umbilical cord perivascular (HUCPV) cells: a source of mesenchymal progenitors. *Stem Cells* **23** (2005) 220–229.
70. Noth U, *et al.* Multilineage mesenchymal differentiation potential of human trabecular bone-derived cells. *J Orthop Res* **20** (2002) 1060–1069.
71. Nakahara H, Goldberg VM, Caplan AI. Culture-expanded human periosteal-derived cells exhibit osteochondral potential *in vivo. J Orthop Res* **9** (1991) 465–476.
72. De Bari C, Dell'Accio F, Tylzanowski P, Luyten FP. Multipotent mesenchymal stem cells from adult human synovial membrane. *Arthritis Rheum* **44** (2001) 1928–1942.
73. De Bari C, *et al.* Mesenchymal multipotency of adult human periosteal cells demonstrated by single-cell lineage analysis. *Arthritis Rheum* **54** (2006) 1209–1221.
74. Seo BM, *et al.* Investigation of multipotent postnatal stem cells from human periodontal ligament. *Lancet* **364** (2004) 149–155.
75. Pierdomenico L, *et al.* Multipotent mesenchymal stem cells with immunosuppressive activity can be easily isolated from dental pulp. *Transplantation* **80** (2005) 836–842.
76. Ame-Thomas P, *et al.* Human mesenchymal stem cells isolated from bone marrow and lymphoid organs support tumor B-cell growth: role of stromal cells in follicular lymphoma pathogenesis. *Blood* **109** (2007) 693–702.
77. Crisan M, *et al.* Purification and Long-Term Culture of Multipotent Progenitor Cells Affiliated with the Walls of Human Blood Vessels: myoendothelial Cells and Pericytes. *Methods Cell Biol* **86C** (2008) 295–309.
78. Crisan M, *et al.* A perivascular origin for mesenchymal stem cells in multiple human organs. *Cell Stem Cell* **3** (2008) 301–313.
79. Zuk PA, *et al.* Human adipose tissue is a source of multipotent stem cells. *Mol Biol Cell* **13** (2002) 4279–4295.
80. Cowan CM, *et al.* Adipose-derived adult stromal cells heal critical-size mouse calvarial defects. *Nat Biotechnol* **22** (2004) 560–567.
81. Simmons PJ, Torok-Storb B. Identification of stromal cell precursors in human bone marrow by a novel monoclonal antibody, STRO-1. *Blood* **78** (1991) 55–62.
82. Gronthos S, *et al.* Molecular and cellular characterisation of highly purified stromal stem cells derived from human bone marrow. *J Cell Sci* **116** (2003) 1827–1835.
83. Dennis JE, Carbillet JP, Caplan AI, Charbord P. The STRO-1+ marrow cell population is multipotential. *Cells Tissues Organs* **170** (2002) 73–82.
84. Shi S, Gronthos S. Perivascular niche of postnatal mesenchymal stem cells in human bone marrow and dental pulp. *J Bone Miner Res* **18** (2003) 696–704.
85. Colter DC, Sekiya I, Prockop DJ. Identification of a subpopulation of rapidly self-renewing and multipotential adult stem cells in colonies of human marrow stromal cells. *Proc Natl Acad Sci USA* **98** (2001) 7841–7845.
86. Tuli R, *et al.* Characterization of multipotential mesenchymal progenitor cells derived from human trabecular bone. *Stem Cells* **21** (2003) 681–693.
87. Haynesworth SE, Baber MA, Caplan AI. Cell surface antigens on human marrow-derived mesenchymal cells are detected by monoclonal antibodies. *Bone* **13** (1992) 69–80.
88. Honczarenko M, *et al.* Human bone marrow stromal cells express a distinct set of biologically functional chemokine receptors. *Stem Cells* **24** (2006) 1030–1041.
89. Etheridge SL, Spencer GJ, Heath, DJ, Genever PG. Expression profiling and functional analysis of want signaling mechanisms in mesenchymal stem cells. *Stem Cells* **22** (2004) 849–860.
90. Gronthos S, *et al.* Surface protein characterization of human adipose tissue-derived stromal cells. *J Cell Physiol* **189** (2001) 54–63.
91. da Silva Meirelles L, Chagastelles PC, Nardi NB. Mesenchymal stem cells reside in virtually all post-natal organs and tissues. *J Cell Sci* **119** (2006) 2204–2213.
92. Peister A, *et al.* Adult stem cells from bone marrow (MSCs) isolated from different strains of inbred mice vary in surface epitopes, rates of proliferation, and differentiation potential. *Blood* **103** (2004) 1662–1668.

93. Reyes M, *et al.* Purification and *ex vivo* expansion of postnatal human marrow mesodermal progenitor cells. *Blood* **98** (2001) 2615–2625.

94. Musina RA, Bekchanova ES, Sukhikh GT. Comparison of mesenchymal stem cells obtained from different human tissues. *Bull Exp Biol Med* **139** (2005) 504–509.

95. Baddoo M, *et al.* Characterization of mesenchymal stem cells isolated from murine bone marrow by negative selection. *J Cell Biochem* **89** (2003) 1235–1249.

96. Murray L, *et al.* Enrichment of human hematopoietic stem cell activity in the CD34+Thy-1+Lin- subpopulation from mobilized peripheral blood. *Blood* **85** (1995) 368–378.

97. Pierelli L, *et al.* CD34+/CD105+ cells are enriched in primitive circulating progenitors residing in the G0 phase of the cell cycle and contain all bone marrow and cord blood CD34+/CD38low/- precursors. *Br J Haematol* **108** (2000) 610–620.

98. Dominici M, *et al.* Minimal criteria for defining multipotent mesenchymal stromal cells. The International Society for Cellular Therapy position statement. *Cytotherapy* **8** (2006) 315–317.

99. Tuan RS, Boland G, Tuli R. Adult mesenchymal stem cells and cell-based tissue engineering. *Arthritis Res Ther* **5** (2003) 32–45.

100. Sakaguchi Y, Sekiya I, Yagishita, K, Muneta T. Comparison of human stem cells derived from various mesenchymal tissues: superiority of synovium as a cell source. *Arthritis Rheum* **52** (2005) 2521–2529.

101. Wang QR, Wolf NS. Dissecting the hematopoietic microenvironment. VIII. Clonal isolation and identification of cell types in murine CFU-F colonies by limiting dilution. *Exp Hematol* **18** (1990) 355–359.

102. Grayson WL, Zhao F, Izadpanah R, Bunnell B, Ma T. Effects of hypoxia on human mesenchymal stem cell expansion and plasticity in 3D constructs. *J Cell Physiol* **207** (2006) 331–339.

103. Covello KL, *et al.* HIF-2alpha regulates Oct-4: effects of hypoxia on stem cell function, embryonic development, and tumor growth. *Genes Dev* **20** (2006) 557–570.

104. Koller MR, Manchel I, Palsson BO. Importance of parenchymal: stromal cell ratio for the *ex vivo* reconstitution of human hematopoiesis. *Stem Cells* **15** (1997) 305–313.

105. Strobel ES, Gay RE, Greenberg PL. Characterization of the *in vitro* stromal microenvironment of human bone marrow. *Int J Cell Cloning* **4** (1986) 341–356.

106. Tavassoli M, Takahashi K. Morphological studies on long-term culture of marrow cells: characterization of the adherent stromal cells and their interactions in maintaining the proliferation of hemopoietic stem cells. *Am J Anat* **164** (1982) 91–111.

107. Tuli R, *et al.* Transforming growth factor-beta-mediated chondrogenesis of human mesenchymal progenitor cells involves N-cadherin and mitogen-activated protein kinase and Wnt signaling cross-talk. *J Biol Chem* **278** (2003) 41227–41236.

108. Nelson WJ and Nusse R. Convergence of Wnt, beta-catenin, and cadherin pathways. *Science* **303** (2004) 1483–1487.

109. Kaigler D, *et al.* Endothelial cell modulation of bone marrow stromal cell osteogenic potential. *FASEB J* **19** (2005) 665–667.

110. Gerstenfeld LC, Barnes GL, Shea CM, Einhorn TA. Osteogenic differentiation is selectively promoted by morphogenetic signals from chondrocytes and synergized by a nutrient rich growth environment. *Connect Tissue Res* **44** Suppl. 1 (2003) 85–91.

111. Francois S, *et al.* Local irradiation not only induces homing of human mesenchymal stem cells at exposed sites but promotes their widespread engraftment to multiple organs: a study of their quantitative distribution after irradiation damage. *Stem Cells* **24** (2006) 1020–1029.

112. Bensidhoum M, *et al.* Homing of *in vitro* expanded Stro-1- or Stro-1+ human mesenchymal stem cells into the NOD/SCID mouse and their role in supporting human CD34 cell engraftment. *Blood* **103** (2004) 3313–3319.

113. Son BR, *et al.* Migration of bone marrow and cord blood mesenchymal stem cells *in vitro* is regulated by stromal-derived factor-1-CXCR4 and hepatocyte growth factor-c-met axes and involves matrix metalloproteinases. *Stem Cells* **24** (2006) 1254–1264.

114. Song L, Tuan RS. Transdifferentiation potential of human mesenchymal stem cells derived from bone marrow. *FASEB J* **18** (2004) 980–982.

115. Tamama K, Fan VH, Griffith LG, Blair HC, Wells A. Epidermal growth factor as a candidate for *ex vivo* expansion of bone marrow-derived mesenchymal stem cells. *Stem Cells* **24** (2006) 686–695.

116. Kang YJ, *et al.* Role of c-Jun N-terminal kinase in the PDGF-induced proliferation and migration of human adipose tissue-derived mesenchymal stem cells. *J Cell Biochem* **95** (2005) 1135–1145.

117. Bianchi G, *et al. Ex vivo* enrichment of mesenchymal cell progenitors by fibroblast growth factor 2. *Exp Cell Res* **287** (2003) 98–105.

118. Solchaga LA, *et al.* FGF-2 enhances the mitotic and chondrogenic potentials of human adult bone marrow-derived mesenchymal stem cells. *J Cell Physiol* **203** (2005) 398–409.

119. Tsutsumi S, *et al.* Retention of multilineage differentiation potential of mesenchymal cells during proliferation in response to FGF. *Biochem Biophys Res Commun* **288** (2001) 413–419.

120. Chiou M, Xu Y, Longaker MT. Mitogenic and chondrogenic effects of fibroblast growth factor-2 in adipose-derived mesenchymal cells. *Biochem Biophys Res Commun* **343** (2006) 644–652.

121. Zaragosi LE, Ailhaud G, Dani C. Autocrine fibroblast growth factor 2 signaling is critical for self-renewal of human multipotent adipose-derived stem cells. *Stem Cells* **24** (2006) 2412–2419.

122. Jian H, *et al.* Smad3-dependent nuclear translocation of beta-catenin is required for TGF-beta1-induced proliferation of bone marrow-derived adult human mesenchymal stem cells. *Genes Dev* **20** (2006) 666–674.

123. Dealy CN, Roth A, Ferrari D, Brown AM, Kosher RA. Wnt-5a and Wnt-7a are expressed in the developing chick limb bud in a manner suggesting roles in pattern formation along the proximodistal and dorsoventral axes. *Mech Dev* **43** (1993) 175–186.

124. Parr BA, McMahon AP. Dorsalizing signal Wnt-7a required for normal polarity of D-V and A-P axes of mouse limb. *Nature* **374** (1995) 350–353.

125. Kengaku M, *et al.* Distinct WNT pathways regulating AER formation and dorsoventral polarity in the chick limb bud. *Science* **280** (1998) 1274–1277.

126. Hartmann C, Tabin CJ. Dual roles of Wnt signaling during chondrogenesis in the chicken limb. *Development* **127** (2000) 3141–3159.

127. Kawakami Y, *et al.* WNT signals control FGF-dependent limb initiation and AER induction in the chick embryo. *Cell* **104** (2001) 891–900.

128. van de Wetering M, *et al.* The beta-catenin/TCF-4 complex imposes a crypt progenitor phenotype on colorectal cancer cells. *Cell* **111** (2002) 241–250.

129. Alonso L, Fuchs E. Stem cells of the skin epithelium. *Proc Natl Acad Sci USA* **100**(Suppl. 1) (2003) 11830–11835.

130. Austin TW, Solar GP, Ziegler FC, Liem L, Matthews W. A role for the Wnt gene family in hematopoiesis: expansion of multilineage progenitor cells. *Blood* **89** (1997) 3624–3635.

131. Reya T, *et al.* A role for Wnt signalling in self-renewal of haematopoietic stem cells. *Nature* **423** (2003) 409–414.

132. Boland GM, Perkins G, Hall, DJ, Tuan RS. Wnt 3a promotes proliferation and suppresses osteogenic differentiation of adult human mesenchymal stem cells. *J Cell Biochem* **93** (2004) 1210–1230.

133. Kleber M, Sommer L. Wnt signaling and the regulation of stem cell function. *Curr Opin Cell Biol* **16** (2004) 681–687.

134. Metcalf D. The unsolved enigmas of leukemia inhibitory factor. *Stem Cells* **21** (2003) 5–14.

135. Pruijt JF, Lindley IJ, Heemskerk DP, Willemze, R, Fibbe WE. Leukemia inhibitory factor induces *in vivo* expansion of bone marrow progenitor cells that accelerate hematopoietic reconstitution but do not enhance radioprotection in lethally irradiated mice. *Stem Cells* **15** (1997) 50–55.

136. Heymann D, Rousselle AV. gp130 Cytokine family and bone cells. *Cytokine* **12** (2000) 1455–1468.

137. Carcamo-Orive I, *et al.* ERK2 protein regulates the proliferation of human mesenchymal stem cells without affecting their mobilization and differentiation potential. *Exp Cell Res* **314** (2008) 1777–1788.

138. Song L, Webb NE, Song Y, Tuan RS. Identification and functional analysis of candidate genes regulating mesenchymal stem cell self-renewal and multipotency. *Stem Cells* **24** (2006) 1707–1718.

139. Le Blanc K, Tammik C, Rosendahl K, Zetterberg E, Ringden O. HLA expression and immunologic properties of differentiated and undifferentiated mesenchymal stem cells. *Exp Hematol* **31** (2003) 890–896.

140. Tse WT, Pendleton JD, Beyer WM, Egalka MC, Guinan EC. Suppression of allogeneic T-cell proliferation by human marrow stromal cells: implications in transplantation. *Transplantation* **75** (2003) 389–397.

141. Le Blanc, K, Ringden O. Immunomodulation by mesenchymal stem cells and clinical experience. *J Intern Med* **262** (2007) 509–525.

142. Uccelli A, Pistoia V, Moretta L. Mesenchymal stem cells: a new strategy for immunosuppression? *Trends Immunol* **28** (2007) 219–226.

143. Aggarwal S, Pittenger MF. Human mesenchymal stem cells modulate allogeneic immune cell responses. *Blood* **105** (2005) 1815–1822.

144. Bartholomew A, *et al.* Mesenchymal stem cells suppress lymphocyte proliferation *in vitro* and prolong skin graft survival *in vivo. Exp Hematol* **30** (2002) 42–48 2002.

145. Zappia E, *et al.* Mesenchymal stem cells ameliorate experimental autoimmune encephalomyelitis inducing T-cell anergy. *Blood* **106** (2005) 1755–1761.

146. Corcione A, *et al.* Human mesenchymal stem cells modulate B-cell functions. *Blood* **107** (2006) 367–372.

147. Beyth S, *et al.* Human mesenchymal stem cells alter antigen-presenting cell maturation and induce T-cell unresponsiveness. *Blood* **105** (2005) 2214–2219.

148. Ramasamy R, *et al.* Mesenchymal stem cells inhibit dendritic cell differentiation and function by preventing entry into the cell cycle. *Transplantation* **83** (2007) 71–76.

149. Jiang XX, *et al.* Human mesenchymal stem cells inhibit differentiation and function of monocyte-derived dendritic cells. *Blood* **105** (2005) 4120–4126.

150. Spaggiari GM, *et al.* Mesenchymal stem cells inhibit natural killer-cell proliferation, cytotoxicity, and cytokine production: role of indoleamine 2,3-dioxygenase and prostaglandin E2. *Blood* **111** (2008) 1327–1333.

151. Sotiropoulou PA, Perez SA, Gritzapis AD, Baxevanis CN, Papamichail M. Interactions between human mesenchymal stem cells and natural killer cells. *Stem Cells* **24** (2006) 74–85.

152. Jiang Y, *et al.* Pluripotency of mesenchymal stem cells derived from adult marrow. *Nature* **418** (2002) 41–49.

153. Harada S, Rodan GA. Control of osteoblast function and regulation of bone mass. *Nature* **423** (2003) 349–355.

154. Madras N, Gibbs AL, Zhou Y, Zandstra PW, Aubin JE. Modeling stem cell development by retrospective analysis of gene expression profiles in single progenitor-derived colonies. *Stem Cells* **20** (2002) 230–240.

155. Chen D, Zhao M, Mundy GR. Bone morphogenetic proteins. *Growth Factors* **22** (2004) 233–241.

156. Friedman MS, Long MW, Hankenson KD. Osteogenic differentiation of human mesenchymal stem cells is regulated by bone morphogenetic protein-6. *J Cell Biochem* **98** (2006) 538–554.

157. Jeon EJ, *et al.* Bone morphogenetic protein-2 stimulates Runx2 acetylation. *J Biol Chem* **281** (2006) 16502–16511.

158. Kaneki H, *et al.* Tumor necrosis factor promotes Runx2 degradation through up-regulation of Smurf1 and Smurf2 in osteoblasts. *J Biol Chem* **281** (2006) 4326–4333.

159. De Boer J, Wang HJ, Van Blitterswijk C. Effects of Wnt signaling on proliferation and differentiation of human mesenchymal stem cells. *Tissue Eng* **10** (2004) 393–401.

160. Gaspar C, Fodde R. APC dosage effects in tumorigenesis and stem cell differentiation. *Int J Dev Biol* **48** (2004) 377–386.

161. Tuan RS. Biology of developmental and regenerative skeletogenesis. *Clin Orthop Relat Res* (2004) S105–S117.

162. Baksh D, Song L, Tuan RS. Adult mesenchymal stem cells: characterization, differentiation, and application in cell and gene therapy. *J Cell Mol Med* **8** (2004) 301–316.

163. Barry F, Boynton RE, Liu, B, Murphy JM. Chondrogenic differentiation of mesenchymal stem cells from bone marrow: differentiation-dependent gene expression of matrix components. *Exp Cell Res* **268** (2001) 189–200.

164. Kato Y, Gospodarowicz D. Sulfated proteoglycan synthesis by confluent cultures of rabbit costal chondrocytes grown in the presence of fibroblast growth factor. *J Cell Biol* **100** (1985) 477–485.

165. Marie PJ, Coffin JD, Hurley MM. FGF, FGFR signaling in chondrodysplasias and craniosynostosis. *J Cell Biochem* **96** (2005) 888–896.

166. Massague J, Blain SW, Lo RS. TGFbeta signaling in growth control, cancer, and heritable disorders. *Cell* **103** (2000) 295–309.

167. Boskey AL, Paschalis EP, Binderman I, Doty SB. BMP-6 accelerates both chondrogenesis and mineral maturation in differentiating chick limb-bud mesenchymal cell cultures. *J Cell Biochem* **84** (2002) 509–519.

168. Sekiya I, Larson BL, Vuoristo JT, Reger RL, Prockop DJ. Comparison of effect of BMP-2, -4, and -6 on *in vitro* cartilage formation of human adult stem cells from bone marrow stroma. *Cell Tissue Res* **320** (2005) 269–276.

169. Gooch KJ, *et al.* Bone morphogenetic proteins-2, -12, and -13 modulate *in vitro* development of engineered cartilage. *Tissue Eng* **8** (2002) 591–601.

170. Nochi H, *et al.* Adenovirus mediated BMP-13 gene transfer induces chondrogenic differentiation of murine mesenchymal progenitor cells. *J Bone Miner Res* **19** (2004) 111–122.

171. Goumans MJ, Mummery C. Functional analysis of the TGFbeta receptor/Smad pathway through gene ablation in mice. *Int J Dev Biol* **44** (2000) 253–265.

172. Abecassis L, Rogier E, Vazquez A, Atfi A, Bourgeade MF. Evidence for a role of MSK1 in transforming growth factor-beta-mediated responses through p38alpha and Smad signaling pathways. *J Biol Chem* **279** (2004) 30474–30479.

173. Kahata K, *et al.* Regulation of transforming growth factor-beta and bone morphogenetic protein signalling by transcriptional coactivator GCN5. *Genes Cells* **9** (2004) 143–151.

174. Li WJ, *et al.* Evaluation of articular cartilage repair using biodegradable nanofibrous scaffolds in a swine model: a pilot study. *J Tissue Eng Regen Med* (2008).

175. Yan H, Yu C. Repair of full-thickness cartilage defects with cells of different origin in a rabbit model. *Arthroscopy* **23** (2007) 178–187.

176. Nuttall ME, Gimble JM. Controlling the balance between osteoblastogenesis and adipogenesis and the consequent therapeutic implications. *Curr Opin Pharmacol* **4** (2004) 290–294.

177. McBeath R, Pirone DM, Nelson CM, Bhadriraju K, Chen CS. Cell shape, cytoskeletal tension, and RhoA regulate stem cell lineage commitment. *Dev Cell* **6** (2004) 483–495.

178. Jakkaraju S, Zhe X, Pan D, Choudhury R, Schuger L. TIPs are tension-responsive proteins involved in myogenic versus adipogenic differentiation. *Dev Cell* **9** (2005) 39–49.

179. Hauner H, Schmid P, Pfeiffer EF. Glucocorticoids and insulin promote the differentiation of human adipocyte precursor cells into fat cells. *J Clin Endocrinol Metab* **64** (1987) 832–835.

180. Wolfman NM, *et al.* Ectopic induction of tendon and ligament in rats by growth and differentiation factors 5, 6, and 7, members of the TGF-beta gene family. *J Clin Invest* **100** (1997) 321–330.

181. Helm GA, *et al.* A light and electron microscopic study of ectopic tendon and ligament formation induced by bone morphogenetic protein-13 adenoviral gene therapy. *J Neurosurg* **95** (2001) 298–307.

182. Aspenberg P, Forslund C. Enhanced tendon healing with GDF 5 and 6. *Acta Orthop Scand* **70** (1999) 51–54.

183. D'Souza D, Patel K. Involvement of long- and short-range signalling during early tendon development. *Anat Embryol (Berl)* **200** (1999) 367–375.

184. Kawai S, *et al.* Mouse smad8 phosphorylation downstream of BMP receptors ALK-2, ALK-3, and ALK-6 induces its association with Smad4 and transcriptional activity. *Biochem Biophys Res Commun* **271** (2000) 682–687.

185. Hoffmann A, *et al.* Neotendon formation induced by manipulation of the Smad8 signalling pathway in mesenchymal stem cells. *J Clin Invest* **116** (2006) 940–952.

186. Altman GH, *et al.* Cell differentiation by mechanical stress. *FASEB J* **16** (2002) 270–272.

187. Juncosa-Melvin N, *et al.* The effect of autologous mesenchymal stem cells on the biomechanics and histology of gel-collagen sponge constructs used for rabbit patellar tendon repair. *Tissue Eng* **12** (2006) 369–379.

188. Hsieh AH, *et al.* Time-dependent increases in type-III collagen gene expression in medical collateral ligament fibroblasts under cyclic strains. *J Orthop Res* **18** (2000) 220–227.

189. Klein MB, Pham H, Yalamanchi N, Chang J. Flexor tendon wound healing *in vitro*: the effect of lactate on tendon cell proliferation and collagen production. *J Hand Surg [Am]* **26** (2001) 847–854.

190. Yalamanchi N, Klein MB, Pham HM, Longaker MT, Chang J. Flexor tendon wound healing *in vitro*: lactate up-regulation of TGF-beta expression and functional activity. *Plast Reconstr Surg* **113** (2004) 625–632.

191. Awad HA, *et al.* Autologous mesenchymal stem cell-mediated repair of tendon. *Tissue Eng* **5** (1999) 267–277.

192. Awad HA, *et al. In vitro* characterization of mesenchymal stem cell-seeded collagen scaffolds for tendon repair: effects of initial seeding density on contraction kinetics. *J Biomed Mater Res* **51** (2000) 233–240.

193. Ouyang HW, Goh JC, Lee EH. Viability of allogeneic bone marrow stromal cells following local delivery into patella tendon in rabbit model. *Cell Transplant* **13** (2004) 649–657.

194. Juncosa-Melvin N, *et al.* Effects of cell-to-collagen ratio in mesenchymal stem cell-seeded implants on tendon repair biomechanics and histology. *Tissue Eng* **11** (2005) 448–457.

195. Young RG, *et al.* Use of mesenchymal stem cells in a collagen matrix for Achilles tendon repair. *J Orthop Res* **16** (1998) 406–413.

196. Ouyang HW, Goh JC, Mo XM, Teoh SH, Lee EH. The efficacy of bone marrow stromal cell-seeded knitted PLGA fiber scaffold for Achilles tendon repair. *Ann N Y Acad Sci* **961** (2002) 126–129.

197. Ouyang HW, Goh JC, Thambyah A, Teoh SH, Lee EH. Knitted poly-lactide-co-glycolide scaffold loaded with bone marrow stromal cells in repair and regeneration of rabbit Achilles tendon. *Tissue Eng* **9** (2003) 431–439.

198. Kryger GS, *et al.* A comparison of tenocytes and mesenchymal stem cells for use in flexor tendon tissue engineering. *J Hand Surg [Am]* **32** (2007) 597–605.

199. Adamson JE, Wilson JN. The history of flexor-tendon grafting. *J Bone Joint Surg Am* **43-A** (1961) 709–716.

200. Mayer L. The physiological method of tendon transplants reviewed after forty years. *Instr Course Lect* **13** (1956) 116–120.

201. Mason ML, Allen HS. The Rate of Healing of Tendons: an experimental study of tensile strength. *Ann Surg* **113** (1941) 424–459.

202. Pulvertaft RG. Repair of tendon injuries in the hand. *Ann R Coll Surg Engl* **3** (1948) 3–14.

203. Pulvertaft RG. Tendon grafts for flexor tendon injuries in the fingers and thumb; a study of technique and results. *J Bone Joint Surg Br* **38-B** (1956) 175–194.

204. Boyes JH, Stark HH. Flexor-tendon grafts in the fingers and thumb. A study of factors influencing results in 1000 cases. *J Bone Joint Surg Am* **53** (1971) 1332–1342.

205. White WL. Secondary restoration of finger flexion by digital tendon grafts; an evaluation of seventy-six cases. *Am J Surg* **91** (1956) 662–668.

206. White WL. Tendon grafts: a consideration of their source, procurement and suitability. *Surg Clin North Am* **40** (1960) 403–413.

207. Tubiana R. Results and complications of flexor tendon grafting. *Orthop Clin North Am* **4** (1973) 877–883.

208. Asencio G, Abihaidar G, Leonardi C. Human composite flexor tendon allografts. A report of two cases. *J Hand Surg [Br]* **21** (1996) 84–88.

209. Bischoff RJ, *et al.* The effects of proximal load on the excursion of autogenous flexor tendon grafts. *J Hand Surg [Am]* **23** (1998) 285–289.

210. Zhang AY, Chang J. Tissue engineering of flexor tendons. *Clin Plast Surg* **30** (2003) 565–572.
211. Campion DR. The muscle satellite cell: a review. *Int Rev Cytol* **87** (1984) 225–251.
212. Qu-Petersen Z, *et al.* Identification of a novel population of muscle stem cells in mice: potential for muscle regeneration. *J Cell Biol* **157** (2002) 851–864.
213. Asakura A, Seale P, Girgis-Gabardo A, Rudnicki MA. Myogenic specification of side population cells in skeletal muscle. *J Cell Biol* **159** (2002) 123–134.
214. Dellavalle A, *et al.* Pericytes of human skeletal muscle are myogenic precursors distinct from satellite cells. *Nat Cell Biol* **9** (2007) 255–267.
215. Blau HM, Webster C. Isolation and characterization of human muscle cells. *Proc Natl Acad Sci USA* **78** (1981) 5623–5627.
216. Webster C, Pavlath GK, Parks DR, Walsh FS, Blau HM. Isolation of human myoblasts with the fluorescence-activated cell sorter. *Exp Cell Res* **174** (1988) 252–265.
217. Ferrari G, *et al.* Muscle regeneration by bone marrow-derived myogenic progenitors. *Science* **279** (1998) 1528–1530.
218. Rodriguez AM, *et al.* Transplantation of a multipotent cell population from human adipose tissue induces dystrophin expression in the immunocompetent mdx mouse. *J Exp Med* **201** (2005) 1397–1405.
219. De Bari C, *et al.* Skeletal muscle repair by adult human mesenchymal stem cells from synovial membrane. *J Cell Biol* **160** (2003) 909–918.
220. Dezawa M, *et al.* Bone marrow stromal cells generate muscle cells and repair muscle degeneration. *Science* **309** (2005) 314–317.
221. Li H, Yu B, Zhang Y, Pan Z, Xu W. Jagged1 protein enhances the differentiation of mesenchymal stem cells into cardiomyocytes. *Biochem Biophys Res Commun* **341** (2006) 320–325.
222. Ortiz LA, *et al.* Interleukin 1 receptor antagonist mediates the antiinflammatory and antifibrotic effect of mesenchymal stem cells during lung injury. *Proc Natl Acad Sci USA* **104** (2007) 11002–11007.
223. Ringden O, *et al.* Mesenchymal stem cells for treatment of therapy-resistant graft-versus-host disease. *Transplantation* **81** (2006) 1390–1397.
224. Giordano A, Galderisi U, Marino IR. From the laboratory bench to the patient's bedside: an update on clinical trials with mesenchymal stem cells. *J Cell Physiol* **211** (2007) 27–35.
225. Koc ON, *et al.* Rapid hematopoietic recovery after coinfusion of autologous-blood stem cells and culture-expanded marrow mesenchymal stem cells in advanced breast cancer patients receiving high-dose chemotherapy. *J Clin Oncol* **18** (2000) 307–316.
226. Lee ST, *et al.* Treatment of high-risk acute myelogenous leukaemia by myeloablative chemoradiotherapy followed by co-infusion of T cell-depleted haematopoietic stem cells and culture-expanded marrow mesenchymal stem cells from a related donor with one fully mismatched human leucocyte antigen haplotype. *Br J Haematol* **118** (2002) 1128–1131.
227. Lazarus HM, *et al.* Cotransplantation of HLA-identical sibling culture-expanded mesenchymal stem cells and hematopoietic stem cells in hematologic malignancy patients. *Biol Blood Marrow Transplant* **11** (2005) 389–398.
228. Guo J, Lin GS, Bao CY, Hu ZM, Hu MY. Anti-inflammation role for mesenchymal stem cells transplantation in myocardial infarction. *Inflammation* **30** (2007) 97–104.
229. Strauer BE, *et al.* Repair of infarcted myocardium by autologous intracoronary mononuclear bone marrow cell transplantation in humans. *Circulation* **106** (2002) 1913–1918.
230. Chen SL, *et al.* Effect on left ventricular function of intracoronary transplantation of autologous bone marrow mesenchymal stem cell in patients with acute myocardial infarction. *Am J Cardiol* **94** (2004) 92–95.
231. Wollert KC, *et al.* Intracoronary autologous bone-marrow cell transfer after myocardial infarction: the BOOST randomised controlled clinical trial. *Lancet* **364** (2004) 141–148.
232. Katritsis DG, *et al.* Transcoronary transplantation of autologous mesenchymal stem cells and endothelial progenitors into infarcted human myocardium. *Catheter Cardiovasc Interv* **65** (2005) 321–329.

233. Urbich C, Dimmeler S. Endothelial progenitor cells: characterization and role in vascular biology. *Circ Res* **95** (2004) 343–353.

234. Walter DH, *et al*. Statin therapy accelerates reendothelialization: a novel effect involving mobilization and incorporation of bone marrow-derived endothelial progenitor cells. *Circulation* **105** (2002) 3017–3024.

235. Assmus B, *et al*. Transplantation of Progenitor Cells and Regeneration Enhancement in Acute Myocardial Infarction (TOPCARE-AMI). *Circulation* **106** (2002) 3009–3017.

236. Stamm C, *et al*. Autologous bone-marrow stem-cell transplantation for myocardial regeneration. *Lancet* **361** (2003) 45–46.

237. Perin EC, *et al*. Transendocardial, autologous bone marrow cell transplantation for severe, chronic ischemic heart failure. *Circulation* **107** (2003) 2294–2302.

238. Perin EC, *et al*. Improved exercise capacity and ischemia 6 and 12 months after transendocardial injection of autologous bone marrow mononuclear cells for ischemic cardiomyopathy. *Circulation* **110** (2004) II213–218.

239. Jori FP, *et al*. Molecular pathways involved in neural *in vitro* differentiation of marrow stromal stem cells. *J Cell Biochem* **94** (2005) 645–655.

240. Mazzini L, *et al*. Stem cell therapy in amyotrophic lateral sclerosis: a methodological approach in humans. *Amyotroph Lateral Scler Other Motor Neuron Disord* **4** (2003) 158–161.

241. Deans, RJ, Moseley AB. Mesenchymal stem cells: biology and potential clinical uses. *Exp Hematol* **28** (2000) 875–884.

242. Chuah MK, *et al*. Long-term persistence of human bone marrow stromal cells transduced with factor VIII-retroviral vectors and transient production of therapeutic levels of human factor VIII in nonmyeloablated immunodeficient mice. *Hum Gene Ther* **11** (2000) 729–738.

243. Schwarz EJ, Alexander GM, Prockop DJ, Azizi SA. Multipotential marrow stromal cells transduced to produce L-DOPA: engraftment in a rat model of Parkinson disease. *Hum Gene Ther* **10** (1999) 2539–2549.

244. Song L, *et al*. Electric field-induced molecular vibration for noninvasive, high-efficiency DNA transfection. *Mol Ther* **9** (2004) 607–616.

245. Peister A, Mellad JA, Wang M, Tucker HA, Prockop DJ. Stable transfection of MSCs by electroporation. *Gene Ther* **11** (2004) 224–228.

246. Vanderbyl S, *et al*. Transfer and stable transgene expression of a mammalian artificial chromosome into bone marrow-derived human mesenchymal stem cells. *Stem Cells* **22** (2004) 324–333.

247. Dezawa M, Hoshino M, Ide C. Treatment of neurodegenerative diseases using adult bone marrow stromal cell-derived neurons. *Expert Opin Biol Ther* **5** (2005) 427–435.

248. Sakurai K, *et al*. Brain transplantation of genetically modified bone marrow stromal cells corrects CNS pathology and cognitive function in MPS VII mice. *Gene Ther* **11** (2004) 1475–1481.

249. Martino S, *et al*. Restoration of the GM2 ganglioside metabolism in bone marrow-derived stromal cells from Tay-Sachs disease animal model. *Neurochem Res* **27** (2002) 793–800.

250. Jin HK, Carter JE, Huntley GW, Schuchman EH. Intracerebral transplantation of mesenchymal stem cells into acid sphingomyelinase-deficient mice delays the onset of neurological abnormalities and extends their life span. *J Clin Invest* **109** (2002) 1183–1191.

251. Jin, HK, Schuchman EH. *Ex vivo* gene therapy using bone marrow-derived cells: combined effects of intracerebral and intravenous transplantation in a mouse model of Niemann-Pick disease. *Mol Ther* **8** (2003) 876–885.

252. Kurozumi K, *et al*. BDNF gene-modified mesenchymal stem cells promote functional recovery and reduce infarct size in the rat middle cerebral artery occlusion model. *Mol Ther* **9** (2004) 189–197.

253. Reiser J, *et al*. Potential of mesenchymal stem cells in gene therapy approaches for inherited and acquired diseases. *Expert Opin Biol Ther* **5** (2005) 1571–1584.

254. Olmsted-Davis E.A, *et al*. Use of a chimeric adenovirus vector enhances BMP2 production and bone formation. *Hum Gene Ther* **13** (2002) 1337–1347.

255. Lou J, Tu Y, Li S, Manske PR. Involvement of ERK in BMP-2 induced osteoblastic differentiation of mesenchymal progenitor cell line C3H10T1/2. *Biochem Biophys Res Commun* **268** (2000) 757–762.

256. Gysin R, *et al. Ex vivo* gene therapy with stromal cells transduced with a retroviral vector containing the BMP4 gene completely heals critical size calvarial defect in rats. *Gene Ther* **9** (2002) 991–999.

257. Le Blanc K, *et al.* Fetal mesenchymal stem-cell engraftment in bone after in utero transplantation in a patient with severe osteogenesis imperfecta. *Transplantation* **79** (2005) 1607–1614.

258. Horwitz E.M, *et al.* Transplantability and therapeutic effects of bone marrow-derived mesenchymal cells in children with osteogenesis imperfecta. *Nat Med* **5** (1999) 309–313.

259. Sotiropoulou PA, Perez SA, Salagianni M, Baxevanis CN and Papamichail M. Characterization of the optimal culture conditions for clinical scale production of human mesenchymal stem cells. *Stem Cells* **24** (2006) 462–471.

260. Murphy JM, *et al.* Reduced chondrogenic and adipogenic activity of mesenchymal stem cells from patients with advanced osteoarthritis. *Arthritis Rheum* **46** (2002) 704–713.

261. Sethe S, Scutt A, Stolzing A. Aging of mesenchymal stem cells. *Ageing Res Rev* **5** (2006) 91–116.

CHAPTER 5

Scaffolds

Wei Liu*,† and Yilin Cao†

Introduction

Treatment of tissue injury and defects remains a major challenge in reconstructive surgery, partly because there is a limited source of autologous tissue grafts. The hand is one of the most important parts of human body and responsible for many important physical activities in daily life. As a branch of plastic and reconstructive surgery, hand surgery deals with finger or thumb reconstruction, nerve and tendon transplantation and repair, soft tissue transfer and repair as well as restoration of vascular supply. Almost all of these clinical therapies involve tissue graft transplantation, thus the availability of an ideal tissue graft has become a key issue for success in tissue repair.

Many of the current procedures in hand surgery attempt to repair tissue defects at the cost of creating another defect at a site which is not functionally as important as the site to be repaired (e.g. toe to thumb transfer, tendon transfer, and nerve graft transplantation). Although allogeneic tissue transplantation may be a viable choice in the future, finding an alternative source for autologous tissue grafts is more likely to produce immediate applications to hand surgery. Tissue engineering is a crucial technology in realizing such a goal.

The first scientific meeting sponsored by the National Science Foundation and devoted solely to tissue engineering was held in 1988 at Lake Tahoe, California, where the definition of tissue engineering was developed: *"Tissue engineering" is the application of the principles and methods of engineering and the life sciences toward the fundamental understanding of structure/function relationships in normal and pathological mammalian tissues and the development of biological substitutes to restore, maintain, or improve functions.*[1] In 1993, Drs. Langer and Vacanti proposed the basic paradigm of tissue engineering, employing a degradable scaffold and seeded cells to develop living tissue either *in vitro* or *in vivo*.[2] Using this principle, Drs. Cao *et al.* published their research on generating human ear shaped cartilage in nude mice in 1997,[3] demonstrating the potential of tissue engineering and its obvious applications to reconstructive surgery. In 2001, Dr. C. Vacanti first reported the clinical application of tissue engineered bone in thumb reconstruction. By seeding autologous periosteal cells on coral scaffold, a cell-scaffold construct was formed and then implanted inside repaired soft tissue, regenerating a distal phalange and thus successfully repairing a thumb defect.[4] This result illustrated the possibility of translating the principle of tissue engineering into clinical reality in hand surgery. Moreover, its application also clearly demonstrated three major components of tissue engineering: seed cells, scaffold and microenvironment. In this chapter, we will introduce

---

*Corresponding author.

†Department of Plastic and Reconstructive Surgery, Shanghai 9th People's Hospital, Shanghai Tissue Engineering Center, Shanghai Jiao Tong University School of Medicine, 639 Zhi Zao Ju Road, Shanghai 200011, P.R. China.

the basic concept of scaffold materials, the principles of scaffold preparation, and their applications in the engineering of various types of tissues related to hand surgery.

Basic Principles

As outlined by Stock and Vacanti,[5] the basic concept of tissue engineering includes a scaffold that provides an architecture on which seeded cells can organize and develop into the desired organ or tissue prior to implantation. The scaffold provides an initial biomechanical profile for the replacement tissue until the cells produce an adequate extracellular matrix. During the formation, deposition, and organization of the newly generated matrix, the scaffold is either degraded or metabolized, eventually leaving a vital organ or tissue that restores, maintains, or improves tissue function (Figure 1).

As one of the key players, an ideal scaffold material should have the following characteristics in order to perform its functions properly. These features include:[6]

(1) Good biocompatibility: besides the general requirements for biomaterials (e.g. non-toxic, non-carcinogenic, non-inflammatory) scaffolds should also be able to support cell attachment, proliferation, matrix production and even differentiation. Particularly, the degraded products should not be harmful to seeded cells.

(2) Suitable biodegradability: after tissue formation, scaffolds should be able to completely degrade. In addition, the degradation rate should match the rate of cell growth and tissue formation. Furthermore, the degradation rate should be able to be controlled according to the requirements of different types of tissues.

(3) Three-dimensional porous structure: generally speaking, the porosity should be above 90 percent with a high surface/volume ratio, so that the scaffolds can provide a large surface area for cell attachment and growth, matrix production and deposition, and for nutrition and waste transportation and the access of neovascularization.

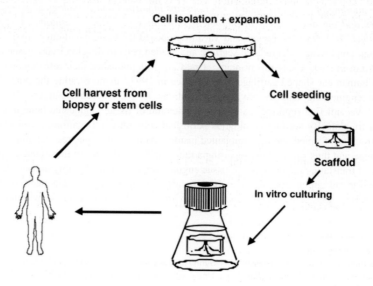

Figure 1.   The basic principle of tissue engineering. Reprint permitted by *Annu. Rev. Med* (Ref. 5).

(4) Good plasticity and mechanical properties: an ideal scaffold should be able to be pre-fabricated into a certain shape and possess a certain level of mechanical strength, so that it can support tissue function before the engineered tissue can be remodeled and become mature enough to regain normal mechanical strength.

(5) Appropriate cell surface properties for cell-scaffold interaction: this is particularly important for maintaining normal cell phenotype, or promoting cell differentiation in addition to cell attachment and growth.

(6) Easy to manufacture: low cost and ease of fabrication is generally required in order to fabricate scaffold in large scale for practical applications.

(7) Sterilization: an ideal material should be easily handled for various types of sterilization without affecting its characteristics.

Based on the source of materials, the scaffolds can be divided into two groups: natural scaffolds and synthetic polymer scaffolds.

Natural scaffolds

Natural scaffolds are generally derived from plant and animal sources and are composed of protein or carbohydrates, including collagen, gelatin, glycosaminoglycans, hyaluronic acid, fibrin or chitosan.

Collagen is the major component of mammal connective tissues and it is estimated to account for about 30 percent of all protein in human body.[7] Although 14 collagen types have been found, collagen type I predominates in almost every major tissue that requires strength and flexibility, particularly tendon, skin, bone and fascia. Thus these tissues also become the convenient and abundant sources for collagen extraction and preparation.[8]

Collagen molecules exhibit a triple-helical structure; posttranslational modification of peptide-bound prolyl residues leads to 4-hydroxyproline, which provides a distinctive serum marker of these molecules.[9] Due to their role as extracellular matrix molecules, collagens exhibit their excellent biocompatibility and cell compatibility and thus are able to support cell growth and differentiation. The weak mechanical properties of individual collagen fibers can be overcome by cross-linking, which is usually mediated through any number of physical or chemical techniques[10] including ultraviolet radiation, dehydration, and administration of glutaraldehyde. Increased intermolecular crosslinking usually leads to: 1) increased biodegradation time by making collagen less susceptible to enzymatic digestion; 2) making collagen less able to absorb water; 3) decreasing collagen solubility and 4) further increasing the tensile strength of collagen fibers.

Gelatin is a derivative of collagen, formed after chemical treatment. It contains the important amino acid components of collagen. Unlike collagen, gelatin is non immunogenic and is more easily degraded. Gelatin is usually used for coating grafts because of its excellent biocompatibility. It can also be mixed with other materials to form a complex scaffold. For example, when gelatin is mixed with chitosan, the resulting scaffold not only provides an ideal 3-D porous structure, but also exhibits excellent biocompatibility for chondrocyte attachment, proliferation and matrix production, and thus has proved to be a good scaffold candidate for cartilage engineering.[11]

Glycosaminoglycans (GAGs) are polysaccharides made of repeating disaccharide units arranged linearly. GAGs usually contain an uronic acid component (such as glucuronic acid) and a hexosamine component (such as *n-acetyl-d*-glucosamine). GAGs are generally present as part of proteoglycan molecules; they are formed by attaching their sugar chains to the core proteins of proteoglycans, such as chondroitin sulfate, dermatan sulfate, keratin sulfate and heparin sulfate. The largest GAG is hyaluronic acid or hyaluronan;

it is the major extracellular matrix molecule in many tissues, particularly in cartilage. It is an anionic polysaccharide with repeating disaccharide units of n-acetylglucosamine and glucuronic acid, with unbranched units ranging from 500 to several thousand in number. Because of this, it has strong capacity for water-binding and is usually very viscous when diluted in solution.

Hyaluronic acid can be either extracted from a natural source such as rooster combs or made from microbial fermentation.[12] The physical properties of hyaluronic acid can also be modified by chemical treatment, such as by esterification of the carboxyl moieties to reduce its water solubility and increase its viscosity.[13] In addition, hyaluronan can also be cross-linked to form a gel. Due to its relative ease of isolation and modification, and its ability to appear in different physical forms such as solution and solid, it can potentially be used to coat scaffolds or be used as a scaffold itself.

Fibrin is a biopolymer composed of monomer fibrinogens. The fibrinogen molecule consists of two sets of three polypeptide chains including α, β, and γ, which are joined together by six disulfide bridges.[14] After thrombin-mediated cleavage of fibrinopeptide A from the α chains and fibrinopeptide B from the β chains, a conformational change leads to the exposure of polymerization sites and thus the formation of fibrin monomer, which has a great tendency to self-associate and form insoluble fibrin.[15] Fibrin hydrogels are the usual physical form for scaffold applications in tissue engineering, and have been successfully used in the engineering of adipose tissue,[16] cardiovascular tissue,[17] muscle,[18] liver,[19] skin,[20] cartilage,[21] and bone.[22] The disadvantages of using fibrin scaffolds for tissue engineering include gel shrinkage, rapid degradation and weak mechanical properties. Therefore fibrin is best combined with other materials to serve as a scaffold for applications in tissue engineering.[22]

Chitosan, the deacylated derivative of chitin, is a naturally occurring polysaccharide which is usually extracted from crustacean exoskeletons or generated by fungal fermentation processes. Chitosan is a β-1,4-linked polymer of 2-amino-2-deoxy-d-glucose, therefore it carries a positive charge from amine groups.[23] The degradation rate of chitosan can be manipulated by varying the amount of residual acetyl content. Additionally, chitosan can also be modified into different physical formats with different mechanical properties, such as film, fiber, powders and hydrogel. While chitosan is widely used in wound dressings due to its excellent compatibility with epithelial cell types,[24] it should be noted that chitosan may not well support mesodermal cell attachment, proliferation and matrix production.[11]

Synthetic scaffolds

The other source of scaffold materials falls into the category of synthetic polymers, which were actually the first scaffolds created during the pioneering research in tissue engineering.[25-27] Unlike natural scaffolds, synthetic polymers can be designed and manufactured to achieve exact degradation properties, pore size and porosity, and other physical and chemical properties. Thus the quality of synthetic scaffolds can be better controlled compared to natural scaffolds. This allows for production of large quantities of scaffold with reliable characteristics. Additionally, there is minimal risk of transferring pathogens to host compared to the use of natural tissues. Synthetic polymers generally include Poly(α-hydroxy acids), Poly(dioxanone) (a polyether-ester), Poly(ε-caprolactone) (PCL), Polyanhydrides, Poly(amino acids), Polyorthoesters, Polyhydroxyalkanoates (PHA), etc.[28] The nomenclature for commonly used synthetic polymers is listed in Figure 2.[28]

Nomenclature	
Abbreviations	
LPLA	poly(L-lactide)
PGA	poly(glycolide)
DLPLA	poly(DL-lactide)
PDO	poly(dioxanone)
LDLPLA	poly(DL-lactide-*co*-L-lactide)
SR	self-reinforced
DLPLG	poly(DL-lactide-*co*-glycolide)
PGA-TMC	poly(glycolide-*co*-trimethylene carbonate)
LPLG	poly(L-lactide-*co*-glycolide)
PCL	poly(ε-caprolactone)

Figure 2. The nomenclature for commonly used synthetic polymers. Reprint permitted by *Biomaterials* (Ref. 28).

The poly (α-hydroxy acids) are hydrolyzed by cleaving their ester bonds in a water containing environment, which results in a reduction of the molecular weight, but not the total mass, of the scaffold. The degradation via hydrolysis usually reduces the molecular weight to around 5000. Afterwards, cellular degradation takes over the degradation process to further degrade the polymer into monomers, and they are finally metabolized into water and CO_2.[29] Poly (α-hydroxy acids) are considered optimal synthetic scaffolds for tissue engineering. In the poly (α-hydroxy acids) category, polyglycolic acid (PGA) is one of the most commonly used scaffolds from the earliest times[30] up to the present day[31] in tissue engineering research. Polylactic acid (PLA) is another polymer that is widely used for orthopedic applications. Generally, PGA degrades faster and becomes mechanically weaker than PLA, while PLA has more acidic degradation products than PGA, which may affect cell compatibility and tissue formation. Thus, the co-polymer poly(lactic acid-co-glycolic acid), or PLGA, was developed to take advantage of the properties of both polymers while minimizing their disadvantages.[32]

Poly (ether-ester) polydioxanone (PDS) is prepared by a ring-opening polymerization of p-dioxanone. PDS degrades into low-toxicity monomers *in vivo* and thus has gained increasing interest in the medical field. Poly(ε-caprolactone) or PCL is an aliphatic polyester. Comparing with PGA and PLA, PCL degradation is significantly slower and is thus most suitable for use in long-term implants. Accordingly, PCL has been used successfully in bone engineering.[33,34] Other synthetic polymers which have been reported for their use in medical devices or medical sutures are far less frequently used for tissue engineering scaffolds compared to PGA, PLA, PLGA or PCL.

Polymers must be processed to possess certain key characteristics required for efficacy as scaffold materials; these desired characteristics include high porosity and surface area, structural strength, specific three-dimensional shape and varied physical forms including fiber, hydrogel, membrane or solid object forms. Scaffold fabrication techniques usually determine whether the fabrication itself will affect the biocompatibility of constructed materials and whether scaffolds with desirable and reproducible characters can be manufactured.

Fiber bonding is a common technique for creating polymer fiber based scaffolds. Because fibers can provide a large surface:volume ratio, they are considered desirable building blocks for scaffolds. Fiber binding is required to hold together individual fibers while maintaining the desired shape. As an example, by dissolving PLA in dichloromethane, the resulting polymer solution can be cast over a nonwoven mesh of PGA fibers to solidify the PGA mesh, and the residual PLA can bind PGA fibers together after evaporation of the solvent. This technique has been used for engineering a human ear-shaped cartilage, which requires a complex 3-D structure.[3]

Mikos *et al.* have developed a solvent casting and particulate leaching technique to generate porous constructs of synthetic biodegradable polymers with specific porosity, surface:volume ratio, pore size and crystallinity for different applications.[35] By dispersing sieved salt particles into a PLLA/chloroform solution and casting the mixture into a container, a polymer construct containing certain sized salts can form after chloroform evaporation. When salts are leached out in water, a PLLA scaffold is formed with pore size and porosity determined by dispersed salt particles that are insoluble in chloroform.

Melting molding is an alternative method for constructing three dimensional scaffolds. For example, a fine PLGA powder could be mixed with sieved gelatin microspheres and poured into a Teflon mold, which is then heated above the glass transition temperature of the polymer to form a PLGA/gelatin construct. After removal of the construct from the mold, it is then placed into water to leach out gelatin that is water soluble, and finally a PLGA porous scaffold can form with a 3-D structure identical to that of the mold.[36]

Another method to fabricate polymer scaffold with variable porosity and pore size is to employ an emulsion/free-drying process. For example, water is mixed with a solution of PLGA dissolved in methylene chloride to create an emulsion. The mixture is then homogenized, poured into a copper mold, and quenched in liquid nitrogen followed by a freezing and drying process to remove water and solvent, resulting in the desired porous structure of the scaffold.[37]

Three dimensional printing is a technique that has been developed to fulfill the requirement of tissue repair and replacement, i.e., the need for creating scaffolds with complex three-dimensional shapes. 3-D printing produces a scaffold by first laying down a thin layer of powder over a building platform, and then an inkjet printer head prints a liquid binder onto the powder bed followed by another layer of powder being laid down. With repeated cycles of printing, the layers of polymers are merged together to form a 3-D structure. The 3-D printing technique belongs to the subset of fabrication techniques known as solid free-form (SFF) methods. The printing process is controlled by a computer-assisted design and manufacture (CAD/CAM) program.[38] As an example, 3-D printing was used for organogenesis of liver tissue[39] due to its advantage in generating fine 3-D complex structures.

Bioceramics are important scaffolds for bone engineering due to their chemical compositional resemblance to bone mineral, and their strong mechanical properties which are required for engineering weight-bearing bone. The commonly used bioceramics include tricalcium phosphate (TCP, $Ca_3[PO_4]$), hydroxyapatite (HA, $Ca_{10}[PO4]_6[OH]_2$), modified calcium phosphate ceramics, carbonate-substituted hydroxyapatite, silicate-substituted hydroxyapatite bioglass, and coral. Besides their strong mechanical properties, the other advantage of bioceramics is that their rich mineral components help to create a microenvironment for bone lineage differentiation when mesenchymal stem cells are used for bone regeneration. For example, β-TCP[40] and coral[41] were used for generating mandibular bone and foot bone tissue for repairing respective tissue defects. However, it is noted that the mechanical property of bioceramics remains less satisfactory when compared to natural cortical bone, particularly due to the low compressive strength and fracture toughness of porous bioceramics. This is due to the high tensile strength and fracture toughness of bone which results from tough and flexible collagen fibers reinforced by HA crystals. With this in mind, porous polymer/inorganic composite scaffolds have been proposed for bone tissue engineering in order to mimic the natural bone composition and structure.[42]

Recent Research in Hand Surgery

The hand is a composite tissue which is composed of skin, tendon, bone, cartilage, nerve and blood vessels. Some tissues such as skin or bone are less of a challenge because the hand is relatively small when compared to the entire human body; therefore tissue sources are readily available from other parts of the body. Nevertheless, tendon defects may become a particular problem to surgeons because the hand is the major tissue source for tendon grafts. Surgeons usually face challenges when searching for available tendon grafts in particularly large avulsion or crush injuries that lead to significant tendon defects. It is a common procedure to transfer tendons from one finger to another in order to restore the function of the more important finger.[43] "Repairing one tissue defect by creating another tissue defect" may be avoided in the future, and thus tissue engineering might become an important approach to further developing surgical techniques for the hand. The application of scaffolds in engineering different types of tissues, as well as *in vivo* tissue repair, will be introduced in the following sections.

Scaffolds for tendon engineering

Tendons are the connective tissue that link muscles to bones, so that the tensile forces created by muscles can be transmitted to bone for body movement. The main tendon extracellular matrix is type I collagen, which is highly organized in a hierarchy of bundles that are aligned in a parallel fashion. This unique structure provides the unique biomechanical properties of tendon tissues. Therefore, parallel alignment structure and strong mechanical properties are necessary for tendon scaffold design. It has been proposed that an ideal tendon scaffold should fulfill the following requirements:[44]

(1) Biodegradability with adjustable degradation rate.
(2) Biocompatibility before, during and after degradation.
(3) Superior mechanical properties and maintenance of mechanical strength during the tissue regeneration process.
(4) Biofunctionality: the ability to support cell proliferation and differentiation, ECM secretion and tissue formation.
(5) Processability: the ability to be processed to form desired constructs of complicated structures and shapes, such as woven or knitted scaffolds etc.

Based on literature review, the major categories of scaffold materials for tendon engineering include: poly (α-hydroxy acids), collagen derivatives, acellular tendon, xenogenic acellular extracellular matrix, silk derivatives, and polysaccharides.

As early as 1994, Cao *et al.* performed tendon engineering by using polyglycolic acid (PGA) fibers as the scaffold for *in vivo* tendon engineering in a nude mouse model.[45] First, unwoven PGA fibers were arranged in a parallel fashion, after which tenocytes isolated from calf tendons were seeded on the scaffold, followed by *in vivo* implantation in the subcutaneous tissue of nude mice. After 12 weeks of implantation, tendon-like tissue formed and revealed longitudinally aligned collagen fibers. Afterwards, the same scaffold was used for engineering tendon in immunocompetent animals.

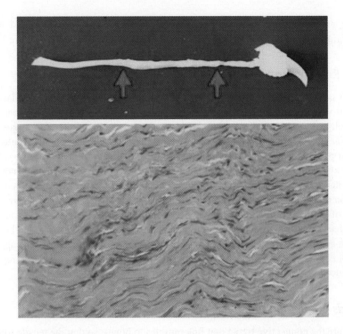

Figure 3. Tissue engineered tendon after 14 weeks of *in vivo* implantation (Ref. 46). Top panel: gross view of engineered tendon (between arrows); Bottom panel: H&E staining of engineered tendon.

In the first experiment, the hen claw was used as a model for tendon regeneration and repair inside a tendon sheath. First, autologous tenocytes were isolated and seeded on the unwoven PGA fibers. Then the cell-scaffold construct was cultured *in vitro* for one week, followed by *in vivo* transplantation to repair a 3 centimeter long defect of flexor digitorum profundus tendon. At 14 weeks postrepair, mature tendon tissue was formed when observed grossly (Figure 3, top panel). Histologically, longitudinally aligned collagen fibers with a curving pattern could be observed as well (Figure 3, bottom panel). More importantly, engineered tendon reached 83% of native tendon's tensile strength.[46] In the second experiment, a porcine model was used to perform the study. However, dermal fibroblasts were used to replace tenocytes as seed cells to engineer tendon and repair tendon defects. As similarly described, dermal fibroblasts were first isolated and *in vitro* cultured and expanded, then the cells were seeded onto unwoven PGA fibers for one week of *in vitro* culture. Scanning electron microscope examination revealed abundant matrix production by both dermal fibroblasts and tenocytes that were seeded onto the scaffold, indicating good cell compatibility between the cells and the scaffold (Figure 4). The pre-fabricated cell-scaffold construct was transplanted *in vivo* to repair a 3 centimeter long defect created on flexor digitorum superficialis tendon. When examined at 26 weeks postrepair, mature tendon tissue was formed, which was similar to the tendon tissue engineered by autologous tenocytes in gross view. Histologically, similar tissue structure was observed among dermal fibroblast and tenocyte engineered tendons and native tendon tissues (Figure 5). Importantly, strong mechanical properties were also achieved in engineered tendons.[47]

These two experiments showed the feasibility of using unwoven PGA fibers as a scaffold for *in vivo* tendon engineering, and the scaffold did fit the requirement of good

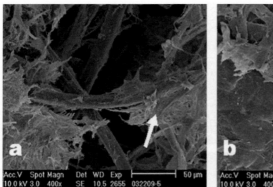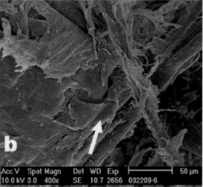

Figure 4. Scanning electron microscopic view of an *in vitro* cultured cell-scaffold construct, which shows adhesion and matrix production of dermal fibroblasts (A) and tenocytes (B) on PGA fibers. Arrow indicates cell attachment and matrix production. Reprint permitted by *Tissue Eng.* (Ref. 47).

biocompatibility, biodegradability with proper degradation rate, biofunctionality for supporting cell growth and matrix production, and processability. However, an obvious disadvantage of unwoven fibers is the lack of proper mechanical support during tendogenesis, therefore an acellular small intestinal submucosa membrane was needed to wrap around the cell loaded PGA fibers in order to enhance the mechanical strength of the cell-scaffold constructs.[46,47]

Compared to PGA, PLGA, the co-polymer of PGA and PLA, can retain good cell compatibility while degrading more slowly, and thus can better maintain its mechanical strength during tendogenesis. In addition, the change in physical form of the polymer will also help to further enhance the mechanical strength of the scaffold to be used. Ouyang *et al.* reported using knitted PLGA and allogeneic bone marrow stromal cells (BMSCs) for *in vivo* tendon engineering and the repair of rabbit Achilles tendon.[48] In the study, the tendon defect was either repaired with BMSC-loaded PLGA scaffold or PLGA scaffold alone as a control. At 2 and 4 weeks postrepair, a higher rate of tissue formation was observed in cell-loaded scaffold comparing to the control. Nevertheless, no significant difference was observed between two groups at 8 and 12 weeks postrepair. The formed tissues contained both collagen type I and III. In addition, the tensile stiffness and modulus of cell-seeded group were 87% and 62.6% of normal tendon respectively, whereas they were about 56.4% and 52.9% of normal tendon in scaffold alone group, respectively. This study also demonstrated that PLGA could also serve as a scaffold for *in vivo* tendon engineering and repair.[48] Similarly, Cooper *et al.* employed three-dimensional braiding technology to fabricate PLGA (10:90 ratio) polymer fibers for ACL ligament replacement and the resulting micro-porous scaffold exhibited optimal pore diameter with a range of 175–233 micron. Additionally, the construct's initial mechanical strength was similar to that of native ligament.[49]

To further improve cell attachment and cell proliferation on knitted PLGA scaffold, Sahoo *et al.* reported creating hybrid polyester scaffolds by coating the knitted scaffolds with a thin film of poly (epsilon-caprolactone) or poly (D, L-lactide-co-glycolide) nanofibers on type 1 collagen. The study showed that the coating of knitted PLGA could modulate the mechanical properties and facilitate cell attachment and proliferation in the hybrid scaffold, and thus could be applied in tendons/ligament engineering.[50]

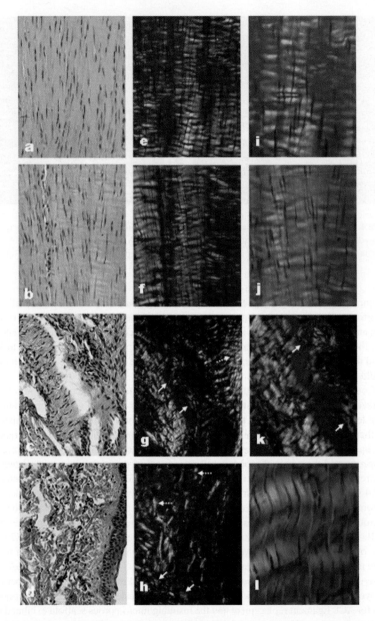

Figure 5. Histological finding of formed tissue at 26 weeks. H&E staining shows histological structures of fibroblast (A), tenocyte (B) engineered tendons, a control tissue in control group 2 (C) and normal pig skin (D). Collagen III (delicate collagen fibers with a light-green color) is detected only in the polarized images of control tissues (G and K) and in normal pig skin (H), as indicated by white arrows. In addition, collagen I (golden color, indicated by white dotted arrows) is also detected in these tissues. In the polarized images of fibroblast (E and I) and tenocyte (F and J) engineered tendons and natural tendons (L), collagen I (golden color) is the predominant collagen type. Original magnification ×400 (I–K); ×200, all others. Reprint permitted by *Tissue Eng.* (Ref. 47).

Although PGA, PLA and PLGA all belong to the group of poly-α-hydroxy acids or poly-a-hydroxyesters, they vary greatly individually in term of their degradation rate and their cell compatibility. Generally, the degradation rate increases from PLA to PLGA to PGA. For example, PGA unwoven fibers usually completely degrade between 8–12 weeks after *in vivo* implantation,[45-47] whereas PLA fibers will not be completely degraded during the first 12 months of implantation according to our own recent study. PLGA degradation rate falls between PGA and PLA, and depends on the PLA percentage in the co-polymer. On the other hand, PGA exhibited the best cell compatibility, whereas PLA revealed the worst compatibility relatively. Similarly, the cell compatibility for PLGA is between PGA and PLA depending on the percentage of PLA in the co-polymer. It is often observed that PGA exhibits a hydrophilic character whereas PLA usually exhibits hydrophobic properties, leading to poor PLA compatibility to cells. The other factor that might affect cell compatibility is the acidic degradation product which can disturb the cell survival environment. This has become a common problem for all poly-α-hydroxyester polymers. Particularly, the accumulated degradation product of lactic acid derived from PLA polymer is harmful to seeded cells. This may explain why PGA and PGLA are often used as tissue engineering scaffolds, whereas PLA is usually used in medical sutures or orthopedic devices that need good biocompatibility, but not necessarily good cell compatibility. Furthermore, the acidic degradation products of PLA induce strong host inflammatory reaction, which further deteriorates the *in vivo* environment for tissue regeneration. This explains why PGA unwoven fibers behave differently *in vivo* environments for tissue formation between immunodeficient and immunocompetent animal models.[51]

One of the dilemmas for using poly-α-hydroxy as a scaffold is how to take the advantage of its good cell compatibility and avoid the accumulation of undesired degradation products during tendogenesis. One approach is to develop a technique to engineer tendon graft *in vitro* for *in vivo* implantation. Cao *et al.* reported a preliminary study on tendon engineering *in vitro* using hen tenocytes and unwoven PGA fibers.[52] First, unwoven PGA fibers were arranged into a cord like construct and fixed on a U-ring with a static tension. Next, tenocytes were extracted from hen tendon tissue by collagenase digestion, expanded *in vitro*, and then seeded onto the PGA scaffold. After culture of the cell-scaffolds for 6 or 10 weeks *in vitro*, a neo-tendon tissue was formed when examined grossly (Figure 6, top panel). Histologically, collagen fibers with longitudinal alignment were observed in the tissue section. Interestingly, PGA fibers were mostly degraded through *in vitro* culture (Figure 6, bottom panel). Apparently, *in vivo* implantation of neo-tendon tissue instead of cell-seeded polymer may avoid the formation of acidic degradation products which adversely affect tissue formation. Importantly, when a properly designed bioreactor is used for *in vitro* tendon engineering, the system can provide perfusion of culture medium to efficiently remove the acidic degradation products, which might further improve the tissue quality of engineered tendon. Recently in our center, PGA unwoven fibers and human dermal fibroblasts were similarly used to generate human neo-tendon tissue after *in vitro* culture for 14 weeks on a U-spring (Figure 7, unpublished data). Regardless of the cell type used, the formed neo-tendon exhibited very weak tensile strength (usually less than 5 N), likely due to the lack of proper mechanical loading that is applied to tendon tissue in daily life. In a recent study performed in our center, we developed a bioreactor to provide dynamic mechanical loading to the *in vitro* engineered tendon at a certain frequency. The resulting tissue quality as well as the mechanical strength of loaded engineered tendon were significantly improved after 10–12 weeks of *in vitro* culture (data not shown). It is noted however, that *in vitro* engineered tendon tissue remains weaker in mechanical strength when compared with the strength of native tendon tissue.

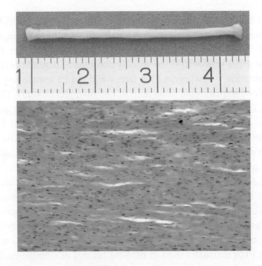

Figure 6. Growth view (top) and histology (bottom) of *in vitro* engineered tendon using hen teno-cytes and unwoven PGA fibers after 10 weeks of *in vitro* culture. Note, PGA fibers were completely degraded. (Ref. 52)

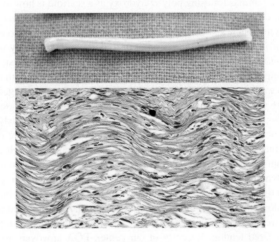

Figure 7. Growth view (top) and histology (bottom) of *in vitro* engineered tendon using human dermal fibroblasts and unwoven PGA fibers after 14 weeks of *in vitro* culture. (Dan Deng's Thesis).

In a recent study performed in our center, we tried to delineate the optimal conditions for tissue engineering mediated tendogenesis.[53] In this study, PGA long fibers and human fetal extensor tenocytes (isolated from 3 month old aborted fetus donated by the patients for research only) were used to engineer extensor tendon equivalents. The PGA long fibers were arranged to mimic the extensor tendon complex like structure (Figure 8) followed by

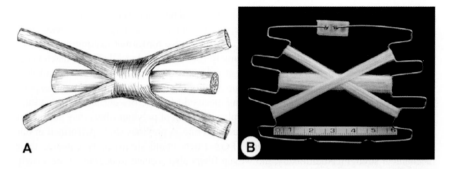

Figure 8. Design of extensor tendon complex scaffold. (A) Schema chart showing the central slip and two lateral bands. (B) PGA scaffold secured on a custom-made spring to mimic the complex structure. Reprint permitted by *Biomaterials*. (Ref. 53).

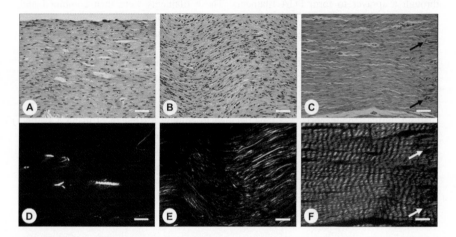

Figure 9. H&E staining (A–C) and polarized microscopic views (D–F) of *in vitro* engineered tendon after total 12 weeks (A and D) and of *in vivo* implanted tendons for 14 weeks without (B and E) and with (C and F) loading. Original magnifications: ×200; bar represents 50 mm for all. Reprint permitted by *Biomaterials*. (Ref. 53).

cell seeding onto the scaffold. After *in vitro* culture for 6 weeks, the cell-scaffold constructs were further divided into three groups: 1) *in vitro* culture with mechanical loading; 2) *in vivo* implantation without mechanical loading; 3) *in vivo* implantation with mechanical loading by suturing the construct to fascia, allowing mouse movement to provide a natural dynamic loading. The results showed that human fetal cells could form an extensor tendon complex structure *in vitro* and become further matured *in vivo* by mechanical stimulation. In contrast to *in vitro* loaded and *in vivo* non-loaded tendons, *in vivo* loaded tendons exhibited bigger tissue volume, better aligned collagen fibers, more mature collagen fibril structure with D-band periodicity, and stronger mechanical properties (Figure 9). These results indicate that *in vivo* mechanical loading via an *ex vivo* approach might be an optimal method for engineering functional tendon tissue. Therefore, a reasonable strategy for engineered functional repair of tendon defect might be to generate a neo-tendon tissue first *in vitro* and then to implant *in vivo* for further maturation.

This leads to the second dilemma, namely finding the balance between mechanical strength and cell compatibility for the use of poly-α-hydroxy acids as polymer scaffolds. PLA fibers usually exhibited poor cell compatibility relative to PGA fibers in our experience. In contrast, PLA could maintain strong mechanical properties during *in vitro* culture compared to PGA fibers. Additionally, the physical form of polymer fibers may also affect its mechanical property and likely its degradation rate. A previous study performed in our center found that once in woven form, PGA fibers could significantly enhance their mechanical strength. Additionally, the woven fibers also seemed to degrade more slowly compared to non-woven fibers at the same time points.[54]

Based on these findings, a new strategy is being developed for using PGA fibers as tendon scaffold (unpublished data). First, we replaced PGA unwoven short fibers with PGA long fibers that were generated using a technique similar to that of weaving medical suture. The PGA raw materials were melted at high temperature (above 200°C), then went through a sprayer to form PGA filaments. These filaments were then combined and stretched to finally form PGA long fibers (Figure 10a). The advantage of using long fibers is that all fibers can be arranged in a longitudinal aligned pattern, which allows collagen fiber formation to be guided into parallel strands (Figure 10b). Unlike unwoven fibers which are short and disconnected, long fibers are continuous though the whole construct and thus provide stronger initial mechanical strength which allow for mechanical loading.

Due to degradation, even long PGA fibers themselves would not be able to maintain a superior mechanical strength, which is essential to ensure the structural integrity of the scaffold during tendogenesis. We thus decided to employ a strategy of using a knitted polymer net wrapping around the neo-tendon in order to strengthen the mechanical properties of engineered neo-tendon, thus allowing for *in vivo* implantation and further maturation (Figure 10c). The idea is that the neo-tendon tissue inside the net will be the main part of engineered tissue, while the net will became the peripheral part of the engineered tissue responsible for enhanced mechanical property (Figure 10d).

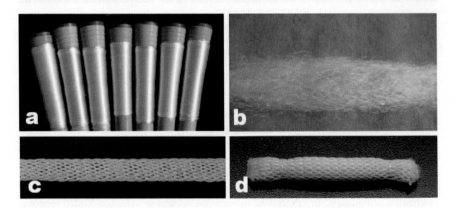

Figure 10. PGA long fibers are manufactured with high temperature melting technique (a). The long fibers are arranged longitudinally for engineering of neo-tendon (b) or knitted to form a net (c), which can wrap around the inserted PGA long fibers as a composite scaffold (d). (Bin Wang's Thesis).

Figure 11. Gross view of 4 sets of polymer nets during hydrolysis assay before (A) and 3 months (B) and 6 months (C) after the hydrolysis. The columns from left to right are in an order of pure PGA, PGA/PLA (4:2), PGA/PLA (2:4) and pure PLA.

In order to minimize degradation and optimize mechanical strength, 4 different sets of nets knitted with pure or combined polymer fibers were designed by varying the ratio of PLA and PGA in the combined polymer nets, including 1) pure PGA fibers; 2) PGA:PLA in 4:2 ratio; 3) PGA:PLA in 2:4 ratio; 4) pure PLA fibers. To characterize these nets, a hydrolysis assay was performed by culturing 4 sets of net scaffolds in phosphate-buffered saline (PBS) at 37°C in an incubator for 12 and 24 weeks to investigate the relationship between degradation and mechanical property change. As shown in Figure 11, gross examination showed that the pure PGA net was partly degraded at 3 months, and mostly degraded at 6 months. In contrast, the main knitted fiber structure remained observable in other sets of nets at both 3 months and 6 months. Examination by scanning electronic microscopy confirmed the changes seen in gross view. As revealed with scanning electronic microscopy examination (Figure 12), pure PGA degraded faster than other groups and lost its knitted structure completely at 6 months. Degradation was also observable in other three groups with the increase of incubation time. Additionally, increasing PLA percentage led to decreased degradation. In contrast to pure PGA group, the knitted structure could be maintained if PLA fibers were incorporated into the net scaffold.

More importantly, pure PGA almost lost all of its mechanical strength at 3 months, when evaluated for maximal strength. Interestingly, the net with PGA/PLA in 2:4 ratio could have around 55 N maximal strength at 3 months. The maximal strength for PGA/PLA in 2:4 ratio and pure PLA was 116 and 149 N respectively. At 6 months, the

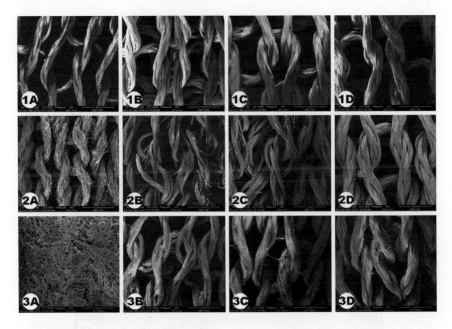

Figure 12. Scanning electron microscopic view of 4 sets of polymer nets during hydrolysis assay before (top panels) and after 3 months (middle panels) and 6 months (bottom panels). The columns of (A), (B), (C) and (D) represent pure PGA, PGA/PLA (4:2), PGA/PLA (2:4) and pure PLA respectively.

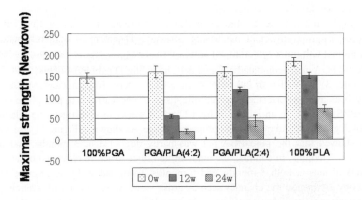

Figure 13. Change of maximal strength in 4 different sets of scaffold nets during a hydrolysis assay for 3 and 6 months.

maximal strength for the three groups of PGA/PLA (4:2), PGA/PLA (2:4) and pure PLA was 18, 43 and 71 N respectively (Figure 13). This result indicated that incorporating part of PLA fibers could significantly enhance the mechanical strength of the net structure.

Next, we investigated the cell compatibility of these knitted nets and how they affect tissue formation. As mentioned previously, we inserted pure PGA long fibers into 4 sets of knitted nets, which were then seeded with human dermal fibroblasts isolated from circumcised foreskin. The cell seeded constructs were then loaded on a bioreactor to give mechanical loading as a way of *in vitro* tendon engineering. After *in vitro* engineering for 4 and 8 weeks, the engineered tissues were evaluated with gross and histological examinations, as well as mechanical testing. The results showed that no mature tissue was formed at 4 weeks. At 8 weeks, poor tissue formation in the scaffolds wrapped with the nets of pure PLA and PGA/PLA in a 2:4 ratio, although the mechanical strength remained good (data not shown). In contrast, good tissue formation was observed in the group with pure PGA net, but with poor mechanical strength (<5 N), possibly due to the repaid degradation of PGA scaffold, even while formed tissue remain immature. Among the four sets of nets, only the net knitted with PGA and PLA in 4:2 ratio seemed to achieve the needed balance between cell compatibility or tissue formation and mechanical strength. After 8 weeks of culture, the engineered tissue was able to bear a tensile strength of 55 N with good tissue formation both grossly and histologically (Figure 14). All of these results indicate that poly-α-hydroxyesters polymers remain an ideal candidate for tendon engineering, given modification of physical form and proper combination of different types of polymers.

In addition to poly-α-hydroxy acid, collagen derivatives have also been intensively investigated for their application in tendon tissue engineering. This is because tendon tissues are mainly composed of type I collagen, and the collagen derivatives would be highly biocompatible to seeded cells, supporting cell adhesion, proliferation and matrix production. Gel is a common form for collagen scaffold. Collagen gels are usually employed for studying the effect of mechanical loading on neo-tendon formation in an *in vitro* culture system. For example, Garvin *et al.* used a bioreactor to stretch a collagen gel construct containing avian flexor tenocytes and they found that mechanical loading could significantly enhance the tensile strength of the construct within a short time period.[55] However, collagen gel is not suitable for clinical application due to its shortcomings of quick degradation and lack of sufficient mechanical strength. Therefore, many combinations of collagen gels with other materials for use in engineering or repair have been attempted. For example, Awad *et al.* applied the gel along with polyglyconate suture for patellar tendon repair.[56] Collagen gel has also been combined with collagen fibers, or sponged, to

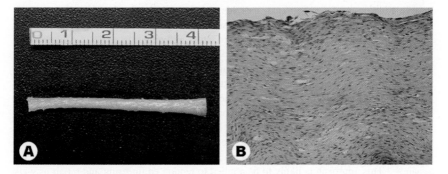

Figure 14. Gross (A) and histological (B) views of neo-tendon tissue engineered *in vitro* for 8 weeks using PGA/PLA (4:2) net along with inserted PGA long fibers as the scaffold.

further enhance its mechanical strength.[57] Cross-linking with chemical agents may also enhance the mechanical strength of collagen based scaffold.

Another approach is the use of acellular collagen matrix. Tischer et al. reported the preparation of acellular tendon scaffold for cruciate ligament reconstructions.[58] They used sodium dodecyl sulfate (SDS) as the main ingredient to decellularize cell components from the tendon tissue. After acellularization, dermal fibroblasts were injected into the tendon and cultured for different time periods at 37°C in 5% CO_2. The results showed that processed acellularized tendon exhibited similar mechanical strength to native tendons including the maximum load to failure, stiffness and elongation. Additionally, cell-seeded graft was able to produce pro-collagen I as revealed by immunohistochemistry. However, it was reported that tenocytes usually failed to penetrate into the decellularized scaffold, although the cells were able to attach to the processed tendon tissues.

Ingram et al. reported the use of ultrasonication to assist recellularization of acellular tissue scaffolds for potential reconstruction of anterior cruciate ligament.[59] Their study showed that ultrasonication plus standard acellularization process could significantly enhance the cell penetration, which allowed extrinsic cells to migrate into the center of the acellular scaffold. This was because the sonication process could produce a more microscopically open porous matrix, but would not damage the overall architecture of the scaffold, and the essential biomechanical characteristics of the native tissue biochemical constituents (collagen, glycosaminoglycans) could be well retained. However, the viability of the cells in the center of the scaffold was compromised. Therefore, promoting cell survival might remain a challenge in the use of acellular dense connective tissue like acellularized tendon graft.

Different from these reports, Badylak et al. reported the application of acellularized small intestinal submucosa (SIS) membrane for tendon repair.[60] Porcine SIS membrane was generated and then used to repair a 1.5 cm segmental defect of the Achilles tendon created in dogs. The animals were sacrificed at different time points and neotendons were evaluated. The results showed that implanted SIS could be penetrated by host cells and the implanted materials could be gradually remodeled into neotendons. More importantly, the remodeled neotendon could achieve strong mechanical strength, indicating a functional repair of the tendon. This result suggests that the implanted SIS biomaterial probably underwent a slow degradation within the first several weeks and served as a temporary scaffold around which the host could deposit appropriate and organized connective tissue. This cell-free approach may also become a promising approach for tendon tissue regeneration and defect repair.

In addition to aforementioned scaffolds, silk fibers have also been used as the scaffold materials for tissue engineering because of their unique mechanical properties and the possibility of modification with growth factors and adhesion molecules on the side chains of the fibers. Particularly, silk fibers will be a good candidate for scaffold materials if sericin (glue-like proteins), a potential antigenic protein, can be completely removed.[61] Fan et al. reported the use of knitted silk mesh incorporated with microporous silk sponges as a scaffold to investigate the regeneration of anterior cruciate ligament using mesenchymal stem cells (MSCs) as the cell source.[62] In addition, knitted silk fiber scaffold was also incorporated with gelatin[63] or collagen matrix[64] for ligament regeneration experiments. It was noted that silk fibers elicited little inflammation and degraded slowly, and thus could provide mechanical support for ligament regeneration and repair.[63,64] This approach is likely to be applied to tendon engineering and repair as well. However, these reports also revealed that silk fibers were usually very slowly degraded and no apparent mass loss could be observed, even at one year postrepair.[64] This may cause a challenge to tendon engineering because long term presence of the scaffold materials may cause foreign body reaction and further fibrosis, and prevent implanted scaffold from tissue regeneration by occupying the space where tissue is intended to regenerate. It is expected that further sophistication of silk process technique might be able to modify silk fibers into a scaffold materials with proper degradation rate and strong mechanical strength.[61]

Polysaccharides might also be a potential material for tendon engineering. Polysaccharides include chitin, chitosan, alginate and agarose, etc. Bagnaninchi *et al.* reported that porous chitosan scaffolds with microchannels were successfully designed to engineer tendon tissues.[65] However, hybridization of chitosan with other polysaccharides seems to be a more practical approach for its application. For example, a hybridized chitosan-hyaluronan scaffold exhibited not only enhanced mechanical strength, but also better bioreactivity with enhanced cell attachment and matrix production, when compared to chitosan alone.[66] More detailed study regarding their cell compatibility as well as their performance in tendon regeneration *in vivo* might provide valuable information to guide their applications.

Before a scaffold material is chosen for tendon engineering, the interaction between cells and the material should be studied. Kryger *et al.* compared the biological behavior of 4 different types of cells *in vitro* and their functions were further evaluated with *in vivo* tendon engineering analysis. The results showed that bone marrow derived mesenchymal stem cells, adipose derived stem cells, tenocytes and sheath fibroblasts could all survive on acellular tendon grafts after *in vivo* implantation.[67] Additionally, growth factor application[68] as well as mechanical stimulation[69] can assist cell proliferation and matrix production as revealed by *in vitro* studies, and their effect on tendon engineering might need to be further investigated based within the scope of cell-scaffold interaction.

Scaffolds for bone and cartilage engineering

Bone is one of the major tissue types in hand, and bone defects are commonly seen in hand injuries. Generally, bone defects of a small size are not a challenge for repair because autologous bone graft is readily available, and harvest of bone graft does not cause significant donor site morbidity. However, large size bone defects remain a clinical challenge. In addition, bone regeneration, as opposed to autologous bone graft, would be a preferred way of bone repair.

Bone engineering and repair in large animal models have already been reported in literature using β-TCP, coral and calcium alginate as the scaffold materials.[40,41,70] Poly-α-hydroxy acid such as PLGA have also been used for regenerating cranial bone tissue in a mouse model.[71] In addition, composite scaffold composed of both polymer and inorganic materials is also a candidate for bone scaffold.[42] Different from large bone tissue, engineering of small size bone tissue like phalangeal bone is not frequently reported in the literature. This is because an exact three-dimensional morphology needs to be replicated in order to successfully engineer a human phalange-like bone tissue.

The hand contains multiple tissue types. Besides bone, cartilage is the major tissue component in interphalangeal joints or metacarpophalangeal joints. There are already several studies reporting the engineering of composite hand tissue. As early as 1999, Isogai *et al.* reported the formation of small phalanges and the whole joints using a tissue engineering approach.[72] They used PLGA or PGA as the scaffold, and fresh bovine periosteum was harvested to wrap around the scaffold for bone engineering.

Additionally, freshly harvested chondrocytes and tenocytes were seeded onto the scaffold in order to form articular cartilage and associated tendon tissue. The different types of cell-seeded constructs were then sutured together to make a composite construct followed by *in vivo* implantation in athymic mice. After two weeks of implantation, a composite tissue formed with the shape and dimensions of human phalanges as well as the joints. In addition, histological structures of mature articular cartilage and subchondral bone as well as tenocapsule were also observed. In 2005, a similar approach was employed to further test the concept of hand composite tissue engineering with a much larger sample size and longer *in vivo* implantation time.[73] In the study, more than 250 samples were tested. Bovine periosteum, cartilage and tendon cells were seeded onto either PGA or PLLA or PCL scaffold to form a middle phalange, or distal interphalangeal joint constructs and then implanted *in vivo* in nude mice for over 60 weeks. The results showed the formation of new cartilage, bone and tendon tissues with a matrix gene expression profile that confirmed the origin as bovine, indicating the potential for clinical application. Recently, Landis *et al.* investigated the effect of periosteum on growth plate cartilage development in a model of human digit engineering, and found that periosteum wrapping around the polymer midshafts promotes the formation of growth plate cartilage at chondrocyte-seeded sheets, directly adjacent to periosteum-wrapped midshaft.[74]

As these studies used poly-α-hydroxyacid polymer for both bone and cartilage engineering, all of them employed periosteum derived osteoblasts and chondrocytes, respectively as the cell source for bone and cartilage engineering. Although it is quite reasonable to use these cells to prove the concept, those cells might be difficult to use for practical application because of donor site morbidity. Instead, there appears to be a trend towards using mesenchymal stem cells as the cell source for both bone and cartilage engineering.[40,41,75] In this case, the scaffolds themselves may play an important role in driving the lineage differentiation of seeded mesenchymal stem cells. Particularly, the inorganic chemical component might be important for osteogenic differentiation of mesenchymal stem cells even without growth factor induction.[76] In contrast, ceramic might not be suitable for chondrogenic differentiation and cartilage formation. One study demonstrated the design of biphasic composite scaffold for osteochondral tissue regeneration in articular tissue repair. This composite scaffold contained a ceramic part made of beta-tricalcium phosphate for bone engineering and a PLGA polymer part for cartilage engineering.[77] This concept might be also applied to the engineering of joint composite tissues in hand when stem cells are considered as the cell source.

As mentioned earlier, the hand is a complex structure, composed of many different types of tissues with complicated anatomical relationships. Although more work needs to be done in order to engineer a truly composite tissue graft for clinical application, the engineering of a single type of tissue for clinical application, such as bone, tendon, or skin alone, may be very close to becoming a clinical reality. Dr. Charles Vacanti first reported clinical application of tissue engineering graft for reconstructing thumb digital bone.[4] A patient with thumb injury was first repaired for soft tissue defect using abdominal skin flap, and then the distal phalangeal bone was reconstructed by a block of specially treated natural coral (porous hydroxyapatite; 500-pore Pro Osteon, Interpore International, Irvine, California) that was seeded with autologous periosteal cells. Six weeks after the implantation, the normal length of the phalange bone was restored. Twenty-eight weeks after the implantation, a good pinch strength of 2.3 kilogram was attained with engineered phalangeal bone. With further sophistication of tissue engineering techniques, more successful clinical application of single type tissues could be expected.

Scaffolds for engineering of skin, peripheral nerves and blood vessels

Like bone tissue, skin defects during hand injury may not necessarily pose a challenge for reconstructive surgeon because of easy availability of skin graft. However, the use of engineered skin graft remains desirable as it can avoid donor site morbidity and scarring. Generally, engineered skin resides in one of two categories: artificial epidermis and full thickness skin. Epidermal graft contracts significantly after transplantation and thus would not be the primary choice for repairing skin defects located at the hand, which usually requires full thickness skin graft to carry out hand function.

Skin engineering research might be more advanced than the research on engineering other types of tissues. Particularly, commercially engineered skin products are now widely available, and many of them have already been applied in clinical treatment. Generally, the scaffolds used in skin engineering include natural extracellular matrix and synthetic materials. The former group includes collagen, glycosaminoglycan, decellularized dermis, hyaluronic acid and fibrin; the latter group contains PGA/PLA, silicone, silicone and nylon mesh.[78] Different scaffolds employed in engineered skin products are listed in Table 1.

Table 1.   Currently commercially available or marketed matrices and products for tissue engineered skin substitutes.

Material	Brand Name	Manufacture
Collagen gel + cult. Allog. HuK + allog. HuFi	Apligraf™ (earlier name: Graftskin™)	Organgenesis, Canton, MA
Cult. Autol HuK	Epicell™	Genzyme Biosurgery, Cambridge, MA
PGA/PLA + ECMP DAHF	Transcyte™	Advanced Tissue LaJolla, CA
Collagen GAG–silicone foil	Integra™	Integra Life Science, Plainsborough, NJ
Acellular dermis	AlloDerm™	Lifecell Corporation, Branchberg, NJ
HAM + cult. HuK	Laserskin™	Fidia Advanced Biopolymers, Padua, Italy
PGA/PLA + allog. HUFi	Dermagraft™	Advanced Tissue Sciences, LaJolla, CA
Collagen + allog HuFi + allog HuK	Orcel™	Ortec International, Inc., New York, NY
Fibrin sealant + cult. Autol HuK	Bioseed™	BioTissue Technologies, Freiburg, Germany
PEO/PBT + autol. HuFi + cult autol HuK	Polyactive™	HC Implants
HAM + HuFi	Hyalograft 3D™	Fidia Advanced Biopolymers, Padua, Italy
Silicone + nylon mesh + collagen	Biobrane™	Dow Hickham/Bertek Pharmac., Sugar Land, TX

Table legend: ECMP = extracellular matrixproteins, DAHF = derived from allog. HuFi, GAG = glycosaminoglycan, PGA = polyglycolic acid (Dexon™), PLA = polylactic acid(Vicryl™), PEO = polyethylen oxide, PBT = polybutyliterephthalate, cult. = cultured; autol. = autologous, allog. = allogeneic, HuFi = human fibroblasts, HuK = human keratinocytes, HAM = microperforated Hyaluronic Acid Membrane (benzilic esters of hyaluronic acid = HYAFF-11®). Modified from and reprint permitted by J Cell Mol Med. (Ref. 78).

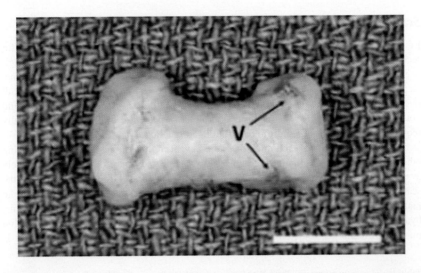

Figure 15. An example of an intact tissue-engineered middle phalange model implanted for 20 weeks in a nude mouse. The construct consists of two chondrocyte-seeded PGA sheets sutured at each end of a central periosteum-wrapped PCL/PLLA copolymer. This macroscopic view shows vascular networks (V) over the surface of the construct. Bar, 1 cm. Reprint permitted by *Cells Tissues Organs*. (Ref. 74).

It is noted, however, that autologous keratinocytes were used mainly for engineering artificial epidermis. However, most of engineered full thickness skin products use allogeneic cells, like Apligraf. In the future, development of more autologous cell based engineered skin products, particularly full thickness skin, might be the focus for clinical treatment of hand soft tissue injury.

Peripheral nerve injury and defect are also common in hand surgery. Particularly, when a large defect is created, tissue engineered nerve repair would be a preferred choice in order to avoid potential donor site morbidity caused by harvesting autologous nerve graft. For tissue engineering mediated peripheral nerve repair, two different approaches are generally employed: cell-free nerve conduit and cell-based engineered nerve.

Among different scaffolds tested for peripheral nerve regeneration, agarose hydrogel was used for its effect on neurite extension on a 3-D hydrogel scaffold. It was found that modification by covalent binding of chitosan and alginate to agarose could significantly enhance neurite growth.[81] Another study showed that agarose hydrogel coated with laminin was able to promote neurite extension when tested with PC 12 cells and ganglia cells.[82] Additionally, acellular muscle basal lamina allograft was also tested for use as a nerve engineering scaffold.[83] In a canine model, a resected recurrent laryngeal nerve could be regenerated when the defect was bridged with a PGA conduit, which was even relatively better than the result of autologous nerve graft repair, indicating that tissue engineering is a promising approach for peripheral nerve repair.[84] In addition, a nerve conduit made of semipermeable polysulfone tubes coated with laminin-1 and nerve growth factors could significantly promote nerve regeneration.[85]

Natural extracellular matrix substance is also considered as one of the ideal materials for nerve guide conduit. For example, cylindrical collagen-based scaffolds with axially oriented pore channels have been designed for peripheral or spinal cord regeneration.[86] In an *in vitro* assay, a longitudinal collagen guidance channel was loaded with Schwann cells and the conduit was found to be able to guide longitudinal alignment of Schwann cells, indicating its potential in peripheral nerve regeneration.[87] It is noted that most of nerve guide conduits are often made with rigid materials which may cause cell loss after *in vivo* implantation due to the lack of physiological local stress during body movement. Therefore, microfiber and nanofiber based conduits may become the new type of scaffold for nerve regeneration. These fiber-based scaffolds are usually manufactured by electro-spinning techniques. These spun fibers could form flexible tubular scaffolds that possess high porosity and remarkable surface/volume ratio, and may be better compatible with the *in vivo* environment. Synthetic materials like PGA, PLGA or a blend of PLGA/PCL were used for flexible conduits. Importantly, these nanofibers usually exhibited excellent com-patibility with seeded cells, including cell attachment and neurite outgrowth, particularly after the modification by coupling laminin onto the fibers.[88–90]

Silk fibroin based scaffolds have also been used for nerve guide conduits. In an *in vitro* assay, Schwann cells were cultured in the medium containing silk fibroin extract fluid and the result showed no significant difference in cell morphology, viability and proliferation comparing to the cells cultured in regular medium, indicating good cell com-patibility of silk based scaffold.[91] Following this, a silk fibroin nerve guidance conduit (NGC) inserted with oriented silk fibroin filaments was used to repair a nerve defect in a rat model, and the reparative result approached that of autologous nerve graft.[92]

In comparison to the cell-free approach, cell-based nerve regeneration has been inten-sively invested and seems to be superior for nerve regeneration. To date, Schwann cells and bone marrow stem cells have been used to repair peripheral nerve defects along with various kinds of scaffolds, such as fibrin, PLGA, and PDLLA.[93–96] In addition to small animal studies, a chitosan/PGA artificial nerve graft was used to repair a 30 mm long nerve defect with success in a canine model.[97] Furthermore, the same conduit made of chitosan/PGA was applied in a clinical a trial to repair a 35-mm-long median nerve defect. During a 3 year follow-up period, an ongoing recovery of both motor and sensory function of the median nerve was observed post-implantation, which could reach M4 and S3+ levels.[98] These reports demonstrate the great potential of engineered nerve repair, and this promising approach is expected to become a clinical reality in patient treatment in the near future, with further advancement in stem cell technology and scaffold design and development.

Like nerve tissue, blood vessel injury or defect is also commonly seen in hand surgery. As with other tissue types, engineered vessel graft would be the preferred choice for repairing vessel defect. Currently, success in engineering small size vessel with a diam-eter between 4 mm and 6 mm has been reported and the engineered grafts have even been used to repair vessel defects in animal studies.[99–101] In addition, clinical trials of engineered vessel have also been reported.[102] The scaffold materials used for blood vessel engineering include collagen gels that were usually wrapped with non-degradable mesh to enhance mechanical strength,[103,104] PGA,[100] PLLA and PLGA,[105] poly 4-hydroxybutyrate (P4HB),[106] co-polymer of PGA and polyhydroxyalkanoate (PHA),[107] decellularized vessel[108,109] and other hybridized materials such as co-electrospun PLGA, gelatin and α-elastin,[110] or co-electrospun PLGA, type I collagen and elastin.[111]

Unlike these engineered small size vessels, clinical requirements in hand surgery may necessitate even smaller size vessels with a diameter between 1 and 2 mm, which can be used for repairing vessel defect in the areas of hand and forearm. Unfortunately, it remains rarely reported in this area. This might partly due to the difficulty of making a thin but mechanically strong vessel wall that has the compliance to adapt itself to *in vivo* environment. Another reason might be the difficulty of maintaining long term patency, especially given the inherent difficulties in epithelialization of small diameter engineered vessels. Lack of good compliance can also lead to vessel stenosis or occlusion after *in vivo* implantation. In the future, a good *in vitro* environment that can mimic the *in vivo* niche and promote vessel maturation before implantation may be critical for the development of engineered vessel prostheses.[112]

References

1. Nerem RM, Sambanis A. Tissue engineering: from biology to biological substitutes. *Tissue Eng* **1** (1995) 3–13.
2. Langer R, Vacanti JP. Tissue engineering. *Science* **260** (1993) 920–926.
3. Cao Y, Vacanti JP, Paige KT, Upton J, Vacanti CA. Transplantation of chondrocytes utilizing a polymer-cell construct to produce tissue-engineered cartilage in the shape of a human ear. *Plast Reconstr Surg* **100** (1997) 297–302.
4. Vacanti CA, Bonassar LJ, Vacanti MP, Shufflebarger J. Replacement of an avulsed phalanx with tissue-engineered bone. *N Engl J Med* **344** (2001) 1547–1548.
5. Stock UA, Vacanti JP. Tissue engineering: current state and prospects. *Annu Rev Med* **52** (2001) 443–451.
6. Freed LE, Vunjak-Novakovic G, Biron RJ, Eagles DB, Lesnoy DC, Barlow SK, Langer R. Biodegradable polymer scaffolds for tissue engineering. *Biotechnology (NY)* **12** (1994) 689–693.
7. Pachence JM, Kohn J. Biodegradable polymers. Chapter 22. In *"Principles of tissue engineering"*. Ed. Robert L, Robert L, Vaccanti J. (Eds.), 2nd Ed. pp. 263–277, Academic Press, San Diego, California (1997).
8. van der Rest M, Dublet B, Champliaud MF. Fibril-associated collagens. *Biomaterials* **11** (1990) 28–31.
9. Grant ME. From collagen chemistry towards cell therapy — a personal journey. *Int J Exp Pathol* **88** (2007) 203–214.
10. Peter SJ, Kim P, Yasko AW, Yaszemski MJ, Mikos AG. Crosslinking characteristics of an injectable poly(propylene fumarate)/beta-tricalcium phosphate paste and mechanical properties of the crosslinked composite for use as a biodegradable bone cement. *J Biomed Mater Res* **44** (1999) 314–321.
11. Xia W, Liu W, Cui L, Liu Y, Zhong W, Liu D, Wu J, Chua K, Cao Y. Tissue engineering of cartilage with the use of chitosan-gelatin complex scaffolds. *J Biomed Mater Res B* **71** (2004) 373–380.
12. Balazs EA. Sodium hyaluronate and viscosurgery. In "Healon (Sodium Hyaluronate): A guide to its use in ophthalmic surgery". Ed. *Miller D and Stegmann R*, pp. 5–28, Wiley, New York (1983).
13. Sung KC, Topp EM. Swelling properties of hyaluronic acid ester membranes. *J Membrane Sci* **92** (1994) 157–167.
14. Mosesson MW. Fibrinogen and fibrin structure and functions. *J Thromb Haemost* **3** (2005) 1894–1904.
15. Mosesson MW, Siebenlist KR, Meh DA. The structure and biological features of fibrinogen and fibrin. *Ann NY Acad Sci* **936** (2001) 11–30.
16. Cho SW, Kim I, Kim SH, Rhie JW, Choi CY, Kim BS. Enhancement of adipose tissue formation by implantation of adipogenic-differentiated preadipocytes. *Biochem Biophys Res Commun* **345** (2006) 588–594.
17. Ye Q, Zünd G, Benedikt P, Jockenhoevel S, Hoerstrup SP, Sakyama S, Hubbell JA, Turina M. Fibrin gel as a three dimensional matrix in cardiovascular tissue engineering. *Eur J Cardiothorac Surg* **17** (2000) 587–591.
18. Nieponice A, Maul TM, Cumer JM, Soletti L, Vorp DA. Mechanical stimulation induces morphological and phenotypic changes in bone marrow-derived progenitor cells within a three-dimensional fibrin matrix. *J Biomed Mater Res A* **81** (2007) 523–530.

19. Bruns H, Kneser U, Holzhüter S, Roth B, Kluth J, Kaufmann PM, Kluth D, Fiegel HC. Injectable liver: a novel approach using fibrin gel as a matrix for culture and intrahepatic transplantation of hepatocytes. *Tissue Eng* **11** (2005) 1718–1726.

20. Hojo M, Inokuchi S, Kidokoro M, Fukuyama N, Tanaka E, Tsuji C, Miyasaka M, Tanino R, Nakazawa H. Induction of vascular endothelial growth factor by fibrin as a dermal substrate for cultured skin substitute. *Plast Reconstr Surg* **111** (2003) 1638–1645.

21. Eyrich D, Brandl F, Appel B, Wiese H, Maier G, Wenzel M, Staudenmaier R, Goepferich A, Blunk T. Long-term stable fibrin gels for cartilage engineering. *Biomaterials* **28** (2007) 55–65.

22. Chung YI, Ahn KM, Jeon SH, Lee SY, Lee JH, Tae G. Enhanced bone regeneration with BMP-2 loaded functional nanoparticle-hydrogel complex. *J Control Release* **121** (2007) 91–99.

23. Taravel MN, Domard A. Relation between the physicochemical characteristics of collagen and its interactions with chitosan: I. *Biomaterials* **14** (1993) 930–938.

24. Liu BS, Yao CH, Fang SS. Evaluation of a non-woven fabric coated with a chitosan bi-layer composite for wound dressing. *Macromol Biosci* **8** (2008) 432–440.

25. Vacanti JP, Morse MA, Saltzman WM, Domb AJ, Perez-Atayde A, Langer R. Cell transplantation using bioabsorbable artificial polymers as matrices. *J Pediatr Surg* **23**(1 Pt 2) (1988) 3–9.

26. Cima LG, Vacanti JP, Vacanti C, Ingber D, Mooney D, Langer R. Tissue engineering by cell transplantation using degradable polymer substrates. *J Biomech Eng* **113** (1991) 143–151.

27. Cima LG, Ingber DE, Vacanti JP, Langer R. Hepatocyte culture on biodegradable polymeric substrates. *Biotechnol Bioeng* **38** (1991) 145–158.

28. Middleton JC, Tipton AJ. Synthetic biodegradable polymers as orthopedic devices. *Biomaterials* **21** (2000) 2335–2346.

29. Commandeur S, Van Beusekom HMM, Van Der Giessen WJ. Polymers, Drug Release, and Drug-Eluting Stents. *J Interven Cardio* **19** (2006) 500–506.

30. Freed LE, Marquis JC, Nohria A, Emmanual J, Mikos AG, Langer R. Neocartilage formation *in vitro* and *in vivo* using cells cultured on synthetic biodegradable polymers. *J Biomed Mater Res* **27** (1993) 11–23.

31. Yan D, Zhou G, Zhou X, Liu W, Zhang WJ, Luo X, Zhang L, Jiang T, Cui L, Cao Y. The impact of low levels of collagen IX and pyridinoline on the mechanical properties of *in vitro* engineered cartilage. *Biomaterials* **30** (2009) 814–821.

32. Kumbar SG, Nukavarapu SP, James R, Nair LS, Laurencin CT. Electrospun poly(lactic acid-co-glycolic acid) scaffolds for skin tissue engineering. *Biomaterials* **29** (2008) 4100–4107.

33. Porter JR, Henson A, Popat KC. Biodegradable poly(epsilon-caprolactone) nanowires for bone tissue engineering applications. *Biomaterials* **30** (2009) 780–788.

34. Schantz JT, Hutmacher DW, Lam CX, Brinkmann M, Wong KM, Lim TC, Chou N, Guldberg RE, Teoh SH. Repair of calvarial defects with customised tissue-engineered bone grafts II. Evaluation of cellular efficiency and efficacy *in vivo*. *Tissue Eng* **9** (Suppl. 1) (2009) S127–139.

35. Mikos AG, Lyman MD, Freed LE, Langer R. Wetting of poly(L-lactic acid) and poly(DL-lactic-co-glycolic acid) foams for tissue culture. *Biomaterials* **15** (1994) 55–58.

36. Thomson RC, Yaszemski MJ, Powers JM, Mikos AG. Fabrication of biodegradable polymer scaffolds to engineer trabecular bone. *J Biomater Sci Polym Ed* **7** (1995) 23–38.

37. Whang K, Thomas H, Healy KE. A novel method to fabricate bioabsorbable scaffolds. *Polymer* **36** (1995) 837–841.

38. Hutmacher DW, Sittinger M, Risbud MV. Scaffold-based tissue engineering: rationale for computer-aided design and solid free-form fabrication systems. *Trends Biotechnol* **22** (2004) 354–362.

39. Griffith LG, Wu B, Cima MJ, Powers MJ, Chaignaud B, Vacanti JP. *In vitro* organogenesis of liver tissue. *Ann NY Acad Sci* **831** (1997) 382–397.

40. Yuan J, Cui L, Zhang WJ, Liu W, Cao Y. Repair of canine mandibular bone defects with bone marrow stromal cells and porous beta-tricalcium phosphate. *Biomaterials* **28** (2007) 1005–1013.

41. Petite H, Viateau V, Bensaid W, Meunier A, de Pollak C, Bourguignon M, Oudina K, Sedel L, Guillemin G. Tissue-engineered bone regeneration. *Nat Biotechnol* **18** (2000) 959–963.

42. Rezwan K, Chen QZ, Blaker JJ, Boccaccini AR. Biodegradable and bioactive porous polymer/inorganic composite scaffolds for bone tissue engineering. *Biomaterials* **27** (2006) 3413–3431.

43. Gelb RI. Tendon transfer for rupture of the extensor pollicis longus. *Hand Clin* **11** (1995) 411–422.

44. Liu Y, Ramanath HS, Wang DA. Tendon tissue engineering using scaffold enhancing strategies. *Trends Biotechnol* **26** (2008) 201–209.

45. Cao Y, Vacanti JP, Ma X, Paige KT, Upton J, Chowanski Z, Schloo B, Langer R, Vacanti CA. Generation of neo-tendon using synthetic polymers seeded with tenocytes. *Transplant Proc* **26** (1994) 3390–3392.

46. Cao Y, Liu Y, Liu W, Shan Q, Buonocore SD, Cui L. Bridging tendon defects using autologous tenocyte engineered tendon in a hen model. *Plast Reconstr Surg* **110** (2002) 1280–1289.

47. Liu W, Chen B, Deng D, Xu F, Cui L, Cao Y. Repair of tendon defect with dermal fibroblast engineered tendon in a porcine model. *Tissue Eng* **12** (2006) 775–788.

48. Ouyang HW, Goh JC, Thambyah A, Teoh SH, Lee EH. Knitted poly-lactide-co-glycolide scaffold loaded with bone marrow stromal cells in repair and regeneration of rabbit Achilles tendon. *Tissue Eng* **9** (2003) 431–439.

49. Cooper JA, Lu HH, Ko FK, Freeman JW, Laurencin CT. Fiber-based tissue-engineered scaffold for ligament replacement: design considerations and *in vitro* evaluation. *Biomaterials* **26** (2005) 1523–1532.

50. Sahoo S, Cho-Hong JG, Siew-Lok T. Development of hybrid polymer scaffolds for potential applications in ligament and tendon tissue engineering. *Biomed Mater* **2** (2007) 169–173.

51. Liu W, Cao Y. Application of scaffold materials in tissue reconstruction in immunocompetent mammals: our experience and future requirements. *Biomaterials* **28** (2007) 5078–5086.

52. Cao D, Liu W, Wei X, Xu F, Cui L, Cao Y. *In vitro* tendon engineering with avain tenocytes and polyglycolic acids: a preliminary report. *Tissue Eng* **12** (2006) 1369–1377.

53. Wang B, Liu W, Zhang Y, Jiang Y, Zhang WJ, Zhou G, Cui L, Cao Y. Engineering of extensor tendon complex by an *ex vivo* approach. *Biomaterials* **29** (2008) 2954–2961.

54. Wei X, Zhang PH, Wang WZ, Tan ZQ, Cao DJ, Xu F, Cui L, Liu W, Cao YL. Use of polyglycolic acid unwoven and woven fibers for tendon engineering *in vitro*. *Key Eng Materials* **288–289** (2005) 7–10.

55. Garvin J, Qi J, Maloney M, Banes AJ. Novel system for engineering bioartificial tendons and application of mechanical load. *Tissue Eng* **9** (2003) 967–979.

56. Awad HA, Boivin GP, Dressler MR, Smith FN, Young RG, Butler DL. Repair of patellar tendon injuries using a cell-collagen composite. *J Orthop Res* **21** (2003) 420–431.

57. Juncosa-Melvin N, Boivin GP, Gooch C, Galloway MT, West JR, Dunn MG, Butler DL. The effect of autologous mesenchymal stem cells on the biomechanics and histology of gel-collagen sponge constructs used for rabbit patellar tendon repair. *Tissue Eng* **12** (2006) 369–379.

58. Tischer T, Vogt S, Aryee S, Steinhauser E, Adamczyk C, Milz S, Martinek V, Imhoff AB. Tissue engineering of the anterior cruciate ligament: a new method using acellularized tendon allografts and autologous fibroblasts. *Arch Orthop Trauma Surg* **127** (2007) 735–741.

59. Ingram JH, Korossis S, Howling G, Fisher J, Ingham E. The use of ultrasonication to aid recellularization of acellular natural tissue scaffolds for use in anterior cruciate ligament reconstruction. *Tissue Eng* **13** (2007) 1561–1572.

60. Badylak SF, Tullius R, Kokini K, Shelbourne KD, Klootwyk T, Voytik SL, Kraine MR, Simmons C. The use of xenogeneic small intestinal submucosa as a biomaterial for Achilles tendon repair in a dog model. *J Biomed Mater Res* **29** (1995) 977–985.

61. Altman GH, Diaz F, Jakuba C, Calabro T, Horan RL, Chen J, Lu H, Richmond J, Kaplan DL. Silk-based biomaterials. *Biomaterials* **24** (2003) 401–416.

62. Fan H, Liu H, Wong EJ, Toh SL, Goh JC. *In vivo* study of anterior cruciate ligament regeneration using mesenchymal stem cells and silk scaffold. *Biomaterials* **29** (2008) 3324–3337.

63. Fan H, Liu H, Toh SL, Goh JC. Enhanced differentiation of mesenchymal stem cells co-cultured with ligament fibroblasts on gelatin/silk fibroin hybrid scaffold. *Biomaterials* **29** (2008) 1017–1027.

64. Chen X, Qi YY, Wang LL, Yin Z, Yin GL, Zou XH, Ouyang HW. Ligament regeneration using a knitted silk scaffold combined with collagen matrix. *Biomaterials* **29** (2008) 3683–3692.
65. Bagnaninchi PO, Yang Y, Zghoul N, Maffulli N, Wang RK, Haj AJ. Chitosan microchannel scaffolds for tendon tissue engineering characterized using optical coherence tomography. *Tissue Eng* **13** (2007) 323–331.
66. Funakoshi T, Majima T, Iwasaki N, Yamane S, Masuko T, Minami A, Harada K, Tamura H, Tokura S, Nishimura S. Novel chitosan-based hyaluronan hybrid polymer fibers as a scaffold in ligament tissue engineering. *J Biomed Mater Res A* **74** (2005) 338–346.
67. Kryger GS, Chong AK, Costa M, Pham H, Bates SJ, Chang J. A comparison of tenocytes and mesenchymal stem cells for use in flexor tendon tissue engineering. *J Hand Surg [Am]* **32** (2007) 597–605.
68. Costa MA, Wu C, Pham BV, Chong AK, Pham HM, Chang J. Tissue engineering of flexor tendons: optimization of tenocyte proliferation using growth factor supplementation. *Tissue Eng* **12** (2006) 1937–1943.
69. Riboh J, Chong AK, Pham H, Longaker M, Jacobs C, Chang J. Optimization of flexor tendon tissue engineering with a cyclic strain bioreactor. *J Hand Surg [Am]* **33** (2008) 1388–1396.
70. Shang Q, Wang Z, Liu W, Shi Y, Cui L, Cao Y. Tissue-engineered bone repair of sheep cranial defects with autologous bone marrow stromal cells. *J Craniofac Surg* **12** (2001) 586–593.
71. Cowan CM, Shi YY, Aalami OO, Chou YF, Mari C, Thomas R, Quarto N, Contag CH, Wu B, Longaker MT. Adipose-derived adult stromal cells heal critical-size mouse calvarial defects. *Nat Biotechnol* **22** (2004) 560–567.
72. Isogai N, Landis W, Kim TH, Gerstenfeld LC, Upton J, Vacanti JP. Formation of phalanges and small joints by tissue-engineering. *J Bone Joint Surg Am* **81** (1999) 306–316.
73. Landis WJ, Jacquet R, Hillyer J, Lowder E, Yanke A, Siperko L, Asamura S, Kusuhara H, Enjo M, Chubinskaya S, Potter K, Isogai N. Design and assessment of a tissue-engineered model of human phalanges and a small joint. *Orthod Craniofac Res* **8** (2005) 303–312.
74. Landis WJ, Jacquet R, Lowder E, Enjo M, Wada Y, Isogai N. Tissue engineering models of human digits: effect of periosteum on growth plate cartilage development. *Cells Tissues Organs* **189** (2009) 241–244.
75. Zhou G, Liu W, Cui L, Wang X, Liu T, Cao Y. Repair of porcine articular osteochondral defects in non-weight bearing areas with autologous bone marrow stromal cells. *Tissue Eng* **12** (2006) 3209–3221.
76. Liu Q, Cen L, Yin S, Chen L, Liu G, Chang J, Cui L. A comparative study of proliferation and osteogenic differentiation of adipose-derived stem cells on akermanite and beta-TCP ceramics. *Biomaterials* **29** (2008) 4792–4799.
77. Jiang CC, Chiang H, Liao CJ, Lin YJ, Kuo TF, Shieh CS, Huang YY, Tuan RS. Repair of porcine articular cartilage defect with a biphasic osteochondral composite. *J Orthop Res* **25** (2007) 1277–1290.
78. Horch RE, Kopp J, Kneser U, Beier J, Bach AD. Tissue engineering of cultured skin substitutes. *J Cell Mol Med* **9** (2005) 592–560.
79. Eaglstein WH, Falanga V. Tissue engineering and the development of Apligraf, a human skin equivalent. *Clin Ther* **19** (1997) 894–905.
80. Priya SG, Jungvid H, Kumar A. Skin tissue engineering for tissue repair and regeneration. *Tissue Eng Part B* **14** (2008) 105–118.
81. Dillon GP, Yu X, Sridharan A, Ranieri JP, Bellamkonda RV. The influence of physical structure and charge on neurite extension in a 3D hydrogel scaffold. *J Biomater Sci Polym Ed* **9** (1998) 1049–1069.
82. Yu X, Dillon GP, Bellamkonda RB. A laminin and nerve growth factor-laden three-dimensional scaffold for enhanced neurite extension. *Tissue Eng* **5** (1999) 291–304.
83. Fansa H, Schneider W, Wolf G, Keilhoff G. Host responses after acellular muscle basal lamina allografting used as a matrix for tissue engineered nerve grafts1. *Transplantation* **74** (2002) 381–387.
84. Kanemaru S, Nakamura T, Omori K, Kojima H, Magrufov A, Hiratsuka Y, Ito J, Shimizu Y. Recurrent laryngeal nerve regeneration by tissue engineering. *Ann Otol Rhinol Laryngol* **112** (2003) 492–498.

85. Yu X, Bellamkonda RV. Tissue-engineered scaffolds are effective alternatives to autografts for bridging peripheral nerve gaps. *Tissue Eng* **9** (2003) 421–430.

86. Madaghiele M, Sannino A, Yannas IV and Spector M. Collagen-based matrices with axially oriented pores. *J Biomed Mater Res A* **85** (2008) 757–767.

87. Bozkurt A, Brook GA, Moellers S, Lassner F, Sellhaus B, Weis J, Woeltje M, Tank J, Beckmann C, Fuchs P, Damink LO, Schügner F, Heschel I, Pallua N. *In vitro* assessment of axonal growth using dorsal root ganglia explants in a novel *Tissue Eng* **13** (2007) 2971–2979.

88. Valmikinathan CM, Tian J, Wang J and Yu X. Novel nanofibrous spiral scaffolds for neural tissue engineering. *J Neural Eng* **5** (2008) 422–432.

89. Koh HS, Yong T, Chan CK, Ramakrishna S. Enhancement of neurite outgrowth using nano-structured scaffolds coupled with laminin. *Biomaterials* **29** (2008) 3574–3582.

90. Panseri S, Cunha C, Lowery J, Del Carro U, Taraballi F, Amadio S, Vescovi A, Gelain F. Electrospun micro- and nanofiber tubes for functional nervous regeneration in sciatic nerve transections. *BMC Biotechnol* **8** (2008) 39.

91. Yang Y, Chen X, Ding F, Zhang P, Liu J, Gu X. Biocompatibility evaluation of silk fibroin with peripheral nerve tissues and cells *in vitro*. *Biomaterials* **28** (2007) 1643–1652.

92. Yang Y, Ding F, Wu J, Hu W, Liu W, Liu J, Gu X. Development and evaluation of silk fibroin-based nerve grafts used for peripheral nerve regeneration. *Biomaterials* **28** (2007) 5526–5535.

93. Kalbermatten DF, Kingham PJ, Mahay D, Mantovani C, Pettersson J, Raffoul W, Balcin H, Pierer G, Terenghi G. Fibrin matrix for suspension of regenerative cells in an artificial nerve conduit. *J Plast Reconstr Aesthet Surg* **61** (2008) 669–675.

94. Hadlock T, Sundback C, Hunter D, Cheney M, Vacanti JP. A polymer foam conduit seeded with Schwann cells promotes guided peripheral nerve regeneration. *Tissue Eng* **6** (2000) 119–127.

95. Rutkowski GE, Miller CA, Jeftinija S, Mallapragada SK. Synergistic effects of micropatterned biodegradable conduits and Schwann cells on sciatic nerve regeneration. *J Neural Eng* **1** (2004) 151–157.

96. Chen X, Wang XD, Chen G, Lin WW, Yao J, Gu XS. Study of *in vivo* differentiation of rat bone marrow stromal cells into schwann cell-like cells. *Microsurgery* **26** (2006) 111–115.

97. Wang X, Hu W, Cao Y, Yao J, Wu J, Gu X. Dog sciatic nerve regeneration across a 30-mm defect bridged by a chitosan/PGA artificial nerve graft. *Brain* **128**(Pt 8) (2005) 1897–1910.

98. Fan W, Gu J, Hu W, Deng A, Ma Y, Liu J, Ding F, Gu X. Repairing a 35-mm-long median nerve defect with a chitosan/PGA artificial nerve graft in the human: a case study. *Microsurgery* **28** (2008) 238–242.

99. L'Heureux N, Paquet S, Labbe R, Germain L, Auger FA. A completely biological tissue-engineered human blood vessel. *FASEB J* **12** (1998) 47–56.

100. Niklason LE, Gao J, Abbott WM, Hirschi KK, Houser S, Marini R, Langer R. Functional arteries grown *in vitro*. *Science* **284** (1999) 489–493.

101. Cho SW, Lim SH, Kim IK, Hong YS, Kim SS, Yoo KJ, Park HY, Jang Y, Chang BC, Choi CY, Hwang KC, Kim BS. Small-diameter blood vessels engineered with bone marrow-derived cells. *Ann Surg* **241** (2005) 506–515.

102. Shin'oka T, Imai Y, Ikada Y. Transplantation of a tissue-engineered pulmonary artery. *N England J Med* **344** (2001) 532–533.

103. Seliktar D, Nerem RM, Galis ZS. Mechanical strain stimulated remodeling of tissue-engineered blood vessel constructs. *Tissue Eng* **9** (2003) 657–666.

104. Isenberg BC, Tranquillo RT. Long-term cyclic distention enhances the mechanical properties of collagen based media-equivalents. *Ann Biomed Eng* **31** (2003) 937–949.

105. Mooney DJ, Mazzoni CL, Breuer C, McNamara K, Hern D, Vacanti JP, Langer R. Stabilized polyglycolic acid fibre-based tubes for tissue engineering. *Biomaterials* **17** (1996) 115–124.

106. Opitz F, Schenke-Layland K, Cohnert TU, Starcher B, Halbhuber KJ, Martin DP, Stock UA. Tissue engineering of aortic tissue: dire consequence of suboptimal elastic fiber synthesis *in vivo*. *Cardiovasc Res* **63** (2004) 719–730.

107. Shum-Tim D, Stock U, Hrkach J, Shinoka T, Lien J, Moses MA, Stamp A, Taylor G, Moran AM, Landis W, Langer R, Vacanti JP, Mayer JE, Jr. Tissue engineering of autologous aorta using a new biodegradable polymer. *Ann Thorac Surg* **68** (1999) 2298–2304.

108. Dahl SL, Koh J, Prabhakar V, Niklason LE. Decellularized native and engineered arterial scaffolds for transplantation. *Cell Transplant* **12** (2003) 659–666.
109. Tamura N, Nakamura T, Terai H, Iwakura A, Nomura S, Shimizu Y, Komeda M. A new acellular vascular prosthesis as a scaffold for host tissue regeneration. *Int J Artif Organs* **26** (2003) 783–792.
110. Li M, Mondrinos MJ, Chen X, Gandhi MR, Ko FK, Lelkes PI. Co-electrospun poly(lactide-co-glycolide), gelatin, and elastin blends for tissue engineering scaffolds. *J Biomed Mater Res A* **79** (2006) 963–973.
111. Stitzel J, Liu J, Lee SJ, Komura M, Berry J, Soker S, Lim G, Van Dyke M, Czerw R, Yoo JJ, Atala A. Controlled fabrication of a biological vascular substitute. *Biomaterials* **27** (2006) 1088–1094.
112. Zhang WJ, Liu W, Cui L, Cao Y. Tissue engineering of blood vessel. *J Cell Mol Med* **11** (2007) 945–957.

CHAPTER 6

Animal Models for Engineering Tissues in the Upper Extremity

Xing Zhao[*,†,‡] *and Mark A. Randolph*[*,†,§]

Introduction

The comparative study of animals to gain a deeper understanding of human anatomy and physiology dates back to the time of ancient Greece and Rome. The involvement of animals and animal models in scientific experimentation probably began in earnest in the 18th and 19th centuries and evolved into common practice in the 20th century. Although the exact number of all animals used in medical research is not known, estimates range between 17 and 23 million vertebrate animals used each year in the United States — most of which are small rodents (i.e., mice).[1] Accurate numbers of mice used are not available, but the United States Department of Agriculture tracks the use of all species covered under the Animal Welfare Act (all vertebrates except rats, mice, birds, amphibians, and fish) and slightly more than one million are used each year in the US.[2] Although *in vitro* studies using mammalian cell cultures can provide valuable molecular and metabolic information in isolated conditions, and *ex vivo* studies of human cadaveric cells, tissues, and organs allow some degree of understanding of human anatomy and physiology, the ethical and appropriate use of animals and animal models fills a critical gap in our understanding of biology in intact organisms. Not only do they allow for greater understanding of normal physiology in an unperturbed, normal state, but they also allow us to delve deeper into diseased states either by the use of knockout or transgenic animal models or by using injury models to study healing, repair, and regeneration. There are groups of many different persuasions that believe under no circumstances should animals be used for the benefit of mankind, whether they be used for a food source, enlisted for work in land cultivation, or subjected to experimentation. On the other hand, a valid argument can be made by the majority of society that the humane use of animals for supporting mankind supersedes all arguments against animal use and is necessary for our survival through safe and nutritious food sources and the prevention of disease and other maladies. From my perspective as a long-time investigator using small and large animal models to study orthopedic injury and repair as well as a co-chairperson of an Institutional Animal Care and Use Committee (IACUC), I am cognizant of the competing demands of the many positive and negative influences that have established our current environment of animal experimentation. The decision to use animals for whatever purpose is shaped by society, but my personal belief

---

*Corresponding authors.

[†]Division of Plastic Surgery, Harvard Medical School, Massachusetts General Hospital, 15 Parkman St., WAC 435, Boston, MA 02114.

[‡]E-mail: marandolph@partners.org

[§]E-mail: zhaoxing614@hotmail.com

is that animal use must be highly justified and conducted in the most humane and ethical manner that can be achieved. The intent of this chapter is to give the reader the tools to make the choice of whether an animal model should be used for a tissue engineering study of tissues in the hand and, if so, how to select the most appropriate animal model and outcome measures for the intended study.

This chapter will not provide a lengthy list of all the animal models and their relative advantages and disadvantages in studying engineering of tissues for the hand and upper extremity. The intent is to provide an algorithm for the ethical and humane use of animals once extensive *in vitro* testing has been completed (Figure 1). Investigators are encouraged to perform very extensive and critical literature search of animal models for their particular study prior to embarking on *in vivo* investigation. We strongly reiterate that one's literature review must be very critical of the pros and cons of the published information so

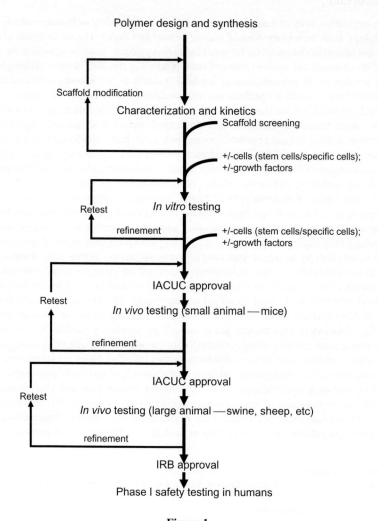

Figure 1.

as to not perform unnecessary animal experimentation that will not add constructive information to our foundation of understanding. All testing that can be performed *in vitro*, such as toxicity and biocompatibility assays, should be completed *in vitro* prior to embarking on animal investigations. Lastly, all animal work must be approved by the IACUC or institutional ethics committee and should adhere to the rules and regulations set forth in the Guide for the Care and Use of Laboratory Animals.[3]

Selection of an Animal Model:

Need for animal model

Several factors need to be considered when selecting a suitable animal model for a tissue engineering study. First and foremost is whether or not an animal should be used to answer the scientific question posed. The hypothesis should be clearly stated and the experimental design critically evaluated by a statistician, team members, and possibly non-team colleagues. Enlisting the opinions of non-team colleagues, who are not competitors to one's own research goals, can be invaluable in assessing the need for utilizing animals and evaluating the choice to use a certain animal model. This broadening of input can serve investigators well if deficiencies in a particular model(s) are identified prior to beginning a study that could ultimately result in a waste of animals if the model is flawed. Once it is determined that an animal model should be used, the next steps include evaluation of animal model, the animal size, the availability of tissues and species-specific reagents, and the relevance to human biology and clinical application. Many answers to these questions relate to the selection of outcome measures that will be discussed in greater detail below.

Suitability

Once the decision to use an animal has been made, the next step is to determine which animal model to use to provide data to support or refute the hypothesis. The model selected for investigation should be suitable and appropriate for the study including size, availability, and institutional constraints. The animal model needs to be carefully selected based on several criteria. Usually the first choice is made on size, which is indirectly related to cost in most cases of mammalian use. The first question often posed by investigators or members of the IACUC is "Can it be done in a mouse?" The question appears somewhat naïve and mistakenly based on the assumption that small, rather distasteful animals are less important or that they suffer less pain and distress than larger species, such as a dog or a horse. This choice is often a financial decision, however investigators have to be careful that large numbers of mice don't suffer or die on unnecessary studies because an investigator can only "afford" to do the experiments in mice. The investigator has to be certain that the selection of animal model is based on scientific criteria and not merely on affordability. Some institutions may not have facilities for the desired species (e.g. horses or NHPs), and this needs to be factored into the choice of model as well. Grant reviewers also should be well informed as to why a particular animal model selected would be the best for a proposed study.

Availability of species-specific reagents

The availability of a vast number of genetically modified mice along with a large catalogue of mice-specific cytokines and growth factors has made the mouse the workhorse in medical scientific investigation. These resources permit careful molecular and cellular study of mice models of human disease. The availability of such a rich resource animal can be quite useful in the engineering of musculoskeletal tissues. For example, one could conceivably introduce an engineered construct to make tissue or modify a microenvironment in a mouse that lacks a critical protein or growth factor for normal tissue generation. Thus, the ability to dissect the effects at the molecular and cellular level in mice is a tremendous asset in understanding normal metabolism and function. The small size of the mouse and the sometimes fatal genetic modifications can limit their use in engineering tissues for musculoskeletal repair and regeneration, however.

In larger animal models, the more limited availability of species-specific reagents may be a hindrance to fully understanding some mechanisms underlying the regeneration or engineering of tissues. However, there are many reagents that are useful across several species. The primary basis for employing large animal models is to closely simulate the defect size and repair mechanisms of humans. Furthermore the anatomical and ultrastructural similarities and differences (e.g. bone size and architecture) need to be considered in choosing the appropriate animal model. Selection of species and defect models will be covered in more detail below.

Scientific and clinical relevance

The selection of an animal model in engineering tissue for the hand and upper extremity should have relevance to the scientific objective and the potential clinical objective. *In vitro* studies should be conducted to confirm biocompatibility and safety prior to embarking on animal studies. These are then followed by studies in rodents, usually mice or rats. New polymer scaffolds, either with or without growth factors are generally placed in rodents to assess *in vivo* biocompatibility and evaluate initial effectiveness for tissue generation. If the intent is to assess the biocompatibility, the rodent model should be immunocompetent to assess any inflammatory response to the scaffold or its components. Due to the frequent need to harvest large volumes of cells from animals or humans in order to seed constructs, many studies employ immunodeficient mice (athymic or severe combined immunodeficient) in order to reduce the immune response to the cellular component of the constructs. In these instances, the athymic nude mouse and the SCID mouse have predominantly been the animal models of choice and the athymic rat has provided a slightly larger, yet similar, animal model. Although these rodent models can provide valuable data on scaffolds and tissue formation, they often are too small to permit clinically relevant engineered solutions in sizes necessary for human application.

The choice for using large animals and the selection thereof should be carefully evaluated in the context of the scientific question being posed. If the anatomical and ultrastructural properties of an animal fail to adequately model the human condition, another species should be selected. Compromises may have to be weighed in selecting one species over another. For example, if only one extremely rare non-human primate species from a remote rain forest perfectly mimicked human bone regeneration in size and mechanism, it is still unlikely that the species will be used for any studies. Clearly, the lack of availability and the political climate by animal activist groups against using NHPs will certainly make investigators choose more common quadruped animal models like dogs or farm animals. This is acceptable since there are numerous similarities to human tissues in many musculoskeletal structures of these quadrupeds. On the other hand, the anatomical dissimilarities quadrupeds can limit their use for exactly replicating the human. For example, the flexed stifle (knee) joint of the hind limb of quadrupeds does not permit evaluation of the biomechanical forces on the ligaments and cartilage of that joint with the same accuracy as the more vertically aligned bipedal human knee. Similarly, there are no animals with an opposing thumb hand that accurately models the human and quadrupeds have entirely different distal limb structures. Nonetheless, quadrupeds can be used for studying engineered constructs such as long bone defects or nerve and tendon deficiencies.

Another consideration is the age of the animal used in studies. Large animals are extremely expensive and even more costly with increasing age (and weight). Thus, with the possible exception of NHPs and equine species, many large animal studies are performed in very young animals — often less than one year old. Although most quadrupeds are skeletally mature (closed physes) by one year, it could be argued that the regenerative capacity of these young (baby) animals is greater than that which would be achieved in older (middle aged) animals. Clearly if regenerative techniques do not work in these young animals, there is strong likelihood that the strategies will not perform in older animals. Conversely, successful regeneration of acute lesions in young healthy animals may not be translated directly to the repair or regeneration of chronic lesions that occur in sometimes unhealthy patients.

Selection of Outcomes Measures

The selection of a suitable animal model relies heavily on the selection of outcome measures that will be used to test the hypothesis. Outcome measures can be qualitative or quantitative measures of structural parameters, scaffold properties, or biological studies of cellular, tissue, or organism responses. Poorly designed studies with unfocused outcome measures will not result in credible, statistically sound data being collected and will likely be difficult to publish. Quantitative outcomes measures that provide discrete data sets, such as the amount of force it takes to break the femur of ten 12-week old wild type C57/Bl6 mice versus ten osteoporotic mice, are easily subjected to statistical comparison. On the other hand, the biological remodeling of a fracture site in ten rat femora can be quite different and sample sizes for mechanically testing comparative groups may require larger numbers to achieve statistical differences or similarities. Many such studies also report on qualitative histological results. Qualitative and often subjective data such as histological scoring may require far larger numbers of specimens in order to be subjected to statistical analysis.

During the experimental design phase investigators should attempt to catalogue the possible outcome measures for collecting data to test their hypothesis without regard to the animal model. Subsequently, a list of animal models that could be used could be developed. For example, if one were to perform a tendon repair to test a particular type of suture technique, the mouse might be the least useful model. On the other hand, if the hypothesis were to examine the knockdown of a particular structural protein in tendon repair and regeneration, the mouse would be the better model since the molecular tools do not exist for using animals such as the horse or goat. If the model chosen cannot provide the requisite data to either support the hypothesis or support the null hypothesis, the investigator has a flawed design and animals could be needlessly used. Certainly, one cannot always predict that a model can provided sufficient statistically sound data, but that is justification for performing carefully designed pilot studies and calculating statistical power of the experimental design. Outcome measures for specific tissues will be covered in the following sections on specific tissues.

Critical Size Defects

In many instances, investigators are engineering tissues to restore integrity to structural tissue deficits due to traumatic loss or degenerative disorders. Studies are commonly designed to make a musculoskeletal defect and grow replacement tissue to place into the defect or encourage new tissue to form in the area of interest through the use of scaffolds and growth factors. In bone tissue engineering, for example, defects are made in the long bones or cranium of animals and filled with the construct or factors under study. Investigators should use caution in the design of such defect models to ensure that the defects are of sufficient size in order to acquire credible data showing that the defect will not spontaneously heal if left alone, or that the regeneration in the defect can be adequately measured to show any differences in healing and regeneration with sufficient statistical probability. That is not to suggest that all studies need to have critical size defects, but that the defect chosen has appropriate sensitivity to accumulate data to test the hypothesis. Critical size defects for bone, articular cartilage, nerve, tendon, and vessels have been determined in many animal models, large and small. A thorough review of the literature is required to ensure that the appropriate defect is made if one claims it is of critical size.

Another concern related to defect models is the nature of the lesion created. In most studies, a defect is made in the tissue of interest and immediately filled with the engineered construct. These defects are acute injuries and generally have little in common the clinical condition where chronic lesions are being treated months or even years after the initial injury. The milieu of an acute would that has healthy surrounding native tissue with good vascularity is, in most cases, not similar to the scarred or degenerated environment in the clinical condition of patients. The generation of chronic wounds in bone, cartilage, or nerve is not often feasible and may be discouraged by an IACUC, but investigators should consider these in designing experiments. The pitfall of making claims drawn from treating acute lesions made in young healthy animals can be a serious flaw in translating the findings to clinical application.

Animal Models of Tissue Engineering

The preceding sections discuss numerous variables to consider when deciding whether or not an animal model can and should be used for study as well as some of the parameters to needed to evaluate various models. This section will provide an overview of possible animal models for studying various tissue types. It is not possible to discuss every feasible animal model for every engineered tissue. Rather, the goal will be to point out some pros and cons of various models in the study of particular tissues. In most cases, the first animal model employed is likely the mouse for screening scaffolds or cells or growth factors and less attention will be paid to the use of mice in this section. Much of the discussion to use mice is covered in preceding sections. As was stated previously, investigators need to perform a thorough and critical review of the published literature when choosing any animal model. If an uninformed decision was made to use an inadequate model because of cost or inconvenience, the data may be flawed and the study deemed useless in the end. If there are established models that have been used to form the foundation of a particular field of interest, it will be very difficult, if not impossible, to sway reviewers that a new or unusual model can be used for comparison. That is not to suggest that reviewers will dismiss new models arrogantly. However, if a certain field of study has long used a particular animal model (e.g. dog), introducing a new untested model that was chosen for cost or political reasons may not be readily received until rigorously tested.

Bone

Generation of bone to fill segmental defects or augment fracture repair has been an area of active investigation for much of the last century. Numerous tissue engineering approaches to generate bone that include cells placed on various types of scaffolds have been described, many of which are discussed in other chapters of this book. In the broader definition of tissue engineering, however, generation of bone can include the delivery of bone morphogenic proteins or other signals that can promote and stimulate bone formation. However broad a definition one accepts, the goal to generate bone tissue to bridge defects of stimulate fracture healing remains the guiding clinical objective. Other chapters of this book will delve deeply into the methods of engineering bone, but thorough *in vitro* testing should be performed prior to embarking on live animal studies.

Many animal models, both large and small, have been developed to study bone healing and repair. Their advantages and disadvantages have been summarized in several publications.[4-6] As stated above, the model needs to be carefully selected to provide data in support of or to refute the hypothesis being tested. Many physiological factors need to be considered such as animal size, gender, the macro and microarchitecture of the bone in the species chosen, similarities in bone composition and remodeling to humans, the availability of species-specific cells and growth factors, and accurate modeling of clinical fractures or the ability to make critical-sized defects. Much can be also learned about bone generation from the dental and craniofacial models of bone defect repair and bone augmentation studies.

Animal selection, size, and macrostructure

Mice and rats have been employed in many studies where they have found utility in early stages of scaffold design and degradation, where recombinant molecules have been specifically generated for the rodent species, and where genetically modified mice have been generated to study specific aspects of bone healing and repair.[7-9] Femoral defect models have been developed in both rats and mice and have some utility in early *in vivo* studies, particularly for initial screening of safety, biocompatibility, and effectiveness. Despite their utility in early phases of testing, rodents cannot model the clinical situation as well as larger species such as dog, sheep, or swine.

Despite their similarities to bipedal humans, very few bone repair studies have been performed in nonhuman primates. Most studies have been performed in quadrupeds and, for many years, dogs were the model of choice to study bone healing and remodeling.[4] Dogs are available in various sizes, yet even the larger breeds commonly used are not of comparable size and load-bearing to that of adult humans. Dogs are generally very docile and tractable making them easy to handle in the research setting. They are also able to ambulate on three legs and tolerate fixation devices, internal as well as external, quite well. Several models of segmental defects in long bones have been reported in dogs including ulnar and radial defects in the forelimb and tibial or femoral defects in the hindlimb. Additionally, surgical equipment along with many fixation devices and implants has been developed for dog studies or treatment in the veterinary field. Dogs and NHPs, however, are politically sensitive species forcing many investigators to turn to farm animals, rabbits, or rodents.

Farm animals or food production species such as sheep, goats, and swine are being used more frequently in all types of research, not only musculoskeletal repair.[2] Although not as variable in size as dog breeds, there is some degree of size variation. Sheep are

generally about the size of adult humans and are being used for studying repair of segmental bone defects, commonly in the tibia.[10] Goats are slightly smaller, and due to their
inquisitive nature may be more difficult to maintain in confinement for long periods. Food
production swine grow very rapidly to extremely large sizes, often reaching 250–300 kilo-
grams in less than a year. However, several smaller or miniature breeds have been
propagated for laboratory investigations including Yucatan miniature swine, Hanford, and
the Massachusetts General Hospital breeds.[11–13] These breeds do not often exceed 100 kg
when grown to adulthood and easily adapt to cage confinement. The short legs with bulky
musculature surrounding the limbs in swine, however, make them less desirable than the
longer and more slender limbs in sheep and goats for long bone repair studies.

The type of defect chosen for repair needs to be seriously considered at the outset of
the experiment. For example, the radius is the primary weight bearing bone in the forelimb
of quadrupeds. The ulna, on the other hand is somewhat protected from full weight bearing and might be selected for this reason. The tibia and the femur also bear full weight in
the hind limb and need protection in the form of internal or external fixation for most
defect repair studies. Unlike compliant patients who follow instructions to not bear full
weight on fractured legs, animals will bear as much as tolerated and can easily destroy the
stability of the fixation devices. A significant shortcoming of ungulate quadruped models
for fracture and segmental bone studies in the upper extremity is the dissimilarity from
humans of the wrist and phalanges. The length and anatomical differences of the wrist and
phalangeal bones in the ungulates does not provide a suitable functional model of the
human hand. Although the hands of nonhuman primate provide a better approximation of
the bones and other tissue structures in the distal extremity, cost and handling difficulties
preclude most use of NHPs for bone repair studies in the hand.

Rabbits are one of the most commonly used animals in musculoskeletal research[4]
probably due to lower costs of procurement and lower per diem fees for care than larger
animal species. Critical-sized defects can be made in the long bones of the rabbit limbs,
but the sizes of these defects are much smaller than those that can be made in large species
and calls into question the relevance of rabbit data to clinical application. Nonetheless, the
rabbit provides a model slightly larger than a rodent in which some valuable data can be
derived on scaffold design and biocompatibility. One significant drawback in rabbit bone
is its lack of similarity to human bone microarchitecture as described below.

Microarchitecture and composition

Whereas the microarchitecture of musculoskeletal tissues like muscle, tendon, and vessels
are similar throughout most laboratory species, the microstructure of the long bones can
be quite different among the species and significantly different than human bone. Adult
human long bones have a secondary osteonal structure with osteons generally greater than
100 microns containing blood vessels and cement lines that form a boundary between
adjacent lamella.[14] Pigs are described as having a similar microstructure to that of the
human.[15] Dogs have a mixed microstructure that includes the secondary osteonal structure
in the center of the cortical bone. However, dogs have a plexiform or laminar type of bone
on the endosteal and periosteal interfaces.[16] Plexiform bone is common in rapidly growing
animals and occasionally in children undergoing periods of rapid growth.[17] Sheep and
goats have predominantly primary bone structure with osteons less than 100 microns and
no cement lines between adjacent lamella.[18,19] Also, the Haversian systems are not distributed uniformly in these species. Rabbit long bones are quite dissimilar in structure to
human bone.[20] They have vascular canals of osteons running parallel to the long axis of

Table 1. Summary of four key features in terms of similarity of animal model to human bone.*

	Rabbit	**Canine**	**Swine**	**Sheep/Goat**
Macrostructure	+	++	++	+++
Microstructure	+	++	++	+
Bone Composition	++	+++	+++	++
Bone Remodeling	+	++	+++	++

+least similar; ++moderately similar; +++most similar

Adapted from Pearce AI, Richards RG, Milz S, Schneider E, Pearce SG. Animal models for implant biomaterial research in bone: a review. Eur Cell Mater. 2007 Mar 2;13:1-10. Review.

the bone on the medullary and periosteal surface. Between these layers is dense Haversian bone. Very dissimilar to human bone, rodents have a lamellar type of long bone structure with large blood vessels passing transversely through the cortex.

There are differences in the composition and density of bone among the various species as well.[21] The dog and the pig have bone density most similar to the human. The composition of dog bone is most similar in terms of water fraction, organic fraction, volatile inorganic fraction and ash. However, these differences in bone composition among the larger species and humans is probably not a significant factor in selecting a model. Table 1 from Pearce *et al.* provides a summary comparison of various attributes of bone canine, pig, rabbit, and sheep and goats.[5]

Bone remodeling and mechanical factors

Bone has tremendous capacity to remodel in the fracture or defect site. Just as important as the selection of the species and defect site is the capacity for remodeling of the new bone in the defect. Remodeling can take months or years in humans and can vary among species and among sites within a single species.[5] Many animal studies, however, are extremely short because of the cost of long-term daily care and full remodeling does not have adequate time to occur. Rabbits tend to have higher remodeling rates than other large animal species. Swine, dogs, and goats generally have a higher remodeling rate than humans, but preclinical data from these animals can still be valuable for predicting clinical outcomes. Remodeling can also depend heavily on the mechanical and micromechanical forces on the bone and the defect site.[22]

Age and gender

The regenerative capacity of young animals is believed to be higher than that of old animals. Age-related differences on the composition of bone are well documented.[23] Additionally, the number of putative stem cells in bone marrow and their regenerative capacity is also believed to decline with increasing age.[24-26] Since many animal models for musculoskeletal research are relatively young, their capacity to repair bone defects may not accurately represent the situation clinically. Unlike most other musculoskeletal tissues, gender has to be considered when performing bone regeneration and tissue engineering. There are validated anatomical and composition differences of the cortical bones between male and female animals[27] that should be considered for regeneration studies. Furthermore, the negative effects of female hormones on bone formation preclude the use of females for most bone formation studies.[28]

Cartilage

Unlike bone that has great capacity for regeneration and remodeling to restore itself to near normal at a fracture or defect site, cartilage has limited capacity for repair and self-regeneration.[29] Cartilage is not innervated, has no lymphatic drainage system, and lacks a direct vascular supply. The lack of vasculature lessens the normal inflammatory response observed in most wound healing responses of other musculoskeletal tissues including the delivery of putative reparative cells to the site of injury. Consequently, injury to cartilage in the articulating joints from trauma, burns, tumors, arthritis, or other metabolic causes results in scar formation that can lead to possible arthritic changes resulting in pain, stiffness, and loss of structure and function. Lesions that penetrate into the subchondral bone can permit inflow of blood and bone marrow cells to the site of injury and forms a repair scar tissue that fills the defect site.[30] In fact, for small isolated lesions in the joint surface a technique known as microfracture involves making multiple small holes through the subchondral bone to promote this influx of bone marrow to fill the lesion.[31] The reparative matrix is not normal hyaline cartilage, however. It is generally of poor quality that is high in type I collagen and lacks the biomechanical strength for long-term durability. Other strategies to deliver cells, either chondrocytes or mesenchymal stem cells, through autologous cell implantation (ACI) without a scaffold[32] or by matrix assisted strategies (MACI)[33] have yielded only moderate results in restoring the joint surface to normal. Not all lesions are confined to the chondral matrix and some treatment approaches involve restoring the subchondral bone as well through the transfer of osteochondral plugs (OATS)[34] within the joint or through engineering multilayer constructs.

Cartilage repair and regeneration has not been extensively studied in the upper extremity (or forelimb) of animal models. The joint spaces of the wrist and phalanges of quadrupeds are difficult to access and manipulate without causing significant additional morbidity. Small rodents are of limited use for *in situ* studies in joint surface repair because of their small joint size and extremely thin cartilage surfaces. The standard models for cartilage repair have been lesions made in the stifle (knee) joint of a multitude of animals including rabbit, dog, sheep, goat, swine, and horse. The focus on the knee joint is probably a reflection of the intense desire to restore function to a joint that is highly susceptible to osteoarthritis and has significant interest from a commercial standpoint. While the knee joint shares may anatomical similarities to diarthrodial joints in the upper extremity in terms of cartilage tissue, the upper extremity of humans is subjected to significantly different biomechanical forces than the forelimb in quadrupeds. Thus, the forelimb of quadrupeds offers little advantage over the hindlimb for articular cartilage repair studies.

Outcome measures

Human articular hyaline cartilage is comprised of chondrocytes embedded in a matrix that is 60–80% water, 4–7% wet weight aggrecan, and 10–20% wet weight collagen, predominantly type II.[29] Water, nutrients, and cellular waste diffuse readily through the matrix. Articular cartilage has a net negative charge due to the glycosaminoglycan (GAG) that plays an important role in resisting compressive mechanical forces. Another key feature of joint cartilage is the low friction surface provided by a tangential layer of cells, including the lamina splendens, which produce the lubricating protein lubricin.[35] Engineered cartilage is generally evaluated biochemically for the total amount of glycosaminoglycan and collagen produced by the cells. Collagen can be further evaluated for the types of collagen produced by biochemical approaches or immunohistochemistry. Microarchitecture is assessed by histological stains including hematoxylin and eosin, Safranin-O, toluidine blue, and trichrome. Since proteoglycan is charged, the intensity of charged dyes like Safranin-O can be correlated to the amount of GAG in the tissue. Several histological scoring methods have been reported including those designed by Mankin,[36] Pineda[37] and O'Driscoll.[38] The ability of joint cartilage to resist compression and cushion the joint, the biomechanical nature of engineered cartilage is extremely important. Although there are mechanical indenters that can be used *in situ* to evaluate some mechanical characteristics of the regenerated cartilage in knee defects, more accurate measures of the compressive modulus can be obtained by compression testing *ex vivo* with mechanical testing devices. While native joint cartilage functions by resisting compressive forces, mechanical forces also have been shown to be important in the generation of engineered cartilage, both *in vitro* and *in vivo*.

Species selection

Initial tissue engineering studies generally have been performed in athymic mice or rats to assess cell-polymer compatibility and the ability to form new cartilage matrix.[39–41] Selection of an immunocompetent animal model, however, relies on the availability of cells and ease with which the scaffold can be implanted and stabilized. Many strategies for autologous chondrocyte implantation involve the harvest of chondrocytes from a joint surface, expanding the cells in culture, and the reimplantation of the cells in or on a scaffold carrier. For approaches using mesenchymal stem cells, the animal needs to have access to sufficient amounts of marrow or source tissue like fat to acquire adequate numbers of cells. Thus, small animals such as rodents are not good models.

At the next stage, the rabbit is often chosen because it is less expensive to procure and has lower daily care charges than larger species. The rabbit knee joint model has been used as a model for articular cartilage repair and regeneration studies for more than 40 years.[42] Seminal studies on chondrocyte implantation by Grande *et al.*[43] and chondrocytes implanted in collagen gel by Wakatani *et al.*[44] established the rabbit stifle joint as a suitable model for tissue engineering and articular surface regeneration. However, the thickness of the cartilage in the stifle joint of the rabbit is about 0.25–0.75 mm compared to 2.2–2.4 mm in the human knee.[45] The thin cartilage makes the rabbit a very difficult model for many tissue engineering approaches confined to chondral lesions. Although it is possible to make a full-thickness osteochondral defect in rabbit knee cartilage, it is not possible to create a partial thickness chondral defect in rabbits.

Commonly used large animal models for joint surface repair in the stifle joints include swine, dog, sheep, goats, and horses.[46–50] There does not appear to be any review providing a head-to-head comparison of the relative merits of each of these large animal models. None have shown any advantage over another because all repair strategies have been largely unsuccessful in restoring the joint surface to the normal histoarchitecture and function in any species. The horse has joint cartilage thickness most closely resembling that of the young adult human. The choice of animal model, however, may rely on the species available and the facilities and experience of an institution. For example, veterinary schools can more easily study horse models than inner city academic medical centers. Other considerations for choosing a model are the tolerance of the animal to the injury, weight-bearing surface or nonweight-bearing site, and manageability of the animal during rehabilitation.

Acute versus chronic lesions

Another limitation of animal models for cartilage repair is the use of acutely made defects in the joint surface. Lesions are generally made by making holes in the joint cartilage that are immediately filled with the engineered construct under study. This does not replicate the clinical condition where chronic and degenerating lesions are not treated for many weeks or months. Furthermore, with probably the exception of the horse, the thickness of the knee joint cartilage is much thinner than human joint cartilage and bleeding from the subchondral bone can often confound reconstructive strategies. A study by Saris *et al.* has demonstrated that early treatment of defects made in the joint of Dutch Milk Goats was significantly better that defects that were treated 10 weeks after the lesions were made.[46] The latter are more indicative of the clinical situation and raise many questions regarding the clinical relevance of experimental models treating acute lesions.

Despite the large investment in articular surface repair, many challenges remain. Engineered constructs have been shown to permit the formation of cartilage matrix in the lesions, but no technique has been successful in restoring all layers of the injured cartilage to normal. Many strategies have been successful in generating cartilage matrix high in type II collagen and others have been only successful in generating fibrocartilage high in type I collagen, which have been shown to break down with time. One aspect that has received little attention during the engineering process is the tangential layer of cells on the surface of cartilage including the laminin splendens and is responsible for producing lubricin, a protein crucial for lubricating the joint and reducing friction.[51] Although many tissue engineering approaches can achieve nearly normal levels of glycosaminoglycan levels, most approaches are ineffective in achieving normal levels of collagen type II. Low levels of type II collagen and limited understanding about the molecular assembly and ultrastructure of the new matrix proteins can possibly lead to inferior biomechanical properties of the engineered cartilage.

Tendon and Ligament

The primary function of tendon and ligaments is to transmit force. For tendons, forces are transmitted from the contraction of muscles to the bones to which they are attached. Ligaments, on the other hand provide stability by attaching to apposing bones in a diarthrodial joint. Tendons and ligaments are commonly grouped together due to the many similarities, yet there are many ways that they differ as well.[52] In the simplest sense, the midsubstance of both tendons and ligaments are linearly aligned, primarily collagen type I fibers with some elastin and cells dispersed throughout. The terminal ends of the tendons connect one end to muscle and the other to bone, whereas ligaments connect to bone at both ends. Tendons can be either intrasynovial (flexors and extensors in the hand) or extrasynovial (Achilles tendon). Ligaments can be intra-articular (anterior cruciate ligament in the knee) or extra-articular. Tendon and ligament injuries are among the most common musculoskeletal problems. Complete or incomplete ruptures of the midsubstance or tears away from their origins and insertion sites as well as inflammation, such as tendonitis, can result severe pain and loss of mobility. Many surgical techniques have been described to suture or stabilize ruptures and many new fixation devices have been introduced to fix ruptured ends back to their bone insertions. Although many of these strategies are successful, however, many injuries cannot be easily repaired with the tissue that remains at the site if injury. Gaps in tendons and severed ligaments are not easily bridged and often require augmentation or replacement such as tendon transfers or replacement as in the case of the anterior cruciate ligament. There are many opportunities for tissue engineering to play a role in repair and regeneration of these structures.

The selection of the animal model will depend on the structure one is investigating and the outcome measures selected for collecting meaningful data. Because the primary function of tendons and ligaments is to bear load and transmit force, the reconstructive goal is to match that ability through replacement or regeneration. Although there have been many advancements in material science over the past three decades, nonbiological replacement materials for tendons and ligaments have not been successful.[53,54] Particularly in the flexor and extensor tendons of the hand, unhindered mobility without any adhesions in the synovial sheath is a critical element for reconstruction and restoration of function.[52] Most models of ligament repair and reconstruction have focused on the stifle (knee) joint in large animals. For example, injury and repair of the medial collateral ligament of the rabbit has been studied extensively.[54,56] Animal models for ligament repair in the hand and upper extremity are poorly developed.

Many biochemical, anatomical, ultrastructural, and some biomechanical studies can be performed in small rodents.[57] The ready availability of genetically manipulated mice make them useful for spontaneous repair of lesions or studies involving the injection of factors into these lesions. The mouse is generally too small, however, for surgical manipulation to study repair techniques or to implant tissue-engineered constructs. The Achilles tendon in rats has been used to evaluate surgical repair and augmentation.[58,59] This model can provide valuable information on extrasynovial tendon repair, but small rodents are not useful for reconstruction of intrasynovial tendon regeneration or ligament reconstruction. The heterotopic placement of tissue-engineered constructs in athymic (nude) animals can be helpful in early studies of cell-polymer interactions and matrix formation.[60]

Very few studies of tendon repair have been performed in nonhuman primates, although their upper extremity anatomy is most similar to humans. The extreme cost of NHPs and difficulty in handling in addition to the challenges of preventing self-mutilation in their manipulated extremities make them a demanding animal model for most investigators. Many studies of flexor tendon repair have been performed in the dog,[61–63] possibly due to a suitable anatomy and the ability of dogs to ambulate on three limbs, thus permitting the protection of any repair. Interestingly, despite the value of the dog model in flexor tendon repair studies, no reported studies involving tissue engineering approaches of tendon have been performed *in vivo* in dogs. Similarly, there have been only a couple of reports if tissue engineered constructs for tendon regeneration placed in other large animals such as sheep, goats, or pigs.[64] The flexor tendons in the feet of chickens have been used widely in the past for studying a functional model of flexor tendon repair,[65,66] but there have only been a couple of reports of tendon tissue engineering in chicken feet tendons.[67] Although the flexor tendons in rabbit feet are quite small and difficult to handle for tissue engineering strategies, the flexor profundus has been used for regenerative strategies with an acellularized tendon scaffold.[68] Thus, the dog forelimb and the chicken foot may the best developed tendon repair models for studying tissue engineered approaches for flexor tendon repair.

Another area of active investigation in the upper extremity is repair of rotator cuff tears or injuries. These injuries are most commonly treated by shoulder and elbow specialists, but deserve some attention here because of the intense interest in surgical repair and tissue augmentation or tissue engineering strategies being studied to improve outcomes. From a mechanical standpoint, the rat rotator cuff has many anatomical similarities to the human.[57] However, the small size of the rat precludes any tissue engineering strategy that requires surgical intervention. The rabbit, dog, sheep (goat), and swine have been studied as models for rotator cuff repair and regeneration.[69,70] Although the sheep model has been used most extensively for rotator cuff repair, none of the large animals have really been established as a model for studying tissue engineering approaches for rotator cuff injuries.

Nerve

Similar to bone repair and regeneration strategies to fill segmental defects or augment fracture healing, the repair and regeneration of peripheral nerves focuses on augmentation of the neurorrhaphy site or the bridging of gaps in peripheral nerves.[71] The primary goal of peripheral nerve repair is the restoration of conduction of electrical impulses to and from the target tissues to permit sensation and promote motor function. As such, a primary outcome measure is the ability of the repaired nerve to conduct the electrical impulse. Disruption of the electrical signal occurs when the nerve is severed, but is slowly regained if the nerve is allowed to regrow to the target tissue through the degenerating distal segment. The speed with which the electrical impulse travels is believed to be related to the quality of the repair and regrowth of the axons. Limb function following nerve regeneration is a function of the severity of the initial injury, the distance the nerve has to regrow (short distances have better functional outcomes), and the amount of innervation achieved at the target muscles. Unlike bone that can repair and remodel itself to near its original strength, peripheral nerves have only limited capacity to recover from injury and often leave patients with measurable sensory of motor deficit.

Species selection

The bulk of studies on repair and regeneration of peripheral nerves in the field of hand surgery has been performed primarily in the sciatic nerve of the rat where there has been a large body of literature generated over the last half century. The sciatic nerve in the rat is a mixed sensory and motor nerve that innervates much of the lower hind limb. It is easily approached through a dorsolateral incision from the pelvis to the stifle (knee) joint followed by gentle dissection of the gluteus muscle. In rats weighing 300–350 grams, this approach provides about three centimeters of sciatic nerve exposure prior to its bifurcation into the common peroneal and tibial branches. Early studies of nerve repair and regeneration involved severing the sciatic and repairing it using a variety of suturing techniques designed to prevent scarring and promote nerve growth into the degenerating distal segment to reinnervate the lower leg muscles and skin. Subsequent studies on the mechanism of nerve repair and regeneration involved using a variety of conduits to bridge the nerve defects and study the growth factor milieu in the nerve gap microenvironment. More recent studies have used conduits with specialized properties engineered into the scaffold or a cellular component to support and promote nerve growth. The athymic rat can also be employed when xenogeneic cells are added to scaffolds to promote peripheral nerve regeneration.[72,73] Thus, the rat sciatic nerve provides a favorable model in which to perform initial tissue engineering studies on new scaffolds and techniques with well established outcome measures, some of which are not available in larger species.

Whereas rats have been the most used model for peripheral nerve repair and have provided valuable insight in the reparative processes, their small size limits the length of nerve regeneration that can be studied and many approaches yield favorable results that might not be achieved in patients. Additionally, dissimilarity to human anatomy limits the utility of rats in translational medicine since there are no forelimb nerve models in rats for comparison to human hand nerve repair. Various large animal species have been employed for nerve repair and regeneration studies that could serve as a foundation for future pre-clinical tissue engineering approaches for hand surgery. It is very important to have a thorough understanding of literature, however, to ensure that the appropriate model is selected and that outcome measures have been substantiated in the model to answer the scientific question being posed. For example, a search of the literature involving peripheral nerve repair and regeneration in goats revealed no peripheral nerve studies and therefore would be a poor model to select. Other, more well-established large animal models for nerve repair would be preferred such as rabbits or dogs.

Large animal species that have reportedly been used in nerve repair and regeneration studies include cats, rabbits, dogs, sheep, swine, and nonhuman primates. Cats and swine frequently have been used for studies in facial nerve studies, and less frequently for nerve studies in the limb. A few reports have employed the median nerve in the sheep forelimb,[74–76] but farm animals have not been used extensively possibly because these species cannot ambulate well on three limbs if one limb is injured or disabled. Although the sciatic nerve has been used occasionally in rabbits,[77] it is unlike the rat sciatic in that this nerve in rabbits branches very proximally into the common peroneal and the tibial nerves. The common peroneal has been used more often for immediate neurorrhaphy or neural gap studies.[78–81] Dogs and macaques have also been used fairly extensively for peripheral nerve repair and regeneration studies. The dog sciatic[82–84] peroneal[85,86] and anterior tibial nerves[87] in the hindlimb can be used as well as the median and ulnar nerves in the forelimb.[88] Dogs ambulate well on three limbs making them a suitable model for studying peripheral nerve injury in limbs. The median[89–91] and less frequently the ulnar[92,93] or radial[94] nerves in the forelimbs of macaques have also been used to study nerve repair and nerve regeneration across gaps, but cost of the animals and high costs for daily care preclude many investigators from using them as an animal model.

Several means for measuring outcomes of nerve repair have been established and Vleggeert-Lankamp has recently published an extensive review of all types of evaluation methods for assessing nerve repair using conduits.[95] Since the ultimate goal in nerve repair and regeneration is reestablishing the nerve fiber channel to conduct electrical impulses, many techniques have been employed or developed to directly or indirectly measure the conduction capacity of the restored nerve. In rats, a well-established indirect means to measure nerve function following repair involves performing a walking track analysis — a technique that cannot be performed in larger species. This is performed by measuring the foot print length and the toe spread to evaluate the return of function in the limb.[96,97] This can be done simply by dipping the rat's feet in ink and allowing it to make prints of the hind feet or digitally with camera recordings of the rat's gait. The rat is unable to walk normally up on its toes after severing the sciatic nerve, yet they still try to walk on all four legs. As a result, the heel drops and forms a long foot print until the return of muscle function that will allow toe walking again. Additionally, the spread of the toes collapses following injury and is slowly restored with increasing nerve regeneration. These parameters can easily be measured and return of function can be calculated using a well-established formula.[97] One noteworthy complication for making accurate measurements with walking track analyses is the problem of automutilation by the rats. Severing the sciatic nerve in rats results in numbness of the foot and the rats often chew off their toes, thus preventing any meaningful walking track data. Sensory tests involving withdrawal reflex in response to various stimuli, such as a hotplate or a pin prick, can add qualitative information to the walking track data as well.

Direct measurements of electrical activity in the injured nerve and limb can be performed as well to assess functional outcomes.[95] Nerve conduction studies can assess the speed of the impulse thus estimating the quality of the repair and regeneration, but is challenging in rats due to the short distances being tested in the hind limb. Measuring the mean conduction velocity (MCV), calculated as the ration of the conducting distance and the latency time to the peak of the maximal action current, involves inserting probes directly into the nerve proximal and distal to the repair. Another technique employs electromyography to acquire an evoked compound muscle action potential (CMAP). The CMAP can be measured by inserting the stimulation probes transcutaneously into the nerve and placing the receiving electrodes into muscles downstream of the nerve repair. Changes in the amplitude and latency of the CMAP can provide useful information on the ability of the repaired nerve to conduct an impulse and is easily performed in larger species such as rabbits and dogs.

One other critical outcome measure is the histoarchitecture of repaired peripheral nerves.[95] Because severed nerve ends have the capacity to sprout from the proximal end of the injured nerve and can grow down the degenerated distal nerve or across a nerve gap in a conduit, histological data can provide valuable insight in the quality of the re-growth. Nerve function, the ability of the injured nerve to transmit pulses as a result of nerve stimuli, is directly related to the number, diameter, and amount of myelination of regenerated nerve fibers. Each of these parameters can be measured in histological sections using well-established nerve histomorphometry methods. From these data, it is possible to calculate the G-ratio, the ration of axon diameter to the entire myelinated fiber diameter, and the N-ratio, the total myelinated fiber are divided by the total tissue area of the nerve cable. It is important to point out, however, that nerve tissue be correctly preserved (generally in a 2% paraformaldehyde, 2% glutaraldehyde solution), postfixed with osmium tetroxide to stain the lipid rich myelin, embedded in materials used for electron microscopy (e.g., Epon), and stained with toluidine blue. Only carefully prepared specimens can be used for accurate histomorphometry and can be applied to nerve repair in all species.

In addition to nerve histomorphometry, many rat studies include data on muscle morphology including weight, circumference, and histomorphometry of cross sections of the muscle. The gastrocnemius muscle is innervated by the sciatic nerve and is easily isolated for direct measurement. Following sciatic nerve transection the gastrocnemius undergoes atrophy and some subsequent degree of restoration if reinnervated. The wet weight of the muscle provides an indirect measure of reinnervation and shows some correlation to return to function measured by walking track analyses. Muscle circumference can also be used, but has been reported less frequently. As the muscle atrophies following nerve transection, the individual muscle fibers also show decreased cross sectional areas. Collectively, nerve and muscle histoarchitecture data in combination with functional and electrical outcome studies can provide insight to the regeneration of injured nerves using conventional repair strategies or tissue engineering approaches.

Vessels and Vasculature

All tissues and organs in the body rely on a vascular supply to deliver oxygen and nutrients and remove cellular waste. Likewise, all engineered tissues depend on capillary systems and arteries and veins for this transfer. Although cartilage is not directly vascularized, it too is reliant on vessels in the surrounding synovium, perichondrium, or subchondral bone for nourishment and survival. All other tissues have highly dispersed networks of vessels of all sizes that provide the requisite circulation. Many strategies, particularly in the fields of cardiovascular and vascular surgery, have focused on engineering large conduits for replacement of primary arteries and veins. Other attempts involve enhancing the microcirculation through neovascularization of engineered tissues or promoting angiogenesis in injured tissues.

The primary goal for large vessel replacement is to generate a conduit that permits blood flow, either arterial or venous, to and from target tissues. It is also important to prevent failure of the conduit by means of aneurysm, dilatation and stretching, or thrombosis. Cell-free approaches have involved using decellularized tissues,[98,99] or synthesis of degradable and nondegradable materials to make implantable tubes.[100–102] Synthetic materials such as expanded polytetrafluoroethylene (ePTFE or Gortex), Dacron, or polyurethane grafts have been somewhat successful for large vessel replacement in patients.[102] Reendothelialization often does not occur throughout these synthetic conduits, especially long ones, and intimal hyperplasia can plague the anastomotic sites.[102] Tissue engineering approaches to enhance endothelialization *in vitro* to prevent thrombus formation and improve performance have been used as well with some success.[103] In hand and upper extremity reconstruction, however, the need is for small and medium sized vessels. Synthetic conduit materials tend to thrombose when used for small diameter vessels and autologous vessel grafts have remained the preferred standard for bridging segmental defects.

The current unavailability of autologous engineered vessels to treat traumatic injury limits their potential use in hand and upper extremity surgery, but advances in the future may open more possibilities for tissue engineered replacements. Biodegradable materials seeded with cells have also been explored for engineering vessels, large and small. Rather than focus on mechanically intact tubes like the ePTFE or Dacron, many approaches have attempted to recreate the layers of the vessels like the intima, media, and adventitia of arteries using different cell populations and materials like polyesters.[104] Another critical element for the normal function of vessels is the production of elastin that has largely been ignored in engineering approaches.[105] One biodegradable material composed of L-lactide and ε-caprolactone in a 50:50 ratio combined with bone marrow cells has been tested for cardiovascular conditions in patients in Japan.[106–108]

Species selection

A comprehensive review by Rashid *et al*, shows that numerous animal models have been utilized for evaluating synthetic conduits and bioengineered vessels.[109] As with many other musculoskeletal tissues, immunodeficient mice can be useful for screening materials and cell-material interactions for forming vessels.[110–112] However, the small size of even their major vessels makes them less desirable for functional studies. The rat and rabbit, although slightly larger, can also be used, but size remains a limitation in these models as well.[113,114] Although there are reports using nonhuman primates, there does not appear to be any advantage of these species over dogs or farm species.[109] The carotid, aortic, and pulmonary arteries have been used for testing large vascular grafts in sheep, dogs, and swine.[100,115–118] The vena cava in dogs and sheep have also been used for evaluating venous grafts.[119,120]

Microcirculation

At the microcirculatory level, new blood-vessel formation is required to supply individual cells with nutrients and oxygen for native tissues to grow beyond 100–200 microns and this is also true for tissue-engineered constructs. After implantation of tissue-engineered constructs, spontaneous vascularization of the implants is usually observed due to an inflammatory wound healing response and endogenous release of angiogenic growth factors stimulated by the hypoxic state created by seeded cells. However, this induced vessel ingrowth is often too slow to provide adequate nutrient transport to the cells in the interior of the implanted engineered construct. Additional strategies for enhancing vascularization are therefore essential to ensure the survival of large tissue-engineered grafts. Several strategies for enhancing vascularization are currently under investigation including novel scaffold design, the inclusion of angiogenic factors, *in vitro* prevascularization, and *in vivo* prevascularization.[121]

There are a couple of common approaches to achieve *in vivo* formation of microvascular networks.[122] One is intrinsic vascularization to spontaneously generate a microcirculatory network from a central macrovascular conduit, such as an arteriovenous loop placed in a protected space, to vascularize an endogenously formed or implanted scaffold. Erol and Spira implanted an arteriovenous shunt loop in the rat groin and later observed the development of new blood vessels originating from the loop that could support engraftment of an overlying skin graft.[123] A number of other experimental studies have used an arteriovenous shunt loop for vascular support of prefabricated skin flaps and even new connective tissue.

Another approach is extrinsic vascularization to produce a microcirculatory network via ingrowth of capillaries from the surrounding recipient vascular bed into the construct tissue. This is a common technique employed in numerous *in vivo* tissue-engineering studies involving implantation of cell-seeded porous scaffolds or hydrogels into already vascularized sites such as the kidney capsule, mesenteric and adipose tissue, and subdermal areas. Mice and rats are commonly used for these types of applications. A couple of commonly used experimental models that allow for detailed analysis of angiogenesis in engineered tissue constructs *in vivo* are the chorioallantoic membrane (CAM) assay in the chick embryo and the dorsal skin fold chamber used in the Syrian golden hamster or mouse.[124]

Although early *in vivo* experiments for many different engineered tissue tissues are performed in the immunodeficient mouse, it is less suitable for vasculature regeneration probably because of their comparatively small size and decreased vasculature distribution. At the other end of the spectrum, little direct angiogenesis and vascularization research has been performed on nonhuman primates, but there has been a report about monitoring vascularization in tissue-engineered bone in rhesus macaques.[125] Other large animals such as rabbit, dog, pig, and sheep also have been used widely in this field. The choice of animal model mainly depends on different research aims in addition to cost, animal behavior, and relevance to the human condition. Lastly, tissue engineering of lymphatic tissues has received very little attention with only a few studies about tissue engineering of lymphatic tissues in mice.[126]

Muscle

Tissue engineering of skeletal muscle has been studied less extensively than the engineering of most other musculoskeletal tissues and heart muscle. Through contractile function, muscles produce forces on tendons and bones to initiate motion of the skeleton. Unlike the relatively low energy needs of the cells in structural tissues such as bone, cartilage, and tendons, this contractile function by muscle cells requires significant amounts of oxygen and energy. Furthermore, the interdependency of muscle with tendons, nerves, and vascular supply for muscle function makes the engineering of muscle much more complex than other musculoskeletal tissues like bone and cartilage.[127–129]

There is a large body of literature on *in vitro* conditions for engineering skeletal muscle, but there have been few *in vivo* studies.[130,131] *In vitro* studies have examined a variety of factors affecting the formation of myoids or organoids as well as the effects of external factors in bioreactors to enhance their formation. For example, Yan *et al* reported that they had successfully constructed a multilayered culture of skeletal muscle cells, derived from neonatal satellite cells that are distributed in a 3D pattern of organization that mimics many of the features of intact muscle tissue.[132] Some approaches involve the use of bioreactors to apply mechanical forces to the formation of muscle *in vitro*.[133–136] Electrophysiological stimulation has also been employed to control the architecture of muscle tissue growth.[137] Scaffolds can also have an effect on the formation of muscle *in vitro*.[138]

The close relationship of muscle with nerves, vessels and tendons has spawned investigations to study the interrelationships of these tissues with muscle. Muscles will have to connect by way of tendons and some studies have begun to evaluate the engineered tendon-muscle interfaces.[139,140] Other studies have begun to study means to promote vascularization of the tissue using recombinant angiogenesis proteins.[141] Another significant obstacle is the formation of neuromuscular junctions and this has been hindered by many of the same problems related to nerve injury and repair where reinnervation is generally poor. Whereas many approaches rare intended to engineer an entire replacement muscle tissue, another approach is to use engineered muscle tissue to deliver genes to diseased muscles.[142,143] For example, engineered muscle tissue modified by gene therapy to produced specific proteins could be used to deliver these proteins to injury or diseased muscle tissue. Despite a growing body of literature on *in vitro* engineering of skeletal muscle tissue, little has been done *in vivo*.

There are only a few reports on *in vivo* studies, and almost all of were performed in nude mice and rats. For example, Okano used rod-shaped hybrid muscular tissues composed of C2C12 cells (skeletal muscle myoblast cell line) and type I collagen, which were prepared using the centrifugal cell-packing method to reconstruct tissue engineered muscle in nude mice.[144] Dhawan *et al* found that neurotization of engineered skeletal muscle significantly increases force generation and caused neuromuscular junctions to develop allowing indirect muscle stimulation in a rat model.[145] One group has implanted myoids in immune competent rats.[146] Because muscle relies so heavily on vasculature and innervation, which have not been conquered in tissue engineering, approaches thus far have not matured sufficiently for further large animal studies. Despite the large foundation of knowledge on muscle physiology and function, outcome measures for tissue-engineered muscle are not well developed and this too hinders the development of suitable animal models.

Skin

Skin is considered as the largest organ in the body and primarily serves as a protective barrier against the environment. Adult skin consists of two tissue layers: a keratinized, stratified epidermis and an underlying thick layer of collagen-rich dermal connective tissue providing support and nourishment.[147] Appendages such as hair and glands are derived from the epidermis and project deep into the dermal layer. Autologous transfer of skin as split- or full-thickness grafts or primarily vascularized free flaps have been very effective in treating complicated wounds (burns) and reconstructing massive defects. In some cases, such as extensive third degree burns, there are a limited number of donor sites from which to harvest grafts and alternatives are needed. Allogeneic[148,149] and xenogeneic tissues[150,151] have been developed as one approach for temporary skin coverage. Collagen-glycosaminoglycan (collagen-GAG) materials, some with a silicone cover, developed in the 1980s have been shown to be effective for burn coverage and providing a substrate to rebuild the dermis prior to skin grafting.[152–154] Keratinocytes and fibroblasts have been added to some of these devitalized tissues or matrices to try and enhance their regenerative capacity.[155–159] Since traumatic injuries and burns to the hand are commonplace, the use of these materials and possible tissue-engineered bioartificial skin substitutes could play an important role in upper extremity reconstruction.

Outcome measures

Unlike muscle, bone, or articular cartilage tissues that are subject to a dynamic physiological environment for growth and function, skin is more like a "static" tissue. The primary goal of engineered substitutes is improved wound healing, which is commonly measured in time to wound closure and epithelialization. Although many strategies can result in faster would healing, unsatisfactory cosmesis and poor pliability of the repair tissue can lead to unfavorable clinical results. Thus, a secondary goal for engineering substitutes should be to restore the wound site to normal with soft, pliable coverage and acceptable cosmetic outcomes.

Species selection

Similarity in structure and composition to the human are the most important elements for considering an animal model. However, the skin of many available animal models is quite different to that of humans. For example, many animals have a muscle layer beneath the skin, the panniculus carnosus, not found in humans that can aid in closing full thickness lesions in the skin through contraction.[160,161] Except swine and some strains hairless mice and guinea pigs, most animals have much more hair than humans that make treatments somewhat difficult. Immunodeficient mice and rats are often the first choice at the early stage for testing and evaluating of different artificial skin composites, particularly if xenogeneic (human) cells or skin are included.[162] Rats are commonly used for burn and excisional wound models.[163] As with other musculoskeletal tissues, the availability of mice transgenic and knockout models make them particularly useful when these models are necessary.[164] All rodents are quite flexible and groom themselves often, so to avoid scratching and biting of the wounds the dorsal region is the most appropriate experimental area.

The domestic pig is the preferred animal for studying the effects of environmental factors on skin and wounds because its integument is more human-like than any other animal.[165,166] The full thickness excision wound model is easier to make on swine for testing dermal and epidermal substitutes.[167–169] Some strategies use isolated keratinocytes and fibroblasts which are mixed with different biodegradable scaffolds to repair the full thickness cutaneous defect.[159] Researchers can also obtain different degrees of burn wound on porcine models by controlling temperature and timing of the burning device such as a heated-brass contact plate. There have been also some reports about full thickness skin defect repair in rabbits and dogs.[160] Whereas there are many similarities between pig and human skin, rabbit and dog skin is quite different from that of humans. However, rabbits are easier to manipulate and cost less making them an optional animal model for skin repair and regeneration. Other animals such as sheep, goats or nonhuman primates have not been used extensively or have no known advantages over rodents, rabbits, and swine.

The Three R's

All investigators using animal models in their research should be well informed of the "three Rs" — reduction, refinement, and replacement — of animal investigation. The origin of this concept is attributed to a Charles Hume, the founder of the Universities Federation for Animal Welfare in the United Kingdom and a subsequent 1959 publication by Russell and Burch entitled "The Principles of Humane Experimental Technique".[170] These principles have been incorporated into laws regarding animal testing by countries around the world and play an important role in protocol review by Institutional Animal Care and Use Committees.

The first tenet is the reduction in the number of animals used for scientific inquiry by suggesting that the number of animals used should be the minimum number necessary to test an experimental hypothesis and give statistically valid results. A basic understanding of experimental design and statistics is a minimum requirement for all scientists and consulting with professional statisticians prior to conducting animal studies is recommended to ensure the appropriate model and minimum animal numbers are used. For tissue engineering studies, extensive *in vitro* evaluation of polymers and other components should be performed prior to embarking on *in vivo* evaluation. Even benchtop cadaveric preparations using limbs or tissues of animals euthanized for other purposes can provide valuable preoperative information prior to embarking on *in vivo* studies and result in reduction of live animals used. The animal experiments need to be carefully designed to provide statistically sound data and avoid inefficient use of animals. For example, investigators could use historical controls (their own or from published works) whenever possible to avoid repeating large numbers of control animals. Another common pitfall is performing negative controls where the expected outcome is failure (e.g., nonimmunosuppressed allotransplant) and the rejected graft will not provide any substantive additional knowledge to the field of study.

Refinement refers to methods to alleviate or minimize potential pain and distress of research animals and enhance their well-being. Many of the animal models discussed above involve injuries to a tissue or organ that can cause pain or distress. It is difficult to objectively assess pain or distress in animals, but one can assume that if the injury could cause pain or distress in humans it can be equally uncomfortable in an animal model. Most pain and distress during and after surgical intervention can be alleviated with appropriate anesthetics and analgesics. Investigators need to have appropriate plans in place to alleviate any postoperative or postprocedural discomfort. Other strategies can be employed to enhance the animals' well-being such as proper and gentle handling, comfortable and appropriately sized cages, and group housing for social species.

Strategies to replace the use of live animals should always be considered. Investigators should be educated in nonanimal methods such as *in vitro* methods, computer or mathematical models, the use of microorganisms and plants, or organisms with limited sentience such as invertebrates. Even thorough evaluation of patients with pathological conditions can provide useful data on etiology of injuries under study to aid in the best possible scientific approach and study design. Thoughtful input and careful design can result in replacing animal experiments with scientifically sound alternatives.

References

1. Office of Technology Assessment, United States Congress. Alternatives to Animal Use in Research, Testing, and Education. *Washington D.C.; Government Printing Office*, OTA–BA–273 (1986).
2. United States Department of Agriculture, Animal and Plant Health Inspection Service. Animal Care Annual Report of Activities, Fiscal Year 2007. *Washington, D.C.; Government Printing Office*, APHIS 41–35–075 (2007).
3. Guide for the Care and Use of Laboratory Animals, National Research Council. *National Academy Press, Washington D.C.* (1996).

4. Neyt JG, Buckwalter JA, Carroll NC. Use of animal models in musculoskeletal research. *Iowa Orthop J* **18** (1998) 118–123.
5. Pearce AI, Richards RG, Milz S, Schneider E and Pearce SG. Animal models for implant bio-material research in bone: a review. *Eur Cell Mater* **13** (2007) 1–10. Review.
6. Reichert JC, Saifzadeh S, Wullschleger ME, Epari DR, Schütz MA, Duda GN, Schell H, van Griensven M, Redl H, Hutmacher DW. The challenge of establishing preclinical models for segmental bone defect research. *Biomaterials* **30**(12) (2009) 2149–2163. Epub 2009 Feb 10. Review.
7. Cheung KM, Kaluarachi K, Andrew G, Lu W, Chan D, Cheah KS. An externally fixed femoral fracture model for mice. *J Orthop Res* **21**(4) (2003) 685–690.
8. Cui Q, Xiao Z, Li X, Saleh KJ, Balian G. Use of genetically engineered bone-marrow stem cells to treat femoral defects: an experimental study. *J Bone Joint Surg Am* **88** (2006) Suppl 3:167–172.
9. Drosse I, Volkmer E, Seitz S, Seitz H, Penzkofer R, Zahn K, Matis U, Mutschler W, Augat P, Schieker M. Validation of a femoral critical size defect model for orthotopic evaluation of bone healing: a biomechanical, veterinary and trauma surgical perspective. *Tissue Eng Part C Methods* **14**(1) (2008) 79–88.
10. Martini L, Fini M, Giavaresi G, Giardino R. Sheep model in orthopedic research: a literature review. *Comp Med* **51**(4) (2001) 292–299. Review.
11. Swindle MM, Smith AC, Hepburn BJ. Swine as models in experimental surgery. *J Invest Surg* **1**(1) (1988) 65–79. Review.
12. Sachs DH, Galli C. Genetic manipulation in pigs. *Curr Opin Organ Transplant* **14**(2) (2009) 148–153. Review.
13. Lee WP, Rubin JP, Bourget JL, Cober SR, Randolph MA, Nielsen GP, Ierino FL, Sachs DH. Tolerance to limb tissue allografts between swine matched for major histocompatibility complex antigens. *Plast Reconstr Surg* **107**(6) (2001) 1482–1490.
14. Kaplan FS, Hayes WC, Keaveny TM, Boskey A, Einhorn TA, Iannotti JP. Form and function of bone. In Orthopaedic Basic Science, Simon SS, ed. *American Academy of Orthopaedic Surgeons, Chicago, IL* (1994), pp. 127–184.
15. Mosekilde L, Kragstrup J, Richards A. Compressive strength, ash weight, and volume of vertebral trabecular bone in experimental fluorosis in pigs. *Calcif Tissue Int* **40**(6) (1987) 318–322.
16. Wang X, Mabrey JD, Agrawal CM. An interspecies comparison of bone fracture properties. *Biomed Mater Eng* **8**(1) (1998) 1–9.
17. Jee WS, Bartley MJ, Cooper R, Dockum N. The beagle as an experimental dog. Ames: *Iowa State University Press* (1970), pp. 162–188.
18. Liebschner MA. Biomechanical considerations of animal models used in tissue engineering of bone. *Biomaterials* **25**(9) (2004) 1697–1714.
19. Newman E, Turner AS, Wark JD. The potential of sheep for the study of osteopenia: current status and comparison with other animal models. *Bone* **16** (1995) (4 Suppl.): 277S–284S.
20. Martiniaková M, Omelka R, Chrenek P, Ryban L, Parkányi V, Grosskopf B, Vondráková M, Bauerová M. Changes of femoral bone tissue microstructure in transgenic rabbits. *Folia Biol (Praha)* **51**(5) (2005) 140–144.
21. Aerssens J, Boonen S, Lowet G, Dequeker J. Interspecies differences in bone composition, den-sity, and quality: potential implications for *in vivo* bone research. *Endocrinology* **139**(2) (1998) 663–670.
22. Liebschner MA, Keller TS. Hydraulic strengthening affects the stiffness and strength of corti-cal bone. *Ann Biomed Eng* **33**(1) (2005) 26–38.
23. Aerssens J, Boonen S, Joly J, Dequeker J. Variations in trabecular bone composition with anatomical site and age: potential implications for bone quality assessment. *J Endocrinol* **155**(3) (1997) 411–421. PubMed PMID: 9487986.
24. Wagner W, Bork S, Horn P, Krunic D, Walenda T, Diehlmann A, Benes V, Blake J, Huber FX, Eckstein V, Boukamp P, Ho AD. Aging and replicative senescence have related effects on human stem and progenitor cells. *PLoS One.* **4**(6) (2009) e5846.

25. Jiang Y, Mishima H, Sakai S, Liu YK, Ohyabu Y, Uemura T. Gene expression analysis of major lineage-defining factors in human bone marrow cells: effect of aging, gender, and age-related disorders. *J Orthop Res* **26**(7) (2008) 910–917.

26. Stolzing A, Scutt A. Age-related impairment of mesenchymal progenitor cell function. *Aging Cell* **5**(3) (2006) 213–224.

27. Martiniaková M, Omelka R, Grosskopf B, Sirotkin AV, Chrenek P. Sex-related variation in compact bone microstructure of the femoral diaphysis in juvenile rabbits. *Acta Vet Scand* **50** (2008) 15.

28. Engin AE, Toney LR, Negulesco JA. Effects of oestrogen upon tensile properties of healing fractured avian bone. *J Biomed Eng* **5**(1) (1983) 49–54.

29. Mankin HJ, Mow VC, Buckwalter JA, Iannotti JP, Ratcliffe A. Form and function of articular cartilage. In Orthopaedic Basic Science, Simon SS, ed. *American Academy of Orthopaedic Surgeons, Chicago, IL* (1994) pp. 1–44.

30. Hunziker EB. Biologic repair of articular cartilage. Defect models in experimental animals and matrix requirements. *Clin Orthop Relat Res* (367 Suppl.) (1999) S135–S146.

31. Steadman JR, Rodkey WG, Rodrigo JJ. Microfracture: surgical technique and rehabilitation to treat chondral defects. *Clin Orthop Relat Res* (391 Suppl.) (2001) S362–S369.

32. Brittberg M, Lindahl A, Nilsson A, Ohlsson C, Isaksson O, Peterson L. Treatment of deep cartilage defects in the knee with autologous chondrocyte transplantation. *N Engl J Med* **331**(14) (1994) 889–895.

33. Behrens P, Bitter T, Kurz B, Russlies M. Matrix-associated autologous chondrocyte transplantation/implantation (MACT/MACI) — 5-year follow-up. *Knee* **13**(3) (2006) 194–202.

34. Hangody L, Feczkó P, Bartha L, Bodó G, Kish G. Mosaicplasty for the treatment of articular defects of the knee and ankle. *Clin Orthop Relat Res* (391 Suppl.) (2001) S328–S336. Review.

35. Jones AR, Gleghorn JP, Hughes CE, Fitz LJ, Zollner R, Wainwright SD, Caterson B, Morris EA, Bonassar LJ, Flannery CR. Binding and localization of recombinant lubricin to articular cartilage surfaces. *J Orthop Res* **25**(3) (2007) 283–292.

36. Mankin HJ, Dorfman H, Lippiello L, Zarins A. Biochemical and metabolic abnormalities in articular cartilage from osteo-arthritic human hips. II. Correlation of morphology with biochemical and metabolic data. *J Bone Joint Surg Am* **53**(3) (1971) 523–537.

37. Pineda S, Pollack A, Stevenson S, Goldberg V, Caplan A. A semiquantitative scale for histologic grading of articular cartilage repair. *Acta Anat (Basel)* **143**(4) (1992) 335–340.

38. O'Driscoll SW, Keeley FW, Salter RB. The chondrogenic potential of free autogenous periosteal grafts for biological resurfacing of major full-thickness defects in joint surfaces under the influence of continuous passive motion. An experimental investigation in the rabbit. *J Bone Joint Surg Am* **68**(7) (1986) 1017–1035.

39. Vacanti CA, Kim W, Schloo B, Upton J, Vacanti JP. Joint resurfacing with cartilage grown in situ from cell-polymer structures. *Am J Sports Med* **22**(4) (1994) 485–488.

40. Paige KT, Cima LG, Yaremchuk MJ, Schloo BL, Vacanti JP, Vacanti CA. De novo cartilage generation using calcium alginate-chondrocyte constructs. *Plast Reconstr Surg* **97**(1) (1996) 168–178.

41. Silverman RP, Passaretti D, Huang W, Randolph MA, Yaremchuk MJ. Injectable tissue-engineered cartilage using a fibrin glue polymer. *Plast Reconstr Surg* **103**(7) (1999) 1809–1818.

42. Meachim G, Roberts C. Repair of the joint surface from subarticular tissue in the rabbit knee. *J Anat* **109**(Pt 2) (1971) 317–327.

43. Grande DA, Pitman MI, Peterson L, Menche D, Klein M. The repair of experimentally produced defects in rabbit articular cartilage by autologous chondrocyte transplantation. *J Orthop Res* **7**(2) (1989) 208–218.

44. Wakitani S, Kimura T, Hirooka A, Ochi T, Yoneda M, Yasui N, Owaki H, Ono K. Repair of rabbit articular surfaces with allograft chondrocytes embedded in collagen gel. *J Bone Joint Surg Br* **71**(1) (1989) 74–80.

45. Reinholz GG, Lu L, Saris DB, Yaszemski MJ, O'Driscoll SW. Animal models for cartilage reconstruction. *Biomaterials* **25**(9) (2004) 1511–1521.

46. Saris DB, Dhert WJ, Verbout AJ. Joint homeostasis. The discrepancy between old and fresh defects in cartilage repair. *J Bone Joint Surg Br* **85**(7) (2003) 1067–1076.

47. Hendrickson DA, Nixon AJ, Grande DA, Todhunter RJ, Minor RM, Erb H, Lust G. Chondrocyte-fibrin matrix transplants for resurfacing extensive articular cartilage defects. *J Orthop Res* **12**(4) (1994) 485–497.

48. Breinan HA, Minas T, Hsu HP, Nehrer S, Sledge CB and Spector M. Effect of cultured autologous chondrocytes on repair of chondral defects in a canine model. *J Bone Joint Surg Am* **79**(10) (1997) 1439–1451.

49. Peretti GM, Xu JW, Bonassar LJ, Kirchhoff CH, Yaremchuk MJ, Randolph MA. Review of injectable cartilage engineering using fibrin gel in mice and swine models. *Tissue Eng* **12**(5) (2006) 1151–1168. Review.

50. Schachar NS, Novak K, Hurtig M, Muldrew K, McPherson R, Wohl G, Zernicke RF, McGann LE. Transplantation of cryopreserved osteochondral Dowel allografts for repair of focal articular defects in an ovine model. *J Orthop Res* **17**(6) (1999) 909–919.

51. Kumar P, Oka M, Toguchida J, Kobayashi M, Uchida E, Nakamura T, Tanaka K. Role of uppermost superficial surface layer of articular cartilage in the lubrication mechanism of joints. *J Anat* **199**(Pt 3) (2001) 241–250.

52. Woo S L-Y , An K-N, Arnoczky SP, Wayne JS, Fithian DC, Myers BS. Anatomy, Biology, and Biomechanics of Tendon, Ligament, and Meniscus. *Orthopaedic Basic Science, Simon SR, ed., American Academy of Orthopaedic Surgeons, Chicago, IL* (1994), pp. 45–87.

53. Jadeja H, Yeoh D, Lal M, Mowbray M. Patterns of failure with time of an artificial scaffold class ligament used for reconstruction of the human anterior cruciate ligament. *Knee* **14**(6) (2007) 439–442.

54. Mäkisalo SE, Visuri T, Viljanen A, Jokio P. Reconstruction of the anterior cruciate ligament with carbon fibres: unsatisfactory results after 8 years. *Knee Surg Sports Traumatol Arthrosc* **4**(3) (1996) 132–136.

55. Frank CB, Loitz B, Bray R, Chimich D, Shrive N. Abnormality of the contralateral ligament after injuries of the medial collateral ligament: an experimental study in rabbits. *J Bone Joint Surg* **76A** (1994) 403–412.

56. Woo SL, Gomez MA, Sites TJ, *et al.* The biomechanical and morphological changes in the MCL of the rabbit after immobilization and remobilization. *J Bone Joint Surg* **69A** (1987) 1200–1211.

57. Carpenter JE, Thomopoulos S, Soslowsky LJ. Animal models of tendon and ligament injuries for tissue engineering applications. *Clin Orthop Relat Res* (367 Suppl.) (1999) S296–S311. Review.

58. Strauch B, Patel MK, Rosen DJ, Mahadevia S, Brindzei N, Pilla AA. Pulsed magnetic field therapy increases tensile strength in a rat Achilles' tendon repair model. *J Hand Surg Am* 31(7) (2006) 1131–1135.

59. Chan BP, Amann C, Yaroslavsky AN, Title C, Smink D, Zarins B, Kochevar IE, Redmond RW. Photochemical repair of Achilles tendon rupture in a rat model. *J Surg Res* **124**(2) (2005) 274–279.

60. Cao Y, Liu Y, Liu W, Shan Q, Buonocore SD, Cui L. Bridging tendon defects using autologous tenocyte engineered tendon in a hen model. *Plast Reconstr Surg* **110** (2002) 1280–1289.

61. Gelberman RH, Woo SL, Amiel D, Horibe S and Lee D. Influences of flexor sheath continuity and early motion on tendon healing in dogs. *J Hand Surg Am* **15**(1) (1990) 69–77.

62. Winters SC, Gelberman RH, Woo SL, Chan SS, Grewal R, Seiler JG 3rd. The effects of multiple-strand suture methods on the strength and excursion of repaired intrasynovial flexor tendons: a biomechanical study in dogs. *J Hand Surg Am* **23**(1) (1998) 97–104.

63. Gelberman RH, Thomopoulos S, Sakiyama-Elbert SE, Das R, Silva MJ. The early effects of sustained platelet-derived growth factor administration on the functional and structural properties of repaired intrasynovial flexor tendons: an *in vivo* biomechanic study at 3 weeks in canines. *J Hand Surg Am* **32**(3) (2007) 373–379.

64. Liu W, Cui L, Cao Y. A closer view of tissue engineering in China: the experience of tissue construction in immunocompetent animals. *Tissue Eng* **9** (Suppl. 1) (2003) S17–30. Review.

65. Strick MJ, Filan SL, Hile M, McKenzie C, Walsh WR, Tonkin MA. Adhesion formation after flexor tendon repair: a histologic and biomechanical comparison of 2- and 4-strand repairs in a chicken model. *J Hand Surg Am* **29**(1) (2004) 15–21.

66. Burt JD, Siddins M, Morrison WA. Laser photoirradiation in digital flexor tendon repair. *Plast Reconstr Surg* **108**(3) (2001) 688–694.

67. Cao Y, Liu Y, Liu W, Shan Q, Buonocore SD, Cui L. Bridging tendon defects using autologous tenocyte engineered tendon in a hen model. *Plast Reconstr Surg* **110**(5) (2002) 1280–1289.

68. Kryger GS, Chong AK, Costa M, Pham H, Bates SJ, Chang J. A comparison of tenocytes and mesenchymal stem cells for use in flexor tendon tissue engineering. *J Hand Surg Am* **32**(5) (2007) 597–605.

69. Dines JS, Grande DA, Dines DM. Tissue engineering and rotator cuff tendon healing. *J Shoulder Elbow Surg* **16**(5 Suppl.) (2007) S204–S207.

70. Rotini R, Fini M, Giavaresi G, Marinelli A, Guerra E, Antonioli D, Castagna A, Giardino R. New perspectives in rotator cuff tendon regeneration: review of tissue engineered therapies. *Chir Organi Mov* **91**(2) (2008) 87–92.

71. Winograd JM, Mackinnon SE. Peripheral nerve injuries: repair and reconstruction. In Plastic Surgery, Second Edition. Mathes SJ, Hentz VR, eds. *Saunder, Elsevier, Philadelphis, PA,* 2006.

72. Santiago LY, Clavijo-Alvarez J, Brayfield C, Rubin JP, Marra KG. Delivery of adipose-derived precursor cells for peripheral nerve repair. *Cell Transplant* **18**(2) (2009) 145–158.

73. Kijima Y, Ishikawa M, Sunagawa T, Nakanishi K, Kamei N, Yamada K, Tanaka N, Kawamata S, Asahara T, Ochi M. Regeneration of peripheral nerve after transplantation of CD133+ cells derived from human peripheral blood. *J Neurosurg* **110**(4) (2009) 758–767.

74. Glasby MA, Fullerton AC, Lawson GM. Immediate and delayed nerve repair using freeze-thawed muscle autografts in complex nerve injuries. Associated arterial injury. *J Hand Surg Br* **23**(3) (1998) 354–359. PubMed PMID: 9665525. median nerve

75. Jeans L, Healy D, Gilchrist T. An evaluation using techniques to assess muscle and nerve regeneration of a flexible glass wrap in the repair of peripheral nerves. *Acta Neurochir Suppl* **100** (2007) 25–28. PubMed PMID: 17985539. median nerves

76. Jeans LA, Gilchrist T, Healy D. Peripheral nerve repair by means of a flexible biodegradable glass fibre wrap: a comparison with microsurgical epineurial repair. *J Plast Reconstr Aesthet Surg* **60**(12) (2007) 1302–1308. Epub 2007 Mar 9.

77. Ding T, Luo ZJ, Zheng Y, Hu XY, Ye ZX. Rapid repair and regeneration of damaged rabbit sciatic nerves by tissue-engineered scaffold made from nano-silver and collagen type I. *Injury* (2009) [Epub ahead of print].

78. Hems TE, Glasby MA. Comparison of different methods of repair of long peripheral nerve defects: an experimental study. *Br J Plast Surg* **45**(7) (1992) 497–502.

79. McCallister WV, McCallister EL, Trumble SA, Trumble TE. Overcoming peripheral nerve gap defects using an intact nerve bridge in a rabbit model. *J Reconstr Microsurg* **21**(3) (2005) 197–206.

80. Choi BH, Zhu SJ, Kim SH, Kim BY, Huh JH, Lee SH, Jung JH. Nerve repair using a vein graft filled with collagen gel. *J Reconstr Microsurg* **21**(4) (2005) 267–272. PubMed PMID: 15971145.

81. Henry FP, Goyal NA, David WS, Wes D, Bujold KE, Randolph MA, Winograd JM, Kochevar IE, Redmond RW. Improving electrophysiologic and histologic outcomes by photochemically sealing amnion to the peripheral nerve repair site. *Surgery* **145**(3) (2009) 313–321.

82. Wang X, Hu W, Cao Y, Yao J, Wu J, Gu X. Dog sciatic nerve regeneration across a 30-mm defect bridged by a chitosan/PGA artificial nerve graft. *Brain* **128**(Pt 8) (2005) 1897–1910.

83. Zhong H, Chen B, Lu S, Zhao M, Guo Y, Hou S. Nerve regeneration and functional recovery after a sciatic nerve gap is repaired by an acellular nerve allograft made through chemical extraction in canines. *J Reconstr Microsurg* **23**(8) (2007) 479–487.

84. Okamoto H, Hata KI, Kagami H, Okada K, Ito Y, Narita Y, Hirata H, Sekiya I, Otsuka T, Ueda M. Recovery process of sciatic nerve defect with novel bioabsorbable collagen tubes packed with collagen filaments in dogs. *J Biomed Mater Res A* (2009) [Epub ahead of print].

85. Matsumoto K, Ohnishi K, Sekine T, Ueda H, Yamamoto Y, Kiyotani T, Nakamura T, Endo K, Shimizu Y. Use of a newly developed artificial nerve conduit to assist peripheral nerve regeneration across a long gap in dogs. *ASAIO J* **46**(4) (2000) 415–420.

86. Nakamura T, Inada Y, Fukuda S, Yoshitani M, Nakada A, Itoi S, Kanemaru S, Endo K, Shimizu Y. Experimental study on the regeneration of peripheral nerve gaps through a polyglycolic acid-collagen (PGA-collagen) tube. *Brain Res* **1027**(1–2) (2004) 18–29.

87. Levinthal R, Brown WJ, Rand RW. Comparison of fascicular, interfascicular and epineural suture techniques in the repair of simple nerve lacerations. *J Neurosurg* **47**(5) (1977) 744–750.

88. Lee WP, Constantinescu MA, Butler PE. Effect of early mobilization on healing of nerve repair: histologic observations in a canine model. *Plast Reconstr Surg* **104**(6) (1999) 1718–1725.

89. Archibald SJ, Shefner J, Krarup C, Madison RD. Monkey median nerve repaired by nerve graft or collagen nerve guide tube. *J Neurosci* **15**(5 Pt 2) (1995) 4109–4123.

90. Krarup C, Archibald SJ, Madison RD. Factors that influence peripheral nerve regeneration: an electrophysiological study of the monkey median nerve. *Ann Neurol* **51**(1) (2002) 69–81. PubMed PMID: 11782986.

91. Zhang P, Yin X, Kou Y, Wang Y, Zhang H, Jiang B. The electrophysiology analysis of biological conduit sleeve bridging rhesus monkey median nerve injury with small gap. *Artif Cells Blood Substit Immobil Biotechnol* **36**(5) (2008) 457–463. PubMed PMID: 18925468.

92. Vsconez LO, Mathes SJ, Grau G. Direct fascicular repair and interfascicular nerve grafting of median and ulnar nerves in the rhesus monkey. *Plast Reconstr Surg* **58**(4) (1976) 482–489.

93. Hu J, Zhu QT, Liu XL, Xu YB, Zhu JK. Repair of extended peripheral nerve lesions in rhesus monkeys using acellular allogenic nerve grafts implanted with autologous mesenchymal stem cells. *Exp Neurol* **204**(2) (2007) 658–666.

94. Ahmed Z, Brown RA, Ljungberg C, Wiberg M, Terenghi G. Nerve growth factor enhances peripheral nerve regeneration in non-human primates. *Scand J Plast Reconstr Surg Hand Surg* **33**(4) (1999) 393–401.

95. Vleggeert-Lankamp CLAM. The role of evaluation methods in the assessment of peripheral nerve regeneration through synthetic conduits: a systematic review. *J Neurosurg* **107** (2007) 1168–1189.

96. de Medinaceli L, Freed WJ, Wyatt RJ. An index of the functional condition of rat sciatic nerve based on measurements made from walking tracks. *Exp Neurol* **77**(3) (1982) 634–643.

97. Bain JR, Mackinnon SE, Hunter DA. Functional evaluation of complete sciatic, peroneal, and posterior tibial nerve lesions in the rat. *Plast Reconstr Surg* **83**(1) (1989) 129–138.

98. Sawyer PN, Fitzgerald J, Kaplitt MJ, Sanders RJ, Williams GM, Leather RP, Karmody A, Hallin RW, R. Taylor R, Fries CC. Ten year experience with the negatively charged glutaraldehyde-tanned vascular graft in peripheral vascular surgery. Initial multicenter trial. *Am J Surg* **154** (1987) 533–537.

99. Uematsu M, Okada M. A modified human ureter graft tanned by a new crosslinking agent polyepoxy compound for small diameter arterial substitutions: an experimental preliminary study. *Artif Organs* **22** (1998) 909–913.

100. Wippermann J, Schumann D, Klemm D, Kosmehl H, Salehi-Gelani S, Wahlers T. Preliminary results of small arterial substitute performed with a new cylindrical biomaterial composed of bacterial cellulose. *Eur J Vasc Endovasc Surg* (2009).

101. Zavan B, Vindigni V, Lepidi S, Iacopetti I, Avruscio G, Abatangelo G, Cortivo R. Neoarteries grown *in vivo* using a tissue-engineered hyaluronan-based scaffold. *FASEB J* **22**(8) (2008) 2853–2861.

102. Zilla P, Bezuidenhout D, Human P. Prosthetic vascular grafts: wrong models, wrong questions and no healing. *Biomaterials* **28**(34) (2007) 5009–5027.

103. Bordenave L, Fernandez P, Rémy-Zolghadri M, Villars S, Daculsi R, Midy D. *In vitro* endothelialized ePTFE prostheses: clinical update 20 years after the first realization. *Clin Hemorheol Microcirc* **33**(3) (2005) 227–234. Review.

104. Schmedlen RH, Elbjeirami WM, Gobin AS, West JL. Tissue engineered small-diameter vascular grafts. *Clin Plast Surg* **30**(4) (2003) 507–517. Review.
105. Patel A, Fine B, Sandig M, Mequanint K. Elastin biosynthesis: The missing link in tissue-engineered blood vessels. *Cardiovasc Res* **71**(1) (2006) 40–49. Epub 2006 Feb 28. Review.
106. Naito Y, Imai Y, Shin'oka T, Kashiwagi J, Aoki M, Watanabe M, Matsumura G, Kosaka Y, Konuma T, Hibino N, Murata A, Miyake T, Kurosawa H. Successful clinical application of tissue-engineered graft for extracardiac Fontan operation. *J Thorac Cardiovasc Surg* **125**(2) (2003) 419–420.
107. Shin'oka T, Matsumura G, Hibino N, Naito Y, Watanabe M, Konuma T, Sakamoto T, Nagatsu M, Kurosawa H. Midterm clinical result of tissue-engineered vascular autografts seeded with autologous bone marrow cells. *J Thorac Cardiovasc Surg* **129**(6) (2005) 1330–1338.
108. Shinoka T, Breuer C. Tissue-engineered blood vessels in pediatric cardiac surgery. *Yale J Biol Med* **81**(4) (2008) 161–166.
109. Rashid ST, Salacinski HJ, Hamilton G, Seifalian AM. The use of animal models in developing the discipline of cardiovascular tissue engineering: a review. *Biomaterials* **25**(9) (2004) 1627–1637. Review.
110. Nelson GN, Mirensky T, Brennan MP, Roh JD, Yi T, Wang Y, Breuer CK. Functional small-diameter human tissue-engineered arterial grafts in an immunodeficient mouse model: preliminary findings. *Arch Surg* (2008).
111. Goyal A, Wang Y, Su H, Dobrucki LW, Brennan M, Fong P, Dardik A, Tellides G, Sinusas A, Pober JS, Saltzman WM, Breuer CK. Development of a model system for preliminary evaluation of tissue-engineered vascular conduits. *J Pediatr Surg* **41**(4) (2006) 787–791.
112. Gong Z, Niklason LE. Blood vessels engineered from human cells. *Trends Cardiovasc Med* **16**(5) (2006) 153–156. Review.
113. Zavan B, Vindigni V, Lepidi S, Iacopetti I, Avruscio G, Abatangelo G, Cortivo R. Neoarteries grown *in vivo* using a tissue-engineered hyaluronan-based scaffold. *FASEB J* **22**(8) (2008) 2853–2861.
114. Sparks SR, Tripathy U, Broudy A, Bergan JJ, Kumins NH, Owens EL. Small-caliber mesothelial cell-layered polytetraflouroethylene vascular grafts in New Zealand white rabbits. *Ann Vasc Surg* **16**(1) (2002) 73–76.
115. Yang D, Guo T, Nie C, Morris SF. Tissue-engineered blood vessel graft produced by self-derived cells and allogenic acellular matrix: a functional performance and histologic study. *Ann Plast Surg* **62**(3) (2009) 297–303.
116. Yokota T, Ichikawa H, Matsumiya G, Kuratani T, Sakaguchi T, Iwai S, Shirakawa Y, Torikai K, Saito A, Uchimura E, Kawaguchi N, Matsuura N, Sawa Y. In situ tissue regeneration using a novel tissue-engineered, small-caliber vascular graft without cell seeding. *J Thorac Cardiovasc Surg* **136**(4) (2008) 900–907.
117. Hoerstrup SP, Cummings Mrcs I, Lachat M, Schoen FJ, Jenni R, Leschka S, Neuenschwander S, Schmidt D, Mol A, Günter C, Gössi M, Genoni M, Zund G. Functional growth in tissue-engineered living, vascular grafts: follow-up at 100 weeks in a large animal model. *Circulation* **114**(1 Suppl) (2006) 1159–1166.
118. Opitz F, Schenke-Layland K, Cohnert TU, Starcher B, Halbhuber KJ, Martin DP, Stock UA. Tissue engineering of aortic tissue: dire consequence of suboptimal elastic fiber synthesis *in vivo*. *Cardiovasc Res* **63**(4) (2004) 719–730.
119. Hibino N, Shin'oka T, Matsumura G, Ikada Y, Kurosawa H. The tissue-engineered vascular graft using bone marrow without culture. *J Thorac Cardiovasc Surg* **129**(5) (2005) 1064–1070.
120. Brennan MP, Dardik A, Hibino N, Roh JD, Nelson GN, Papademitris X, Shinoka T, Breuer CK. Tissue-engineered vascular grafts demonstrate evidence of growth and development when implanted in a juvenile animal model. *Ann Surg* **248**(3) (2008) 370–377.
121. Rouwkema J, Rivron NC, van Blitterswijk CA. Vascularization in tissue engineering. *Trends Biotechnol* **26**(8) (2008) 434–441.
122. Lokmic Z, Mitchell GM. Engineering the microcirculation. *Tissue Eng Part B Rev* **14**(1) (2008) 87–103.

123. Erol OO, Spira M. New capillary bed formation with a surgically constructed arteriovenous fistula. *Surg Forum* **30** (1979) 530–531.

124. Laschke, MW, Harder Y, Amon M, Martin I, Farhadi J, Ring A, Torio-Padron N, Schramm R, Rücker M, Junker D, Häufel JM, Carvalho C, Heberer M, Germann G, Vollmar B, Menger MD. Angiogenesis in tissue engineering: breathing life into constructed tissue substitutes. *Tissue Eng* **12**(8) (2006) 2093–2104.

125. Wang XM, Pei GX, Jin D, Wei KH, Jiang S, Tang GH. Perfusion-weighted magnetic resonance imaging for monitoring vascularization in tissue-engineered bone in rhesuses. *Nan Fang Yi Ke Da Xue Xue Bao* **26**(7) (2006) 931–935.

126. Hitchcock T, Niklason L. Lymphatic tissue engineering: progress and prospects. *Ann NY Acad Sci* **1131** (2008) 44–49. Review.

127. Payumo FC, Kim HD, Sherling MA, Smith LP, Powell C, Wang X, Keeping HS, Valentini RF, Vandenburgh HH. Tissue engineering skeletal muscle for orthopaedic applications. *Clin Orthop Relat Res* (403 Suppl) (2002) S228–S242. Review.

128. Koning M, Harmsen MC, van Luyn MJ, Werker PM. Current opportunities and challenges in skeletal muscle tissue engineering. *J Tissue Eng Regen Med* **3**(6) (2009) 407–415.

129. Liao H, Zhou GQ. Development and Progress of Engineering of Skeletal Muscle Tissue. *Tissue Eng Part B Rev* (2009) [Epub ahead of print].

130. Levenberg S, Rouwkema J, Macdonald M, Garfein ES, Kohane DS, Darland DC, Marini R, van Blitterswijk CA, Mulligan RC, D'Amore PA, Langer R. Engineering vascularized skeletal muscle tissue. *Nat Biotechnol* **23**(7) (2005) 879–884.

131. Scime A, Caron AZ, Grenier G. Advances in myogenic cell transplantation and skeletal muscle tissue engineering. *Front Biosci* **14** (2009) 3012–3023. Review.

132. Yan W, George S, Fotadar U, Tyhovych N, Kamer A, Yost MJ, Price RL, Haggart CR, Holmes JW, Terracio L. Tissue engineering of skeletal muscle. *Tissue Eng* **13**(11) (2007) 2781–2790.

133. Powell CA, Smiley BL, Mills J, Vandenburgh HH. Mechanical stimulation improves tissue-engineered human skeletal muscle. *Am J Physiol Cell Physiol* **283**(5) (2002) C1557–1565.

134. Chromiak JA, Shansky J, Perrone C, Vandenburgh HH. Bioreactor perfusion system for the long-term maintenance of tissue-engineered skeletal muscle organoids. *In vitro Cell Dev Biol Anim* **34**(9) (1998) 694–703.

135. Bian W, Bursac N. Engineered skeletal muscle tissue networks with controllable architecture. *Biomaterials* **30**(7) (2009) 1401–1412.

136. Moon du G, Christ G, Stitzel JD, Atala A, Yoo JJ. Cyclic mechanical preconditioning improves engineered muscle contraction. *Tissue Eng Part A* **14**(4) (2008) 473–482.

137. Serena E, Flaibani M, Carnio S, Boldrin L, Vitiello L, De Coppi P, Elvassore N. Electrophysiologic stimulation improves myogenic potential of muscle precursor cells grown in a 3D collagen scaffold. *Neurol Res* **30**(2) (2008) 207–214.

138. 146: Huang YC, Dennis RG, Larkin L, Baar K. Rapid formation of functional muscle *in vitro* using fibrin gels. *J Appl Physiol* **98**(2) (2005) 706–713.

139. Larkin LM, Calve S, Kostrominova TY, Arruda EM. Structure and functional evaluation of tendon-skeletal muscle constructs engineered *in vitro*. *Tissue Eng* **12**(11) (2006) 3149–3158.

140. Kostrominova TY, Calve S, Arruda EM, Larkin LM. Ultrastructure of myotendinous junctions in tendon-skeletal muscle constructs engineered *in vitro*. *Histol Histopathol* **24**(5) (2009) 541–550.

141. Lu Y, Shansky J, Del Tatto M, Ferland P, Wang X, Vandenburgh H. Recombinant vascular endothelial growth factor secreted from tissue-engineered bioartificial muscles promotes localized angiogenesis. *Circulation* **104**(5) (2001) 594–599.

142. Vandenburgh H, Del Tatto M, Shansky J, Lemaire J, Chang A, Payumo F, Lee P, Goodyear A, Raven L. Tissue-engineered skeletal muscle organoids for reversible gene therapy. *Hum Gene Ther* **7**(17) (1996) 2195–2200.

143. Lu Y, Shansky J, Del Tatto M, Ferland P, McGuire S, Marszalkowski J, Maish M, Hopkins R, Wang X, Kosnik P, Nackman M, Lee A, Creswick B, Vandenburgh H. Therapeutic potential of implanted tissue-engineered bioartificial muscles delivering recombinant proteins to the sheep heart. *Ann NY Acad Sci* **961** (2002) 78–82.

144. Okano T, Matsuda T. Muscular tissue engineering: capillary-incorporated hybrid muscular tissues *in vivo* tissue culture. *Cell Transplant* **7**(5) (1998) 435–442.

145. Dhawan V, Lytle IF, Dow DE, Huang YC, Brown DL. Neurotization improves contractile forces of tissue-engineered skeletal muscle. *Tissue Eng* **13**(11) (2007) 2813–2821.

146. Borschel GH, Dow DE, Dennis RG, Brown DL. Tissue-engineered axially vascularized contractile skeletal muscle. *Plast Reconstr Surg* **117**(7) (2006) 2235–2242.

147. 25: MacNeil S. Progress and opportunities for tissue-engineered skin. *Nature* **445**(7130) (2007) 874–880. Review.

148. Hermans, MH. Clinical experience with glycerol-preserved donor skin treatment in partial thickness burns. *Burns Incl. Therm Inj* **15** (1989) 57–59.

149. Wainwright DJ. Use of an acellular allograft dermal matrix (AlloDerm) in the management of full-thickness burns. *Burns* **21**(4) (1995) 243–248.

150. Chiu T, Burd A. "Xenograft" dressing in the treatment of burns. *Clin Dermatol* **23**(4) (2005) 419–423. Review.

151. Jarman-Smith ML, Bodamyali T, Stevens C, Howell JA, Horrocks M, Chaudhuri JB. Porcine collagen crosslinking, degradation and its capability for fibroblast adhesion and proliferation. *J Mater Sci Mater Med* **15**(8) (2004) 925–932.

152. Burke JF, Yannas IV, Quinby WC Jr, Bondoc CC, Jung WK. Successful use of a physiologically acceptable artificial skin in the treatment of extensive burn injury. *Ann Surg* **194**(4) (1981) 413–428.

153. Stern R, McPherson M, Longaker MT. Histologic study of artificial skin used in the treatment of full thickness thermal injury. *J. Burn Care Rehabil* **11** (1990) 7–13.

154. Heimbach D, Luterman A, Burke J, Cram A, Herndon D, Hunt J, Jordan M, McManus W, Solem L, Warden G, et al. Artificial dermis for major burns. A multi-center randomized clinical trial. *Ann Surg* **208**(3) (1988) 313–320.

155. Bello, YM, Falabella, AF. The role of Graftskin (Apligraf®) in difficult-to-heal venous leg ulcers. *J. Wound Care* **11** (2003) 182–183.

156. Lipkin S, Chaikof E, Isseroff Z, Silverstein P. Effectiveness of bilayered cellular matrix in healing of neuropathic diabetic foot ulcers: results of a multicenter pilot trial. *Wounds* 15 (2003) 230–236.

157. Bannasch H, Unterberg T, Föhn M, Weyand B, Horch RE, Stark GB. Cultured keratinocytes in fibrin with decellularised dermis close porcine full-thickness wounds in a single step. *Burns* **34**(7) (2008) 1015–1021.

158. Matousková E, Broz L, Pokorná E, Königová R. Prevention of burn wound conversion by allogeneic keratinocytes cultured on acellular xenodermis. *Cell Tissue Bank* **3**(1) (2002) 29–35.

159. van den Bogaerdt AJ, Ulrich MM, van Galen MJ, Reijnen L, Verkerk M, Pieper J, Lamme EN, Middelkoop E. Upside-down transfer of porcine keratinocytes from a porous, synthetic dressing to experimental full-thickness wounds. *Wound Repair Regen* **12**(2) (2004) 225–234.

160. Greenhalgh DG. Models of wound healing. *J Burn Care Rehabil* **26**(4) (2005) 293–305. Review.

161. Lindblad WJ. Considerations for selecting the correct animal model for dermal wound-healing studies. *J Biomater Sci Polym Ed* **19**(8) (2008) 1087–1096. Review.

162. Geer DJ, Swartz DD, Andreadis ST. *In vivo* model of wound healing based on transplanted tissue-engineered skin. *Tissue Eng* **10**(7–8) (2004) 1006–1017.

163. Dorsett-Martin WA. Rat models of skin wound healing: a review. *Wound Repair Regen* **12**(6) (2004) 591–599. Review.

164. Grose R, Werner S. Wound-healing studies in transgenic and knockout mice. *Mol Biotechnol* **28**(2) (2004) 147–166. Review.

165. Sullivan TP, Eaglstein WH, Davis SC, Mertz P. The pig as a model for human wound healing. *Wound Repair Regen* **9**(2) (2001) 66–76. Review.

166. Sullivan TP, Eaglstein WH, Davis SC, Mertz P. The pig as a model for human wound healing. *Wound Repair Regen* **9**(2) (2001) 66–76. Review.

167. Markowicz MP, Steffens GC, Fuchs PC, Pallua N. Enhanced dermal regeneration using modified collagen scaffolds: experimental porcine study. *Int J Artif Organs* **29**(12) (2006) 1167–1173.

168. Corr DT, Gallant-Behm CL, Shrive NG, Hart DA. Biomechanical behavior of scar tissue and uninjured skin in a porcine model. *Wound Repair Regen.***17**(2) (2009) 250–259.

169. Middelkoop E, van den Bogaerdt AJ, Lamme EN, Hoekstra MJ, Brandsma K, Ulrich MM. Porcine wound models for skin substitution and burn treatment. *Biomaterials* **25**(9) (2004) 1559–1567.

170. Russell WMS, Burch RL. *The principles of humane experiemental technique*, Methuen, London, 1959.

CHAPTER 7

Guidance Strategies in Hand Tissue Engineering: Manipulating the Microenvironment Through Cellular and Material Cues

Harvey Chim and Arun K. Gosain[*,†]

Introduction

The field of tissue engineering has seen tremendous exponential growth in recent years. The initial concept of tissue engineering as defined by Langer and Vacanti[1] was "an interdisciplinary scientific field that combines the principles of life sciences and engineering toward the development of biologic substitutes that will serve to restore, maintain, or improve tissue function". This concept relied on the combined use of donor-derived cells, a biodegradable scaffold and possibly a bioreactor[2] to engineer a cell-scaffold construct with potential clinical application as a tissue or organ substitute.

In the special context of hand surgery, a number of tissues are of interest for engineering components of the upper extremity. These include nerve, bone, tendon, skin, vessel, cartilage and composite tissue.[3] While much progress has been made in the field, a number of limitations have prevented widespread clinical application of tissue engineering aside from a number of skin substitutes which have been approved for use by the Food and Drug Administration (FDA).[4,5]

One major limitation has been the issue of donor site morbidity from cell harvest. The major issue is whether cells are derived from end organs or from more multipotent sources such as bone marrow derived mesenchymal stem cells. While groups have explored the use of embryonic stem cells (ESCs) for use in musculoskeletal tissue engineering,[6] the use of ESCs is fraught with ethical and moral issues. An alternative cellular approach to tissue engineering would be to induce cell migration into an implanted scaffold through creation of a biomimetic environment by selective constant release of cytokines. We would term this approach "cell guidance".[7]

The use of cells and scaffold materials to replace diseased or injured tissue, as per the traditional paradigm of tissue engineering, was also found to be ineffective for purposes of nerve tissue engineering, with a consensus emerging that it will ultimately require the coordinated presentation of multiple permissive signals to be incorporated into tissue engineered biomaterial platforms for regrowth of tissues. Efforts in this area to recreate the complex neural microenvironment have been termed "axon guidance".[8]

*Corresponding author.

[†]Arun K. Gosain, M.D., Department of Plastic Surgery, Case Western Reserve University, University Hospitals of Cleveland, MS 5044, 11100 Euclid Avenue, Cleveland, Ohio 44106, E-mail address: arun.gosain@uhhospitals.org

Another problem that has emerged with translation of traditional tissue engineering techniques has been inadequate vascularization, and hence perfusion of engineered constructs. In the hand, this will potentially pose a major problem in engineering of bone and composite tissue. Inherent limitations with osseous tissue only forming on the surface of a polymeric scaffold were noted early on,[9] and continue to limit engineering of large vascularized bone constructs.[10] By patterning biomimetic vascular channels in a polymer scaffold, induced vascularization of constructs would potentially allow fabrication of much larger constructs than that currently possible following the traditional tissue engineering paradigm. This approach has been termed "vascular guidance".[11]

In this chapter, we describe recent progress in "Guidance" strategies, which aim at manipulating the microenvironment around cells and scaffolds to achieve a biomimetic milieu. These strategies offer an alternative and novel approach to overcoming limitations inherent to the traditional paradigm of tissue engineering.

Cell Guidance: Inducing Site-Specific Homing of Cells Through Creation of a Biomimetic Microenvironment

SDF-1α/CXCR4 interaction in homing of mesenchymal stem cells

In bone tissue engineering, mesenchymal stem cells (MSCs) derived from bone marrow are commonly used as a cell source. MSCs are capable of differentiating into a variety of non-hematopoietic cells such as osteoblasts, adipocytes, chondrocytes and myoblasts under appropriate conditions[12,13] (Figure 1). These primitive progenitors exist postnatally and exhibit stem cell characteristics with extensive renewal potential. Our interest in finding an alternative source of cells for tissue engineering aside from currently available donor-derived sources led us to look further into chemokine interactions in MSCs.

Stromal cell derived factor-1 alpha (SDF-1α), a chemokine, is produced mainly by immature bone forming osteoblasts and is highly expressed by bone marrow endothelial cells.[12] SDF-1α is primarily involved in the trafficking of $CD34^+$ hematopoietic stem cells (HSCs) and elevation of plasma SDF-1α induces HSC mobilization. During conditions of stress and inflammation, the expression of SDF-1α is increased in injured organs and consequently in the peripheral blood. Circulating SDF-1α is internalized by CXCR4 across the physiologic blood vessel barrier into the bone marrow where it binds CXCR4 in the stem cell niche. Presentation of translocated SDF-1α by bone marrow endothelial and other stromal cells recruits circulating $CXCR4^+$ hematopoietic stem and progenitor cells into the bone marrow, which proliferate and differentiate into immature and mature cells, subsequently mobilize into the damaged organ, and also serve for host defense and organ repair.[14]

SDF-1α is critical in promoting the migration of stem cells to bone marrow via its specific receptor, CXCR4 which is expressed on MSCs.[15] Bone marrow derived MSCs were also shown experimentally to migrate in response to SDF-1α using a microwell chemotaxis chamber assay.[16] The mechanism of the SDF-1α/CXCR4 interaction has been shown to be through JAK2 and JAK3, which associate with CXCR4 and are activated, probably by transphosphorylation, in a Gα(i)-independent manner.[17]

Inducing homing of cells using SDF-1α[18]

Given the ability of SDF-1α to induce migration of MSCs, it serves as an ideal candidate for studies into "cell guidance". We were able to demonstrate migration of MSCs towards

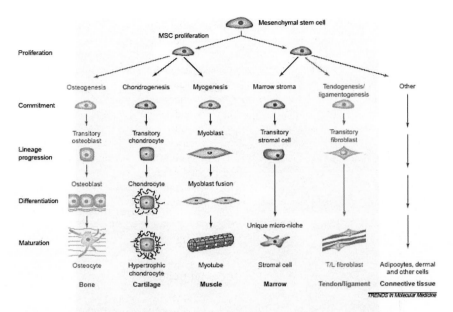

Figure 1. Differentiation pathways of mesenchymal stem cells. (Reproduced from: Trends in Molecular Medicine Vol. 7 No. 6 June 2001 259.)

a SDF-1a stimulus *in vitro* using a chemotaxis assay, with demonstration of preferential homing towards the strongest stimulus, equating to the highest concentration of SDF-1α. In a subsequent study using MSCs seeded into three-dimensional polycaprolactone (PCL) scaffolds, MSCs migrated against gravity in response to a SDF-1 stimulus (Figure 2), showing the feasibility of this approach in the context of the traditional approach to tissue engineering, where cells are seeded onto scaffolds to achieve a functional tissue equivalent.

We subsequently examined the effect of sequential delivery of VEGF, SDF-1, and BMP-6 in a subcutaneous rat model so as to demonstrate the feasibility of a tissue engineering approach whereby a cytokine cocktail is used to stimulate cell migration and proliferation within a three-dimensional polymeric scaffold environment, in order to test the utility of "cell guidance" as an alternative source of cells for tissue engineering. A cytokine microdelivery system was fabricated to ensure constant sequential delivery of these cytokines, and placed on top of a PCL scaffold devoid of any seeded cells. The assembled apparatus was implanted subcutaneously over the panniculosus carnosus in a Wistar rat model. A syringe was used to inject infusate into the pump using aseptic technique every 10 days, with VEGF injected first to induce vascularization, followed by SDF-1 for chemotaxis of native MSCs, and finally BMP-6 to induce osteogenic differentiation and mineralization. Animals were sacrificed after 30 days and scaffolds explanted for histological analysis.

There was evidence of marked vascularization deriving from the animal's panniculosus carnosus obvious on gross examination of the experimental scaffolds, with a dominant vascular pedicle evident (Figure 3). In the control group, no dominant pedicle was observed with minimal ingrowth of tissue into the scaffold. Histological staining with hematoxylin and eosin (H&E) showed evidence of dense vascularized tissue with cellular migration throughout the scaffold in response to the SDF-1α

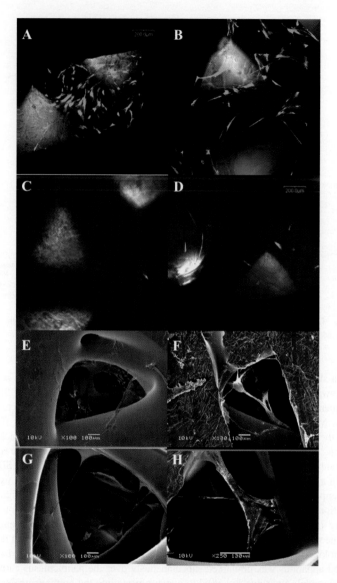

Figure 2. Mesenchymal stem cells (MSCs) showed chemotaxis in response to SDF-1 within 3D scaffolds. Confocal laser microscopy demonstrated a large number of cells for both control (A) and experimental (B) scaffolds, away from the chemotactic SDF-1 stimulus. At the top of the scaffold, closest to the site of SDF-1 release, no cells were present for control scaffolds (C), but sizable numbers of cells were observed for experimental scaffolds (D), indicating that MSCs had migrated from the bottom to the top of the scaffold against gravity in response to the chemotactic stimulus of SDF-1. Scanning electron micrographs confirmed these findings, with large numbers of cells at the bottom of the scaffold observed for control (E) and experimental (F) groups. At the top of the scaffold, no cells were seen for the control (G) group; however experimental scaffolds (H) showed evidence of cell migration in response to SDF-1, with cells bridging adjacent struts of the scaffold. Original magnification: (A-D) × 50, (E-G) × 100 and (H) × 250.

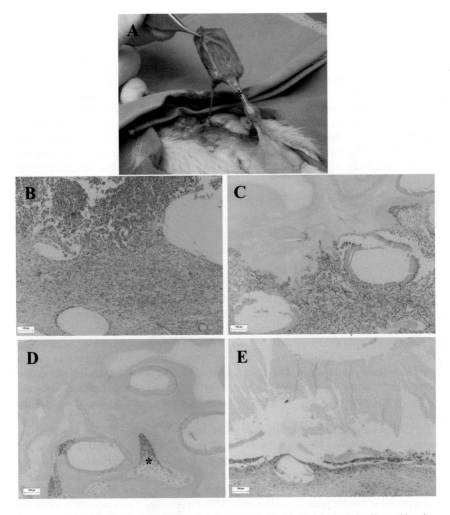

Figure 3. Explanted scaffolds from the *in vivo* rat model after 30 days showed site directed homing of cells following implantation of an unseeded PCL scaffold exposed sequentially to VEGF, SDF-1 and BMP-6. (A) Gross examination of the explanted scaffold showed evidence of a vascular pedicle (marked with *). Vertical sections through the explanted scaffold were obtained (B, bottom [closest to rat wound bed]; C, middle; D, top [closest to cytokine release from microneedle apparatus]). A clear transition in tissue structure was observed (B), with organized precursor tissue formation nearest the rat wound bed and less dense cellular infiltration towards the top of the scaffold. Angiogenesis was evidenced by formation of blood vessels (red areas). (C) The middle section of the scaffold demonstrated the advancing tissue front of cellular infiltration in response to the SDF-1 chemotactic stimulus. (D) Migrating cell ridge islands, representing polarized clusters of cells moving against gravity towards the SDF-1 chemotactic stimulus were observed ahead of the main bulk of tissue infiltration. Cells can be seen to be concentrated in the advancing front of a migrating cell island ridge (marked with *) demonstrating chemotactic cellular infiltration against gravity. (E) In contrast, for control scaffolds there was negligible cell migration and tissue formation, with no evidence of angiogenesis. Clear acellular areas represent scaffold material dissolved during processing for histology. Original magnification: (A-E) × 40.

chemotactic stimulus. Vertical sections through the explanted scaffold were obtained, with evidence of a clear transition in tissue structure. Dense organized tissue formation was found closest to the rat wound bed, and less-dense cellular infiltrate toward the upper extent of the scaffold. In addition, angiogenesis was apparent, with blood vessels observed throughout the bottom and middle sections of the scaffold. The transition in tissue structure was a direct result of cell migration from the rat wound bed toward the microneedle apparatus. Cell migration occurred by two mechanisms, with an advancing tissue front and migrating cell island ridges moving toward the chemotactic stimulus ahead of the main bulk of tissue formation, against gravity. Conversely, for control scaffolds, there was negligible cell migration and tissue ingrowth from the wound bed, with no evidence of vascularization or mineralization. In effect, tissue formation was achieved without the requirement for cells to be seeded prior to implantation, demonstrating an alternative approach compared to the use of donor-derived cells for tissue engineering.

SDF-1α as an agent for cell guidance

A major problem with the traditional paradigm in tissue engineering is the requirement for a source of donor-derived cells. With advances in stem cell technology, an increasing variety of cell sources are available. However, a bone marrow aspirate to obtain MSCs, for example, is still an invasive procedure, with risks of infection and bleeding; as well as causing a significant degree of pain and discomfort to the patient. Use of cells from end organs, which were the first source of cells described for tissue engineering, such as differentiated bone and cartilage, still remains an option, but carries the problem of patient morbidity during cell harvest. The development of such advances as induced pluripotent stem cells[19,20] offers the advantage of a pluripotent cell source without resorting to the use of embryonic stem cells, but ultimately still requires a source of cells harvested from the donor.

Cell guidance offers an alternative option, based on site-directed homing of native stem cells to cytokine-loaded scaffolds, so as to induce tissue formation without the needed for a prior seeded cell source. This approach may be especially promising for tissue engineering and regenerative medicine, as it allows a scaffold to be implanted immediately into the site of a clean tissue defect requiring reconstruction, without the need for a period of culture prior to implantation. SDF-1α provides a very attractive option for cell guidance, due to its ability to direct migration of MSCs towards itself.

The use of a cytokine microdelivery system with a novel microneedle apparatus (Figure 4) allows specific control of the "macroanatomy" and "microanatomy" of the total in vivo tissue engineered construct. Control of the "macroanatomy" of a tissue engineered construct allows selective homing of desired cells to a tissue defect within a scaffold, such as for regeneration of a segmental long bone defect. Control of the "microanatomy" of a construct would entail differential release of cytokines within specific areas of a scaffold, so as to engineer, for example, an osteochondral construct for joint resurfacing by release of osteogenic and chondrogenic differentiation factors at the polar opposite ends of a scaffold implanted into a joint defect.

In tissue engineering of the hand, where all major structures are relatively superficial, cell guidance would be a particularly attractive approach, as the housing and delivery apparatus used to induce cell migration and tissue formation could be placed externally, to allow controlled constant delivery of varying cytokines for regeneration of various tissues, with the aim of achieving composite tissue regeneration.

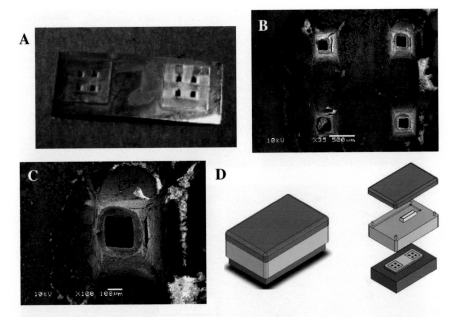

Figure 4. A novel cytokine micro-delivery system comprised of a reservoir housing unit and microneedle apparatus was fabricated for controlled delivery of cytokines to specific areas of the scaffold. (A) The microneedle apparatus is rectangular in shape and composed of two arrays of four needles each. (B, C) Scanning electron micrographs showing microneedles at high magnification. (D) The reservoir housing unit comprises three components, with the top piece designed to fit onto the bottom piece to prevent leakage of the chemokine from the system. The pump and tubing for delivery of cytokines fits within a groove in the middle piece, and leads to a reservoir where the cytokine is channeled into the microneedles for sustained and controlled delivery throughout the scaffold. The bottom piece is specially designed to accommodate the microneedles within its interior as well as being able to fit onto a scaffold. Original magnification: (B) × 35 and (C) × 100. (Figures 2, 3 and 4 reproduced from Schantz JT, Chim H, Whiteman M. Cell guidance in tissue engineering: SDF-1 mediates site-directed homing of mesenchymal stem cells within three-dimensional polycaprolactone scaffolds. *Tissue Eng.* **13**(11) (2007) 2615–2624.)

Axonal Guidance as a Strategy for Nerve Tissue Engineering

Nerve regeneration through molecular cues, micropatterns and electromagnetic cues

A number of cues have been described which are useful in manipulating the microenvironment to promote axonal regeneration. Among the most basic include the use of trophic factors such as brain-derived neurotrophic factor (BDNF), insulin-like growth factor (IGF), fibroblast growth factor (FGF) and nerve growth factor (NGF) to direct turning of growth cones towards the molecule of interest.[21,22] Concentration gradients of soluble factors have also been created through engineering techniques such as micropatterning, microfluidics and self-assembly of monolayers[23] for covalent binding of peptides. These techniques have application in fabrication of scaffolds for nerve tissue engineering. As an example, dorsal root ganglia (DRG) explants were shown to respond to immobilized gradients of the IKVAV peptide (Figure 5). Through micropatterning of laminins onto line

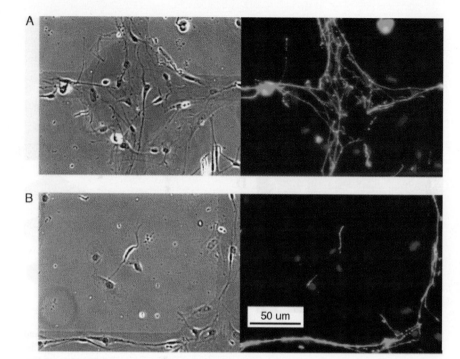

Figure 5. Dorsal root ganglia neurons follow micropatterns of poly-L-lysine–laminin immobilized on polyglutamic acid–doped polypyrrole network at nodes (A) and on channels (B). Neurites stained positive for GAP-43 (green fluorescence). Cell nuclei were labeled with 40,6-diamidino-2-phenylindole (blue fluorescence). Neurons preferentially adhere and neurites preferentially extend on micropatterns. (Reproduced from Biomaterials, Vol. 27, No. 3, Song *et al.* Micropatterns of positive guidance cues anchored to polypyrrole doped with polyglutamic acid: A new platform for characterizing neurite extension in complex environments, pp. 473–484.)

or grid patterned microcontact printed substrates, cellular adhesion and neurite extension was observed in dissociated neurons.[24]

The application of anisotropy for presenting chemical cues through gradients or stripes to create directional bias has been largely successful *in vitro*. Recent advances in microfabrication techniques in microcontact printing and microfluidics have increased our capacity to present directional information on a biologically relevant scale. Optimal dimensions of features range from tens to hundreds of microns[25] and can be used together with other types of guidance cues for synergy in a complex microenvironment.

Electric fields have been shown to reversibly influence the growth and length of neurites,[26] as well as induce neurite growth towards the cathodes.[27] Electric gradients have also been shown to direct neural crest cell migration toward the cathode.[28] Magnetic fields, in contrast, have been shown to enhance neuronal differentiation in PC12 cells, while inhibiting neurite growth.[29]

Use of topographical cues

Neurons have the ability to respond to topographical features in their microenvironment, such as grooves in micro and nano scales, and have been shown to adhere, migrate and orient their axons in response to these cues. Studies[30] have shown that microchannels of 20 to 30 μm were most effective at neurite direction. Polypropylene filaments have also been shown to direct neurite and Schwann cell growth along the long axis of the filaments, with fibronectin and laminin coatings leading to greater maximal neurite lengths.[31] The phenomenon of neurite "bridging", where neurites have been shown to span grooves coated with laminin, has also been demonstrated,[32] with neuritis observed to climb up groove walls to generate these bridges.

The use of nanotopography to generate nanoscale etches onto silicon wafers,[33] carbon nanotubes on flat surfaces[34] and nanofibers on the surface of scaffolds[35] have been reported, and is an area of intense research. Methods of fabrication include chemical etching with hydrofluoric acid to create unordered nanoscale grooves and electrospinning with polymeric materials such as PLA and PLGA. The creation of biomaterial platforms with nanoscale guidance cues, together with the use of electrical and molecular cues, is an area of much promise for axon guidance.

Cellular topographical cues are another area that plays a factor in axon guidance. Neurites have been found to respond to Schwann cell topography by responding to the direction of Schwann cell alignment[36] (Figure 6). Micromolded polymeric templates of aligned astrocyte monolayers have also been fabricated, and found to direct neurite growth according to the direction of replica astrocyte alignment.[8]

Nerve guidance channels for axon guidance

Based on lessons learned from studies into molecular, electromagnetic and topographic cues, studies are being carried out on the use of nerve guidance channels (NGCs). NGCs are biomaterial conduits which aim to provide a biomimetic channel mimicking the *in vivo* environment of a nerve autograft, so as to provide a pathway for axon guidance and nerve regeneration.[37,38] The typical animal model is that of a rat sciatic nerve, a well established model for peripheral nerve injury.

The tubular structure of the NGC itself provides a major directional cue, serving to direct regeneration of growing neurites. Microfilaments also provide topographical guidance, with PLA microfilament containing conduits found to promote cable tissue formation and promote Schwann cell migration *in vivo*.[39] The packing density of PLA microfilaments has been found to correlate with the number of successfully regenerated nerves.[40] In combination with molecular cues such as heregulin-β1,[41] microfilaments have been found to show a synergistic effect in inducing directional regeneration of nerves.

Chemical cues have been incorporated into the walls of NGCs, such as nerve growth factor (NGF), and found to support the growth of nerve cables, myelinated axons and neurons *in vivo*. Various materials, such as alginate, collagen, fibrin sponges and gels have also been used to fill the inner volume of NGCs.[42] These were found to increase the number of myelinated axons and promote Schwann cell migration from the distal nerve stump. Electrical guidance cues have also been incorporated into NGCs electrically conductive biodegradable polymers such as pPy,[43] which were found to enhance neurite outgrowth in a current-dependent fashion.

Through using a tissue engineering approach in fabrication of various NGCs, biocompatible, biodegradable conduits may potentially be available for repair of motor and sensory nerves in the hand. A recent example of a NGC undergoing a prospective trial is Neurotube,[44] a polyglycolic acid bioresorbable conduit which was found to result in better recovery of sensory function than autografts for nerve gaps less than 3 cm in length. Another example of a NGC in clinical use is Neurogen, a collagen matrix tube which has been used in such applications as repair of digital nerves and brachial plexus injury.[45]

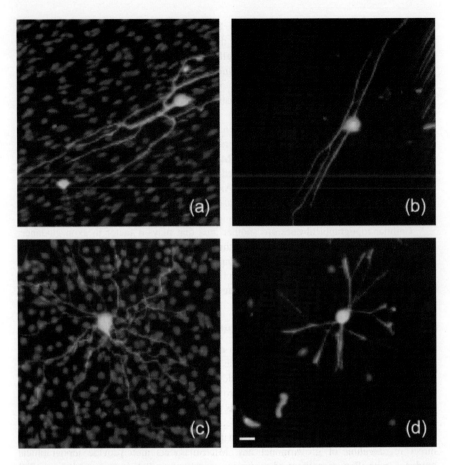

Figure 6. Dorsal root ganglia cultured on aligned underlying Schwann cell (SC) monolayer culture are directed in the orientation of SCs. Fluorescent images of 40,6-diamidino-2-phenylindole–stained SCs and neurofilament-stained DRGs on aligned (A, B) and unaligned (C, D) SC monolayers. (Reproduced from Thompson and Buettner, Annals of Biomedical Engineering Vol. 34, No. 1, 2006, pp. 161–168.)

Vascular Guidance: Microstructural Patterning of Polymeric Scaffolds

Overcoming limitations of currently available scaffolds

The traditional approach to tissue engineering entails seeding cells onto scaffolds to create a functional tissue equivalent. Unfortunately, beyond a certain size, vascularization, and hence perfusion of these constructs becomes a limiting factor in how large these constructs can be. An analogy can be made to the presence of capillaries in human tissue. Just as microscopic capillaries deriving from arterioles, which arise from arteries, are essential to perfuse microscopic blocks of tissue; an engineered cell-scaffold construct cannot conceivably survive solely by diffusion of nutrients beyond a certain size limit.

This limitation of currently available polymeric scaffolds was noted early on in the bone tissue engineering field with poly-L-glycolic acid (PLGA) foams,[9] where penetration of osseous tissue was restricted to the surface of the tissue engineered construct. Subsequent efforts in developing scaffolds with true material interconnectivity using rapid prototyping techniques such as fused deposition modeling[46] resulted in tissue formation throughout the scaffold structure in benchtop and animal experiments. However, this has only been demonstrated for small constructs. The requirement for true angiogenesis and vasculogenesis *in vivo* for formation of a vessel network has so far prevented engineering of large constructs which might be used in conceivable clinical applications such as a skull flap in calvarial reconstruction, a long bone segment for reconstruction after trauma or oncologic surgery, or even a vascularized osteochondral joint segment.

A conceivable solution would be to create polymeric scaffolds through patterning of a branching vessel network mimicking that found in the natural state of a tissue, so as to create a natural pathway for vasculogenesis and resultant perfusion of tissue. This could be done similar to the manner of a prefabricated free flap. In this case, a ligated vascular pedicle or arteriovenous loop is implanted into the proximal egress of the scaffold channels, and subsequently allows ingrowth of vessels into the scaffold architecture, thereby facilitating vasculogenesis. The use of cytokines such as vascular endothelial growth factor (VEGF) to induce directional vasculogenesis and promote angiogenesis at the microscopic level is another adjunct that can be used with this technique. We describe here our experiments into "vascular guidance", as we would term this approach.

Modeling the perfusion of a calvarial bone flap

We chose to use a frontal-parietal calvarial bone flap as a template for modeling a vascular network, as the branching network of vessels originating from the anterior branch of the middle meningeal artery that supplies the flap is easily observed. Grooves resulting from the pulsating course of the artery on the internal aspect of the calvarial bone flap result in an arborizing vascular pattern which is visible to the naked eye.

Using computer-aided design (CAD) software such as the Pro/ENGINEER software (Parametric Technology Corporation, Needham, MA), various models of vascular branching were conceived in 3D. A model that most closely resembled that observed on the calvarial bone flap was eventually chosen, and flow modeled through the potential vascular network.

We subsequently used a well-established rapid prototyping technology, fused deposition modeling to fabricate a scaffold from polycaprolactone (PCL) and hydroxyapatite (HA) with full interconnectivity based on a filament lay-down pattern of 0, 60 and 120 degrees. Data on the modeled vascular network was obtained from computer software, and used to fabricate a series of branching channels within the polymer scaffold (Figure 7).

Experimental studies into vascular guidance

We used a groin nude rat model to investigate the feasibility of vascular guidance *in vivo*. The ligated inferior epigastric pedicle was inserted into the proximal egress of the channel network in the scaffold, and placed in a subcutaneous pocket. Following explantation of specimens after 3 weeks, vasculogenesis was seen following the framework of fabricated channels, as well as angiogenesis throughout the scaffold architecture. In control scaffolds, in contrast, only granulation tissue was observed without evidence of angiogenesis or vasculogenesis. We also transferred this construct as a prefabricated composite osteocutaneous groin flap to the neck using microsurgical technique, with survival of all flaps following free tissue transfer.

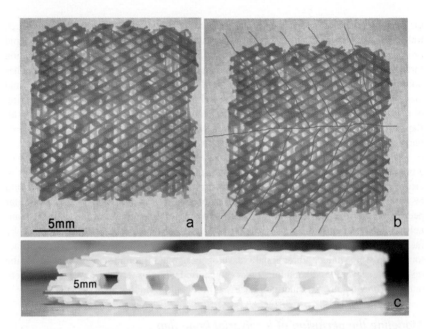

Figure 7. Fabricated PCL scaffold with incorporated vascular channels in arborizing pattern based on course of middle meningeal artery in the calvarium. (A) Top view. (B) Red markings show branching vascular channels. (C) Side view.

Application of vascular guidance in hand tissue engineering

The hand is a singularly unique and complex structure, with an arborizing network of arteries originating from the radial and ulnar arteries, ending in individual digital arteries that perfuse the fingers. If tissue engineering techniques were to progress to the extent that fabrication of a composite tissue allograft, or put simply, a complete hand replacement were possible, vascular guidance would provide a technique to recreate the complex vascular architecture of the hand. This would not be possible using only the traditional tissue engineering paradigm of cells seeded onto scaffolds.

While engineering of individual phalanges is not a technical challenge, and has been achieved *in vivo*[47] and applied clinically[48] to reconstruct the distal phalanx and interphalangeal joint of a thumb after an amputation, the available armamentarium of surgical techniques that can be used instead, provide a potent argument against the tremendous cost of widespread clinical use of tissue engineered constructs as a first line option. Rather, it is in the reconstruction of complex and extensive hand injuries that tissue engineering, and vascular guidance, may offer a solution.

Guidance Strategies in Hand Tissue Engineering — Looking Forward

The hand is one of the most complex structures in the human body. A large variety of tissues, comprising bone, cartilage, tendon, nerve, vessels and skin come together to create a functional composite in a small space that is elegant in form and function, but extremely complex in its composition and arrangement of different tissues. A large number of surgical options are available for reconstruction of any conceivable injury to the hand, ranging from flaps for digital coverage, to pollicization for a nonreconstructable thumb amputation, to vascularized joint or toe-thumb/finger transfer for reconstruction of digits. All these options are very effective in recovery of function, though some may result in a less pleasing aesthetic result. The engineering of such discrete structures as individual phalanges or small areas of skin substitutes is feasible and has been used clinically in select patients. However, the prohibitive cost of engineering a custom construct, as well as other factors such as time and requirement for lab facilities, suggests that for the foreseeable future tissue engineering will not be the first-line option for treatment of most hand defects requiring reconstruction. Rather, surgical options are more than adequate for treatment of most common hand injuries.

In more complex mutilating hand injuries where there is loss of multiple digits or in even more proximal injuries where the hand is essentially nonsalvageable, there are no good clinical solutions available currently. While hand transplantation remains under investigation, this alternative introduces a new set of problems including the need for chronic immunosuppression. Severe congenital defects, or those resulting after oncologic surgery, present a similar set of problems. It is in these patients that tissue engineering and "guidance" strategies may offer a solution. The fabrication of a composite tissue allograft will inherently be limited functionally and aesthetically by its vascular and nervous supply. These tissues are the most difficult to engineer in the laboratory, and efforts to do so following traditional tissue engineering techniques will probably be unsuccessful, particularly in the context of the complex and arborizing network of vessels and nerves found in the hand. Rather, vascular and axon guidance would be an effective means of ensuring accurate patterning of an anatomically correct and functioning neurovascular network. The ability of computer-aided design (CAD) software and various rapid prototyping and microfabrication techniques to obtain data from imaging modalities such as computed tomography (CT) and magnetic resonance imaging (MRI) may allow replication of the entire neurovascular network of the contralateral uninjured hand. This could serve as the framework for engineering of a synthetic, functioning hand.

In engineering of large quantities of tissue, such as that envisioned for large defects in the hand or even an entire hand by itself, an enormous quantity of donor cell material will be required. Rather, cell guidance may offer a solution in this context, by attracting autologous cells and inducing subsequent tissue formation within a pre-existing implanted skeletal framework comprised of polymeric scaffolds, nerve guidance conduits, vascular channels and synthetic functional joints.

While tissue engineering may be nowhere near widespread clinical application in the hand at this juncture, it is vital that pre-existing problems and limitations inherent to the traditional approach to tissue engineering be identified and addressed. Guidance strategies may provide a feasible route towards increasing clinical use of engineered tissue components, with the ultimate aim of a composite tissue allograft that is able to replicate the function, structure, and aesthetic appearance of a human hand.

References

1. Langer R, Vacanti JP. Tissue engineering. *Science* **260** (1993) 920.
2. Griffith LG, Naughton G. Tissue engineering-current challenges and expanding opportunities. *Science* **295** (2002) 1009.
3. Chong AKS, Chang J. Tissue engineering for the hand surgeon: a clinical perspective. *J Hand Surg* **31A** (2006) 349.

4.   Balasubramani M, Kumar TR, Babu M. Skin substitutes: a review. *Burns* **27** (2001) 534.
5.   Kearney JN. Clinical evaluation of skin substitutes. *Burns* **27** (2001) 545.
6.   Tromoleda JL, Forsyth NR, Khan NS, Wojtacha D, Christodoulou I, Tye BJ, Racey SN, Collishaw S, Sottile V, Thomson AJ, Simpson AH, Noble BS, McWhir J. Bone tissue formation from human embryonic stem cells *in vivo*. *Cloning Stem Cells* **10** (2008) 119.
7.   Schantz JT, Chim H, Whiteman M. Cell guidance in tissue engineering: SDF-1 mediates site-directed homing of mesenchymal stem cells within three-dimensional polycaprolactone scaffolds. *Tissue Eng* **13** (2007) 2615.
8.   Li GN, Hoffman-Kim D. Tissue-engineered platforms of axon guidance. *Tissue Eng Part B: Reviews* **14** (2008) 33.
9.   Ishaug-Riley SL, Crane-Kruger GM, Yaszemski MJ, Mikos AG. Three-dimensional culture of rat calvarial osteoblasts in porous biodegradable polymers. *Biomaterials* **19** (1998) 405.
10. Holy CE, Fialkov JA, Davies JE, Shoichet MS. Use of a biomimetic strategy to engineer bone. *J Biomed Mater Res A* **15** (2003) 447.
11. Muller D, Chim H, Bader A, Schantz JT. Vascular guidance: microstructural scaffold patterning for inductive angiogenesis. *Cells Tissues Organs, forthcoming*.
12. Prockop DJ. Marrow stromal cells as stem cells for nonhematopoietic tissues. *Science* **276** (1997) 71.
13. Caplan AI. Mesenchymal stem cells. *J Orthop Res* **9** (1991) 641.
14. Dar A, Kollet O, Lapidot T. Mutual, reciprocal SDF-1/CXCR4 interactions between hematopoietic and bone marrow stromal cells regulate human stem cell migration and development in NOD/SCID chimeric mice. *Exp Hematol* **34** (2006) 967.
15. Wakitani S, Saito T, Caplan AI. Myogenic cells derived from rat bone marrow mesenchymal stem cells exposed to 5-azacytidine. *Muscle Nerve* **18** (1995) 1417.
16. Sordi V, Malosio ML, Marchesi F, Mercalli A, Melzi R, Giordano T, Belmonte N, Ferrari G, Leone BE, Bertuzzi F, Zerbini G, Allavena P, Bonifacio E, Piemonti L. Bone marrow mesenchymal stem cells express a restricted set of functionally active chemokine receptors capable of promoting migration to pancreatic islets. *Blood* **106** (2005) 419.
17. Vila-Coro AJ, Rodríguez-Frade JM, Martín De Ana A, Moreno-Ortíz MC, Martínez-A C, Mellado M. The chemokine SDF-1alpha triggers CXCR4 receptor dimerization and activates the JAK/STAT pathway. *FASEB J* **13** (1999) 1699.
18. Schantz JT, Chim H, Whiteman M. Cell guidance in tissue engineering: SDF-1 mediates site-directed homing of mesenchymal stem cells within three-dimensional polycaprolactone scaffolds. *Tissue Eng* **13** (2007) 2615.
19. Yu J, Vodyanik MA, Smuga-Otto K, Antosiewicz-Bourget J, Frane JL, Tian S, Nie J, Jonsdottir GA, Ruotti V, Stewart R, Slukvin II, Thomson JA. Induced pluripotent stem cell lines derived from human somatic cells. *Science* **318** (2007) 1917.
20. Takahashi K, Yamanaka S. Induction of pluripotent stem cells from mouse embryonic and adult fibroblast cultures by defined factors. *Cell* **126** (2006) 663.
21. Boyd JG, Gordon T. Neurotrophic factors and their receptors in axonal regeneration and functional recovery after peripheral nerve injury. *Mol Neurobiol* **27** (2003) 277.
22. Kato AC, Lindsay RM. Overlapping and additive effects of neurotrophins and cntf on cultured human spinal cord neurons. *Exp Neurol* **130** (1994) 196.
23. Dertinger SKW, Jiang X, Li Z, Murthy VN, Whitesides GM. Gradients of substrate-bound laminin orient axonal specification of neurons. *Proc Natl Acad Sci USA* **99** (2002) 12542.
24. Vogt AK, Wrobel G, Meyer W, Knoll W, Offenhausser, A. Synaptic plasticity in micropatterned neuronal networks. *Biomaterials* **26** (2005) 2549.
25. Yeung CK, Lauer L, Offenhausser A, Knoll W. Modulation of the growth and guidance of rat brain stem neurons using patterned extracellular matrix proteins. *Neurosci Lett* **301** (2001) 147.
26. Patel N, Poo MM. Orientation of neurite growth by extracellular electric fields. *J Neurosci* **2** (1982) 483.
27. Wood M, Willits RK. Short-duration, DC electrical stimulation increases chick embryo DRG neurite outgrowth. *Bioelectromagnetics* **27** (2006) 328.
28. Stump RF, Robinson KR. Xenopus neural crest cell migration in an applied electrical field. *J Cell Biol* **97** (1983) 1226.

29. Morgado-Valle C, Verdugo-Díaz L, García DE, Morales-Orozco C, Drucker-Colín R. The role of voltage-gated Ca^{2+} channels in neurite growth of cultured chromaffin cells induced by extremely low frequency (ELF) magnetic field stimulation. *Cell Tissue Res* **291** (1998) 217.

30. Mahoney MJ, Chen RR, Tan J, Saltzman WM. The influence of microchannels on neurite growth and architecture. *Biomaterials* **26** (2005) 771.

31. Wen, X, Tresco, PA. Effect of filament diameter and extracellular matrix molecule precoating on neurite outgrowth and Schwann cell behavior on multifilament entubulation bridging device *in vitro*. *J Biomed Mater Res A* **76** (2006) 626.

32. Goldner JS, Bruder JM, Li, G, Gazzola D, Hoffman-Kim D. Neurite bridging across micropatterned grooves. *Biomaterials* **27** (2006) 460.

33. Fan YW, Cui FZ, Hou SP, Xu QY, Chen LN, Lee IS. Culture of neural cells on silicon wafers with nano-scale surface topograph. *J Neurosci Methods* **120** (2002) 17.

34. Lovat V, Pantarotto D, Lagostena L, Cacciari B, Grandolfo M, Righi M, Spalluto G, Prato M, Ballerini L. Carbon nanotube substrates boost neuronal electrical signaling. *Nano Lett* **5** (2005) 1107.

35. Ahmed I, Liu HY, Mamiya PC, Ponery AS, Babu AN, Weik T, Schindler M Meiners S. Three dimensional nanofibrillar surfaces covalently modified with tenascin-C-derived peptides enhance neuronal growth *in vitro*. *J Biomed Mater Res A* **76** (2006) 851.

36. Bruder JM, Lee AP, Hoffman-Kim D. Biomimetic materials replicating Schwann cell topography enhance neuronal adhesion and neurite alignment *in vitro*. *J Biomater Sci Polym Ed* **18** (2007) 967.

37. Schmidt CE, Leach JB. Neural tissue engineering: strategies for repair and regeneration. *Annu Rev Biomed Eng* **5** (2003) 293.

38. Battiston B, Geuna S, Ferrero M, Tos P. Nerve repair by means of tubulization: literature review and personal clinical experience comparing biological and synthetic conduits for sensory nerve repair. *Microsurgery* **25** (2005) 258.

39. Cai J, Peng X, Nelson KD, Eberhart R, Smith GM. Permeable guidance channels containing microfilament scaffolds enhance axon growth and maturation. *J Biomed Mater Res A* **75** (2005) 374.

40. Ngo TTB, Waggoner PJ, Romero AA, Nelson KD, Eberhart RC, Smith GM. Poly(L-Lactide) microfilaments enhance peripheral nerve regeneration across extended nerve lesions. *J Neurosci Res* **72** (2003) 227.

41. Cai J, Peng X, Nelson KD, Eberhart R, Smith GM. Synergistic improvements in cell and axonal migration across sciatic nerve lesion gaps using bioresorbable filaments and heregulin-beta1. *J Biomed Mater Res A* **69** (2004) 247.

42. Hashimoto T, Suzuki Y, Kitada M, Kataoka K, Wu S, Suzuki K, Endo K, Nishimura Y, Ide C. Peripheral nerve regeneration through alginate gel: analysis of early outgrowth and late increase in diameter of regenerating axons. *Exp Brain Res* **146** (2002) 356.

43. Zhang Z, Rouabhia M, Wang Z, Roberge C, Shi G, Roche P, Li J, Dao LH. Electrically conductive biodegradable polymer composite for nerve regeneration: electricity-stimulated neurite outgrowth and axon regeneration. *Artif Organs* **31** (2007) 13.

44. Weber RA, Breidenbach WC, Brown RE, Jabaley ME, Mass DP. A randomized prospective study of polyglycolic acid conduits for digital nerve reconstruction in humans. *Plast Reconstr Surg* **106** (2000) 1036.

45. Ashley WW Jr, Weatherly T, Park TS. Collagen nerve guides for surgical repair of brachial plexus birth injury. *J Neurosurg* **105** (2006) 452.

46. Hutmacher DW, Schantz T, Zein I, Ng KW, Teoh SH, Tan KC. Mechanical properties and cell cultural response of polycaprolactone scaffolds designed and fabricated via fused deposition modeling. *J Biomed Mater Res* **55** (2001) 203.

47. Isogai N, Landis W, Kim TH, Gerstenfeld LC, Upton J, Vacanti JP. Formation of phalanges and small joints by tissue-engineering. *J Bone Joint Surg* **81A** (1999) 306.

48. Vacanti CA, Bonassar LJ, Vacanti MP, Shufflebarger J. Replacement of an avulsed phalanx with tissue-engineered bone. *N Engl J Med* **344** (2001) 1511.

CHAPTER 8

Bioreactors

Laurence A. Galea*,† and Wayne A. Morrison*,‡

Introduction

Tissue and organ development occurs by self-organization of large cell populations into specific spatial patterns. This is driven by vital genetic, biochemical and biomechanical signals. Bioreactors can be defined as tissue-culture devices that provide a controllable, mechanically active environment in which biological and/or biochemical processes develop.[1] In tissue engineering, bioreactors are used to study and potentially improve 3D engineered tissue structure and properties from isolated and proliferated cells *in vitro*.[2] They do so by attempting to replicate the biochemical and biomechanical environment necessary for the development of tissue and eventually organs.

The term 'bioreactor' can be loosely applied to any mechanical aid that might influence cell growth and behaviour. It may incorporate biochemical as well as static or active biomechanical stimuli and include bioactive matrices, growth factors and cytokines as well as exogenously applied cells. It is clear that within this broad definition there are myriad permutations that could be construed as bioreactors. We have elected to limit our discussion to the principles underlying bioreactor design and describe our *in vivo* tissue engineering chamber as one representative example. Specific applications to tissues relevant to hand tissue engineering are also discussed as illustrative of the concepts but individual chapters by others in this book bring appropriate focus to these endeavours. Bioreactors can be broadly subdivided into *ex-vivo* and *in-vivo*.

Ex-vivo Bioreactors

The first step in establishing a 3D tissue construct is cell proliferation. Usually, only a small number of cells can be obtained from a biopsy specimen and expansion up to several orders of magnitude is required. The simplest and most widely used bioreactors for cell proliferation today are culture dishes and flasks.[3] These provide an environment that is sterile, simple, easy to use and economical to manufacture. However, these bioreactors require individual manual handling for cell seeding and medium exchange amongst others and therefore limit their usefulness when large numbers are required. Another drawback of these so called static bioreactors is that for example for cartilage, the increase in cell number is limited and several sub-cultivations are required. This is thought to be one of

*Corresponding author.

†Laurence A Galea MD, MRCS (Ed), Microsurgery Research Fellow, Bernard O'Brien Institute of Microsurgery, Fitzroy, Melbourne, Australia.

‡Wayne A Morrison MBBS, MD, FRACS, Hugh Devine Professor of Surgery and Head, University Department of Surgery, St Vincent's Hospital, Fitzroy, Melbourne, Australia. Head, Department of Plastic Surgery, St Vincent's Hospital, Fitzroy, Melbourne, Australia. Director, Bernard O'Brien Institute of Microsurgery, Fitzroy, Melbourne, Australia.

the major factors that lead to dedifferentiation of cells, a further disadvantage of these bioreactors.[4,5]

An important step in establishing 3D cultures is the cell seeding of expanded cells onto scaffolds.[6] Although seeding cells at high densities has been associated with enhanced tissue formation in 3D constructs such as increased bone mineralization and high rates of cartilage matrix production, the distribution pattern of cells within the scaffold is essential. It has been shown that the initial distribution of cells within the scaffold after seeding is related to the distribution of tissue subsequently formed within engineered constructs. This suggests that uniform cell-seeding can establish the basis of uniform tissue generation.[7,8]

Significant improvement in seeding efficiency and uniform cell distribution, as opposed to manual loading has been achieved by the introduction of dynamic bioreactors.[9,10] One of these is the spinner flask where a dilute cell suspension is mixed around a stationary scaffold and cells are transported to and into the scaffold by convection. However, although cell seeding is improved, medium stirring generates turbulent eddies and sheer stresses that are thought to cause dedifferentiation of cells.[11] For example seeded chondrocytes show a decreased expression of collagen type II and increased expression of collagen type I.[5] Also, there is non-uniform distributions of cells with a higher density of cells lining the scaffold surface.[12]

Reducing sheer stresses on the cells was diminished by the introduction of rotating wall vessel bioreactors. The vessel walls are rotated at a rate such that forces acting on the construct are balanced, enabling the latter to remain in a state of free fall through the culture medium. Cartilaginous constructs had superior biochemical and biomechanical properties than those grown in static and stirred flasks.[13,14]

Hollow-fibre bioreactors and direct perfusion bioreactors perfuse medium around semi-permeable hollow fibres and directly through the pores of the scaffold respectively. In the latter, cells are transported directly into the scaffold pores, yielding a highly uniform cell distribution. It is imperative that perfusion of the bioreactor is optimized in order to achieve a balance between the mass transfer of nutrients and waste products to and from the cells, the retention of newly synthesized extracellular matrix components within the construct and the fluid-induced sheer stresses within the scaffold pores.[11]

Increasing evidence in the literature suggests that mechanical signals applied to cells are able to modulate cell physiology and may increase the proliferative activity of cells seeded in scaffolds and therefore improve the quality and/or accelerate end product development.[1,11] This led to the development of bioreactors that apply controlled mechanical forces to engineered constructs, particularly musculoskeletal tissue.[15] Cyclical mechanical stretch increased tissue organization and expression of elastin by smooth muscle cells.[16] It also improved the mechanical properties of tissues generated by skeletal muscle cells.[15] Dynamic loading of chondrocytes embedded in a 3D scaffold stimulated glycosaminoglycan synthesis and increased the mechanical properties of the resulting tissue. Mechanical forces applied to engineering constructs have also been shown to direct the differentiation of multi-potent cells. Strain applied to mesenchymal progenitor cells embedded in collagen gel induced cell alignment, formation of oriented collagen fibres and upregulation of ligament-specific genes.[11] However, despite several studies there is still a long way to go in terms of the specific mechanical forces and regimens of application that are stimulatory for a particular tissue. Also, engineered tissue at different stages of development may require different regimens of mechanical stimuli due to increasing accumulation of extracellular matrix and developing structural organization. In the context of this highly complicated field of mechanotransduction, bioreactors have an important role to play by providing reproducible environments for accurate application of specific mechanical forces to 3D tissue constructs.[1]

In-vivo Bioreactors

In vivo tissue engineering relies on the body's capacity to act as a bioreactor to create a functional construct. One of the main problems of *ex vivo* bioreactors is that the size of most 3D engineered constructs is limited as oxygen and nutrients can only be supplied by diffusion up to a thickness of 100–200 microns. On the contrary, *in vivo* bioreactors have their own blood supply in addition to specific cell populations and scaffolds. Therefore they have the potential of producing much bigger constructs.

There are two approaches to *in vivo* vascularization of 3D constructs, extrinsic and intrinsic vascularization. The former relies on the sprouting of microvessels from the surrounding recipient vascular bed into the construct tissue. This type of vascularization is commonly employed in tissue engineering where cell-seeded porous scaffolds are implanted into vascularized beds.[17–19] Although this technique has had some success, there is a delay in recipient blood vessel growth into the scaffold, resulting in limited blood perfusion and oxygen supply to implanted tissues, making it suitable for tissue thickness of approximately 1 to 3 mm only.[20] Our explorations with intrinsic vascularization of 3D tissue engineered constructs derives from experiences with prefabrication of tissue flaps and grafts.[21] Prefabrication is a technique of vascularizing tissues or grafts by implanting an arteriovenous loop or a vascular pedicle underneath or within the tissue. This results in spontaneous angiogenic sprouting from the loop or pedicle and subsequent invasion of new blood vessel growth into the tissue graft thereby establishing a dependence of that tissue on the implanted blood supply and allowing it to be transferred as a separate living entity.[21,22] This phenomenon forms the basis of engineering chambers we have designed in the rat, mouse and pig which have enabled us to produce volumes of vascularized 3D tissues especially fat which is several times larger than any current alternative techniques.[23–28]

The forerunner to *in vivo* bioreactors appeared as a skeletal application device for bone lengthening. The Ilizarov limb[29–32] and later digital ray[33] thumb lengthening[34–36] concept involved osteotomy within a subperiosteal sleeve, followed by progressive distraction of the bone ends, resulting in spontaneous regeneration of new bone tissue within the distracted space. The growth is presumably driven by a combination of mechanical and biochemical signals that are difficult to quantify and the precise protocols for clinical applications have been largely empirical.

Maxillofacial surgeons introduced the concept of 'guided tissue generation' for mandibular bone augmentation where synthetic gutters are moulded over the defect and closed with intact mucoperiosteal flaps.[37] The gutter protects the space from closing and acts as a guiding conduit for new bone to migrate across the gap and repair the defect with bone rather than scar. The process may be enhanced when bone grafts[38] or platelet-rich plasma are added.[39]

The same bioreactor phenomenon seems to occur naturally in clinical hand surgery settings. We have observed spontaneous repair in cases of closed rupture of the extensor and flexor tendons following varying pathologies but when they were later explored the tendon continuity was seen to be intact. Similarly, in the senior author's personal observations, a palmaris longus tendon previously taken for tendon grafting regenerated within its remnant synovial sheath. This is the same phenomenon that allows closed mallet tendon ruptures to repair spontaneously with splinting. Fibular bones can also regenerate in children after harvest provided the periosteal sleeve remains or when beta-tricalcium phosphate is used as a bone graft substitute.[40] Younger tissues are known to have greater regenerative capacity[41] and this suggests that stem cell sources play a role in this predominantly regenerative compared to fibrotic contractile process typical of adult healing.

It seems that the common denominator for tissue growth in these seemingly different scenarios is angiogenesis within a defined space that supports cell proliferation and migration and a specific cellular and extracellular matrix environment which determines the specific tissue that will form within this space. Migration of cells including stem cells to the space follows cytokine and chemokine gradients produced by inflammatory cells recruited to the area of interest.[42,43] Cells also sense their physical environment and are able to migrate down a rigidity gradient from soft to hard substrates, a phenomenon called durotaxis.[44] They also maintain their appropriate cell/cell and cell/extracellular matrix force relationship through a variety of molecules, cellular components and extracellular structures. These include cadherins, gap junctions, integrins, focal adhesions, ion channels, caveolae, surface receptors, and cytoskeletal components including microfilaments, microtubules and intermediate filaments.[45,46] When these forces change to a different modulus the cells through a process of mechanotransduction induce molecular signalling through the Roc/Rho pathways to change their phenotype and proliferate, migrate and differentiate or conversely cluster to form balls of cells and even apoptose.[47,48] For instance myoblasts when cultured on gels of varied elasticity will differentiate into striated myotubes when the stiffness of the gel approaches that of normal muscle.[49] Axial forces across wrist and finger joints no doubt play a part in cell growth and direction during the tendon regeneration observed clinically.

We have developed an *in vivo* tissue engineering chamber that mimics the essential elements of nature's model. It consists of a perforated non collapsible box or chamber that prevents the surrounding tissue from closing down the space. An arteriovenous pedicle or fistula is inset into the chamber via a side hole and the altered forces in association with surgical trauma ischaemia and inflammation stimulate the pedicle tissue inside the chamber to proliferate in an attempt to heal the space from within.[23,24] Although angiogenesis is the primary response as the arteriovenous loop or pedicle is the predominant tissue within the space.[27,28] If the chamber is open or if it abuts adjacent living tissue[50] or if a tissue flap is included with the pedicle this tissue will also proliferate into the space.[51] Systemic endothelial precursors and mesenchymal stem cells can be shown to migrate to the chamber environment and incorporate into the new tissue.[52] When matrix materials, cultured specialized cells or cytokines are added to the chamber the tissue that forms can be manipulated[53] and using these methods we have successfully grown fat,[50,54,55] skeletal muscle,[56] beating heart tissue,[26] insulin secreting pancreatic tissue, and other organ tissues. When the chamber is filled with Matrigel and adjacent inguinal fat pad abuts an opening on the chamber, not only does new fat migrate into the chamber and replace the Matrigel but breast glands that are in the vicinity also proliferate and appear in the new chamber tissue.[50]

This paradigm of intrinsic tissue engineering where tissue grows concurrent with a developing vasculature is essentially different to the traditional extrinsic vascularization model where scaffolds are seeded *ex vivo* with cells and implanted into the tissue with the expectation that these cells will connect to a blood supply that invades the scaffold from the periphery. In the interim cells must survive by diffusion and consequently cell containing constructs greater than a few millimetres in thickness rarely survive.[57]

Bioreactor Models in Tissue Engineering

Our initial design for a tissue engineering chamber was in the rat. It is formed by creating an arteriovenous shunt between the right femoral artery and vein using an interposition vein graft from the left femoral vein. The shunt is then placed onto the base of a polycarbonate cylindrical chamber with an internal diameter of 1.4 cm, the lid closed and the construct sutured to the groin musculature with the aid of small holes on the base of the chamber (Figure 1).[23] This model is able to develop spontaneous *de novo* vascularised fibrous connective tissue which maintains its volume for at least 16 weeks (Figure 2).[28]

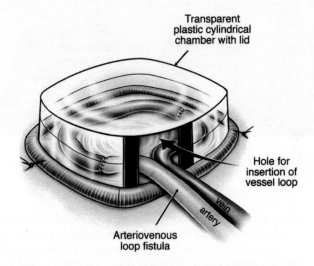

Figure 1.   The rat tissue engineering chamber.

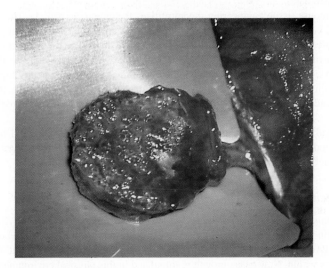

Figure 2.   Tissue construct at four weeks being harvested from the rat chamber.

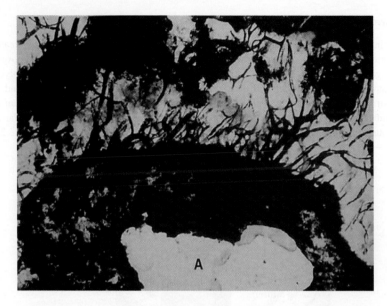

Figure 3.   India ink injection showing new blood vessels developing from the arteriovenous loop in the rat tissue engineering model.

A perfused capillary network that remodels to generate arterioles, postcapillary venules and venules develops (Figure 3).[28] Di Gennaro from our department has shown that using a novel modification of the rat chamber that applies negative pressure its contents is able to produce bigger and more vascular constructs. This is probably related to changes in mechanical transduction forces described above acting on the cells within the chamber (unpublished data).

Our mouse tissue engineering model utilizes the epigastric pedicle. The epigastric artery and vein in the groin are stripped of their surrounding fat and a 5 mm long piece of split silicone tubing is placed around the pedicle. The proximal end and the longitudinal split of the silicone tubing are sealed with bone wax. The chamber is then filled with matrix, the distal end wax sealed and the wound closed. Spontaneous angiogenesis occurs from this pedicle to vascularize the matrix (Figure 4).[50] Although the mouse model broadens experimental applications, the vascularization is not as robust as the rat model.

Several models of tissue engineering specific to tissue type that could be broadly included under the umbrella of bioreactors have been described in the literature by others and are briefly discussed here.

Non specific tissue growth

Tissue expanders not only stretch tissues but also induce a mitogenic response. Rather than a distraction force these devices induce compressive loads which influence cell behaviour.[58]

The *VAC dressing* applies vacuum forces which compress the tissue but also create fluid forces within. New tissue growth is stimulated in this physical environment and this device can clearly be classified as a bioreactor.

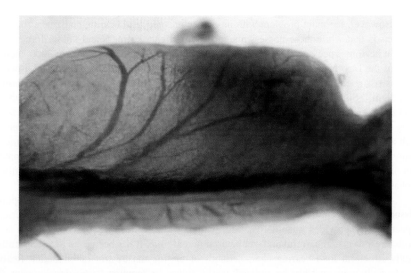

Figure 4. Angiogenesis from the epigastric pedicle in the mouse tissue engineering chamber (Figure provided by Dr Filip Stillaert).

Skin

Skin expansion *ex vivo* is a long established process that was at the vanguard of tissue engineering and led to subsequent efforts to develop skin and dermal substitutes. Integra (Johnson and Johnson) and Alloderm (LifeCell) are prominent examples that have clinical acceptance and fundamentally are a form of *in vivo* bioreactor once implanted into the body. The scaffold matrix orchestrates a cytokine response that leads to revascularization and cell invasion that approximates the missing tissue.

Fat

The BRAVA device described by Roger Khouri for fat augmentation of the breast employs an external vacuum device applied to the skin that distracts the tissues and incites tissue growth.[59] This bioreactor is probably replicating the circumstances of our chamber concept where a space is created in association with microtrauma, ischaemia and inflammation. Cellular and matrix injection into this environment may magnify and direct the growth potential of the response.

Tendon

Fibroblasts from tendon and skin have been utilized in tendon engineering.[60] However, they have low mitotic activity, and bone marrow-derived mesenchymal stem cells (BMMC) are being explored as an alternative. Human bone marrow stromal cells mixed with liquid fibrin glue were injected into standardized full-thickness window defects of the patellar tendon of immunodeficient rats. Twenty days later dense collagen fibres and spindle-shaped cells developed that were mainly orientated along the loading axis.[61] The same cells have been seeded on PLGA scaffolds to repair the Achilles tendon defects in rabbits with limited success.[62]

However, the main problem with *in vivo* engineered load bearing constructs such as tendons is that while these possess a significantly higher stiffness and failure force than natural healing processes, they are quite weak compared to normal tendon. In 2000, members of the U.S. National Committee on Biomechanics addressed the mechanical aspects of this tissue engineering design challenge by proposing a functional tissue engineering paradigm.[63] Some of the principles they came up with included measuring *in vivo* stress and strain histories in normal tissues, determining what signals cells experience *in vivo* as they interact with the extracellular matrix and establishing how physical factors influence cell activity in bioreactors and whether cell-matrix implants can benefit from mechanical stimulation before surgery. This strategy lead to work that improved construct outcome, for instance mechanically stimulating these constructs in modified bioreactors further enhanced repair biomechanics compared to normal.[64]

An important aspect of *in vivo* tissue engineering is *in situ* monitoring of developing tissue in a non destructive and non- or minimally invasive approach. Elastic scattering spectroscopy provides quantitative information, which reflects tissue composition and structure.[65] It is non harmful and can be used in real-time. It has been used to successfully study structural load bearing characteristics of *in situ* and *ex situ* rabbit digital flexor tendons. It may possibly be utilized in the future to monitor tissue development in bioreactors.[65]

Muscle

Primary myoblasts harvested from rat soleus suspended in fibrinogen hydrogel were placed in a silicone chamber around the femoral vessels as a flow-through model in a similar manner to the mouse tissue engineering model described above.[66] Harvesting of tissue constructs at three weeks resulted in 3D skeletal muscle that produced longitudinal contractile force when electrically stimulated. Length-tension, force-voltage, and force-frequency relationships were similar to those found in developing skeletal muscle. Also individual myoblasts had undergone fusion to form multinucleated myotubes and intense angiogenesis had occurred in the construct.[66] Contractile forces of this tissue engineered contractile muscle improved with neurotization of the femoral nerve.[67] Independently we used essentially the same model and found that the myotubes that had formed by three weeks spontaneously disappeared by six weeks suggesting the need for a functional purpose for tissue maintenance. When primary male myoblasts were injected with fibrin into muscle defects in a female animal they showed increasing integration into the host muscle fibres at the injection site with resolution of the fibrin matrix over a period of 12 weeks and no indication of an inflammatory reaction.[68]

Mechanical preconditioning of muscle constructs potentiates skeletal muscle engineering. When human muscle precursor cells seeded on collagen-based acellular tissue scaffolds were subjected to cyclical strain in a computer-controlled bioreactor system and then implanted on the latissimus dorsi of athymic mice, they generated twitch contractile responses which reached 10% of native latissimus dorsi. Although still poor in comparison to normal muscle, the force was substantially higher than in statically cultured control tissue.[69]

Bone

The ideal bone graft substitute requires osteoinductive, osteoconductive and osteogenic components.[70] In one *in vivo* bioreactor model of bone tissue engineering cylindrical scaffolds derived from trabecular exoskeleton of marine coral and chemically converted to a mixture of hydroxyapatite and calcium carbonate/calcium triphosphate were used as the osteoconductive component. They were impregnated with an osteoinductive agent, bone morphogenic protein-2 and sealed with an outer sheet of silicone tubing. This construct was then implanted on the abdominal wall of rats, and the superficial inferior epigastric pedicle was threaded through a hole in the central core of the scaffold. This produced neovascularisation of the bioreactor and acted as a conduit for the recruitment of osteogenic mesenchymal stem cells from the circulation and bone marrow to generate bone.[71] This concept is the natural progeny of bone graft prefabrication where free iliac crest bone grafts in the rat were harvested, bivalved and placed heterotopically in the rat around the mobilized inferior epigastric vascular bundle.[72] Another *in vivo* bone bioreactor was developed by wrapping demineralized bone matrix impregnated in gelatine sponge sheets around a vascular loop in the rat.[73] Bone formation was observed at two, four and six weeks radiographically and histologically and blood supply was considered to be a key factor in the success of this bioreactor. Similar results were obtained using a porous processed bovine cancellous bone matrix.[74]

Another *in vivo* bone bioreactor design was developed in rabbits where the antero-medial tibial periosteum was separated from the underlying bone and alginate gel-supplemented with growth factors transforming growth factor-β1 and fibroblast growth factor-2 or hyaluronic acid-based gel was injected into the space. The new bone formation followed predominantly an intramembranous path, with woven bone matrix subsequently maturing into fully mineralized compact bone exhibiting all of the histological markers and mechanical properties of native bone. Further, neobone was harvested after six weeks and transplanted into contralateral tibial defects, resulting in complete integration after six weeks with no apparent morbidity at the donor site.[75]

In vivo bioreactors for bone development have also been developed in sheep.[76] Hollow rectangular chambers were fabricated from dental-grade polymethyl methacrylate with a polyethylene cuff bonded around the perimeter of the open side to allow suture fixation *in vivo*. When these chambers containing morcellized bone were implanted onto rib periosteum, active endochondral, direct and appositional bone formation was seen for up to 24 weeks. The tissue underwent neovascularisation from vessels supplying the rib periosteum.[77,78]

The idea of using mesenchymal and bone marrow-derived stem cells together with scaffolds and enhancing protein has been used to augment *in vivo* bone bioreactor efficiency.[79] Adipose-derived stem cells from human subjects were transfected with an adenovirus carrying the cDNA for bone morphogenic protein-2 and seeded onto collagen sponges and implanted on the lateral aspect of the femur in severe-combined immunodeficient mice. New bone was produced as quantified by radiographic and histologic studies. Also when an expanded mixed population of bone marrow-derived stem and progenitor cells were seeded on fibronectin-coated porous calcium phosphate ceramic blocks into the backs of severe combined immunodeficient mice they led to the *in vivo* bone formation measured by histology.[80] Di Bella in our department showed that fibronectin scaffolds seeded with predifferentiated osteoblasts derived from lipoaspirate mesenchymal stem cells repaired critical size skull defects in the rabbit much more rapidly than mesenchymal stem cells or differentiated osteoblasts alone. It seems that fibronectin is key to stem cell success.[81]

Nerve

Nerve tissue engineering is focused mainly on developing nerve conduits that bridge nerve gaps in essence, another form of bioreactor. These strategies focused on either natural or synthetic bioreactors. They not only act to guide axon sprouting but also to provide a conduit for neurotrophic and neurotropic factors secreted by the injured nerve stump and help to protect from infiltration of fibrous tissue.[82] Biological conduits include autologous collagen, arteries, veins and acellular muscle grafts with or without cultured Schwann cells. Synthetic substances include lactate polymer, polygalactin mesh, polyethylene and silicone and silicone-polymer tubes.[83] With the recent proliferation of available nerve-growth stimulating factors, nerve repair is now focusing more on molecular biological manipulation of the internal milieu of the bioreactors.[84,85]

Cartilage

Autologous chondrocyte grafting has a well established clinical place[86] and it and skin cell culture are the only non haematological cell based products that are widely approved for human use. Biopsied cartilage chondrocytes are expanded in *ex vivo* bioreactors and reimplanted into focal articular cartilage defects. However, chondrocytes are difficult to isolate in humans, replicate slowly and are prone to phenotypic dedifferentiation in culture.[11] Adult mesenchymal stem cells are attractive cell alternatives as they are highly replicative and can differentiate along the chondrocyte lineage. These are accessible from many tissues including bone marrow, adipose tissue, umbilical cord blood, synovium and blood. There is an expanding body of knowledge on the characteristics, expansion and specific stimuli for chondro-induction of these cells *in vitro* and what is the safest and most functional biomaterial for the delivery of these cells for *in vivo* cartilage formation.[87,88]

The Future

Both *ex vivo* and *in vivo* bioreactors have the potential to improve current therapeutic strategies in hand surgery and therefore improve the quality of life. *In vivo* bioreactor technologies in broad principle try to mimic the tissue specific environment. Tissue growth involves cascades of cellular, growth factor and physical responses which change according to the developmental stage of the tissue. The challenge is to determine and replicate what these parameters are at any one point in time and the pattern of their changes. The next major challenge is to translate research-scale models into applicable manufacturing designs that are reproducible, clinically effective, economically acceptable, as well as complying with the Food and Drug Administration.

References

1. Portner R, Nagel-Heyer S, Goepfert C, Adamietz P, Meenen NM. Bioreactor design for tissue engineering. *J Biosci Bioeng* **100** (2005) 235.
2. Ratcliffe A, Niklason LE. Bioreactors and bioprocessing for tissue engineering. *Ann NY Acad Sci* **961** (2002) 210.
3. Domm C, Schunke M, Christesen K, Kurz B. Redifferentiation of dedifferentiated bovine articular chondrocytes in alginate culture under low oxygen tension. *Osteoarthritis Cartilage* **10** (2002) 13.
4. Holy CE, Shoichet MS, Davies JE. Engineering three-dimensional bone tissue *in vitro* using biodegradable scaffolds: investigating initial cell-seeding density and culture period. *J Biomed Mater Res* **51** (2000) 376.
5. Malda J, van Blitterswijk CA, van Geffen M, Martens DE, Tramper J, Riesle, J. Low oxygen tension stimulates the redifferentiation of dedifferentiated adult human nasal chondrocytes. *Osteoarthritis Cartilage* **12** (2004) 306.

6. Kim BS, Putnam AJ, Kulik TJ, Mooney DJ. Optimizing seeding and culture methods to engineer smooth muscle tissue on biodegradable polymer matrices. *Biotechnol Bioeng* **57** (1998) 46.
7. Carrier RL, Papadaki M, Rupnick M, Schoen FJ, Bursac N, Langer R, Freed, LE, Vunjak-Novakovic G. Cardiac tissue engineering: cell seeding, cultivation parameters, and tissue construct characterization. *Biotechnol Bioeng* **64** (1999) 580.
8. Li Y, Ma T, Kniss DA, Lasky LC, Yang ST. Effects of filtration seeding on cell density, spatial distribution, and proliferation in nonwoven fibrous matrices. *Biotechnol Prog* **17** (2001) 935.
9. Gooch KJ, Kwon JH, Blunk T, Langer R, Freed LE, Vunjak-Novakovic G. Effects of mixing intensity on tissue-engineered cartilage. *Biotechnol Bioeng* **72** (2001) 402.
10. Martin I, Wendt D, Heberer M. The role of bioreactors in tissue engineering. *Trends Biotechnol* **22** (2004) 80.
11. Vunjak-Novakovic G, Obradovic B, Martin I, Bursac PM, Langer R, Freed, LE. Dynamic cell seeding of polymer scaffolds for cartilage tissue engineering. *Biotechnol Prog* **14** (1998) 193.
12. Freed LE, Hollander AP, Martin I, Barry JR, Langer R, Vunjak-Novakovic G. Chondrogenesis in a cell-polymer-bioreactor system. *Exp Cell Res* **240** (1998) 58.
13. Obradovic B, Martin I, Padera RF, Treppo S, Freed LE, Vunjak-Novakovic G. Integration of engineered cartilage. *J Orthop Res* **19** (2001) 1089.
14. Freed LE, Guilak F, Guo XE, Gray ML, Tranquillo R, Holmes JW, Radisic M, Sefton MV, Kaplan D, Vunjak-Novakovic G. Advanced tools for tissue engineering: scaffolds, bioreactors, and signaling. *Tissue Eng* **12** (2006) 3285.
15. Powell CA, Smiley BL, Mills J, Vandenburgh HH. Mechanical stimulation improves tissue-engineered human skeletal muscle. *Am J Physiol Cell Physiol* **283** (2002) C1557.
16. Kim BS, Nikolovski J, Bonadio J, Mooney DJ. Cyclic mechanical strain regulates the development of engineered smooth muscle tissue. *Nat Biotechnol* **17** (1999) 979.
17. Peters MC, Polverini PJ, Mooney DJ. Engineering vascular networks in porous polymer matrices. *J Biomed Mater Res* **60** (2002) 668.
18. Perets A, Baruch Y, Weisbuch F, Shoshany G, Neufeld G, Cohen S. Enhancing the vascularization of three-dimensional porous alginate scaffolds by incorporating controlled release basic fibroblast growth factor microspheres. *J Biomed Mater Res A* **65** (2003) 489.
19. Elcin AE, Elcin YM. Localized angiogenesis induced by human vascular endothelial growth factor-activated PLGA sponge. *Tissue Eng* **12** (2006) 959.
20. Cassell OC, Hofer SO, Morrison WA, Knight KR. Vascularisation of tissue-engineered grafts: the regulation of angiogenesis in reconstructive surgery and in disease states. *Br J Plast Surg* **55** (2002) 603.
21. Morrison WA, Dvir E, Doi K, Hurley JV, Hickey MJ, O'Brien BM. Prefabrication of thin transferable axial-pattern skin flaps: an experimental study in rabbits. *Br J Plast Surg* **43** (1990) 645.
22. Erol OO, Spira M. New capillary bed formation with a surgically constructed arteriovenous fistula. *Surg Forum* **30** (1979) 530.
23. Tanaka Y, Tsutsumi A, Crowe DM, Tajima S, Morrison WA. Generation of an autologous tissue (matrix) flap by combining an arteriovenous shunt loop with artificial skin in rats: preliminary report. *Br J Plast Surg* **53** (2000) 51.
24. Mian R, Morrison WA, Hurley JV, Penington AJ, Romeo R, Tanaka Y, Knight KR. Formation of new tissue from an arteriovenous loop in the absence of added extracellular matrix. *Tissue Eng* **6** (2000) 595.
25. Tanaka Y, Sung KC, Tsutsumi A, Ohba S, Ueda K, Morrison WA. Tissue engineering skin flaps: which vascular carrier, arteriovenous shunt loop or arteriovenous bundle, has more potential for angiogenesis and tissue generation? *Plast Reconstr Surg* **112** (2003) 1636.
26. Morritt AN, Bortolotto SK, Dilley RJ, Han X, Kompa AR, McCombe D, Wright CE, Itescu S, Angus JA, Morrison WA. Cardiac tissue engineering in an *in vivo* vascularized chamber. *Circulation* **115** (2007) 353.
27. Lokmic Z, Mitchell GM. The source and commencement of angiogenesis from the arteriovenous loop model. *Microvasc Res* **75** (2008) 142.

28. Lokmic Z, Stillaert F, Morrison WA, Thompson EW, Mitchell GM. An arteriovenous loop in a protected space generates a permanent, highly vascular, tissue-engineered construct. *FASEB J* **21** (2007) 511.

29. Ilizarov GA, Deviatov AA. [Surgical elongation of the leg]. *Ortop Travmatol Protez* **32** (1971) 20.

30. Ilizarov GA, Deviatov AA, Trokhova VG. [Surgical lengthening of the shortened lower extremities]. *Vestn Khir Im I I Grek* **108** (1972) 100.

31. Ilizarov GA. The tension-stress effect on the genesis and growth of tissues. Part I. The influence of stability of fixation and soft-tissue preservation. *Clin Orthop Relat Res*, **249**, 1989.

32. Ilizarov GA. The tension-stress effect on the genesis and growth of tissues: Part II. The influence of the rate and frequency of distraction. *Clin Orthop Relat Res*, **263**, 1989.

33. Kessler I, Baruch A, Hecht O. Experience with distraction lengthening in digital rays in congenital anomalies. *J Hand Surg [Am]* **2** (1977) 394.

34. Matev IB. Thumb reconstruction in children through metacarpal lengthening. *Plast Reconstr Surg* **64** (1979) 665.

35. Matev IB. Thumb reconstruction through metacarpal bone lengthening. *J Hand Surg [Am]* **5** (1980) 482.

36. Mulliken JB, Curtis RM. Thumb lengthening by metacarpal distraction. *J Trauma* **20** (1980) 250.

37. Lang NP, Hammerle CH, Bragger U, Lehmann B, Nyman SR. Guided tissue regeneration in jawbone defects prior to implant placement. *Clin Oral Implants Res* **5** (1994) 92.

38. Donos N, Kostopoulos L, Karring T. Augmentation of the rat jaw with autogeneic corticocancellous bone grafts and guided tissue regeneration. *Clin Oral Implants Res* **13** (2002) 192.

39. Wojtowicz A, Chaberek S, Urbanowska E, Ostrowski K. Comparison of efficiency of platelet rich plasma, hematopoieic stem cells and bone marrow in augmentation of mandibular bone defects. *NY State Dent J* **73** (2007) 41.

40. Arai E, Nakashima H, Tsukushi S, Shido Y, Nishida Y, Yamada Y, Sugiura, H, Katagiri H. Regenerating the fibula with beta-tricalcium phosphate minimizes morbidity after fibula resection. *Clin Orthop Relat Res*, **233**, 2005.

41. Fischgrund J, Paley D, Suter C. Variables affecting time to bone healing during limb lengthening. *Clin Orthop Relat Res*, **31**, 1994.

42. Zvaifler NJ, Marinova-Mutafchieva L, Adams G, Edwards CJ, Moss J, Burger JA, Maini RN. Mesenchymal precursor cells in the blood of normal individuals. *Arthritis Res* **2** (2000) 477.

43. Han CI, Campbell GR, Campbell JH. Circulating bone marrow cells can contribute to neointimal formation. *J Vasc Res* **38** (2001) 113.

44. Lo CM, Wang HB, Dembo M, Wang YL. Cell movement is guided by the rigidity of the substrate. *Biophys J* **79** (2000) 144.

45. Ingber DE. Cellular mechanotransduction: putting all the pieces together again. *FASEB J* **20** (2006) 811.

46. Discher DE, Janmey P, Wang YL. Tissue cells feel and respond to the stiffness of their substrate. *Science* **310** (2005) 1139.

47. McBeath R, Pirone DM, Nelson CM, Bhadriraju K, Chen CS. Cell shape, cytoskeletal tension, and RhoA regulate stem cell lineage commitment. *Dev Cell* **6** (2004) 483.

48. Yeung T, Georges PC, Flanagan LA, Marg B, Ortiz M, Funaki M, Zahir N, Ming W, Weaver V, Janmey PA. Effects of substrate stiffness on cell morphology, cytoskeletal structure, and adhesion. *Cell Motil Cytoskeleton* **60** (2005) 24.

49. Engler AJ, Griffin MA, Sen S, Bonnemann CG, Sweeney HL, Discher DE. Myotubes differentiate optimally on substrates with tissue-like stiffness: pathological implications for soft or stiff microenvironments. *J Cell Biol* **166** (2004) 877.

50. Cronin KJ, Messina A, Knight KR, Cooper-White JJ, Stevens GW, Penington AJ, Morrison WA. New murine model of spontaneous autologous tissue engineering, combining an arteriovenous pedicle with matrix materials. *Plast Reconstr Surg* **113** (2004) 260.

51. Dolderer JH, Abberton KM, Thompson EW, Slavin JL, Stevens GW, Penington AJ, Morrison WA. Spontaneous large volume adipose tissue generation from a vascularized pedicled fat flap inside a chamber space. *Tissue Eng* **13** (2007) 673.

52. Simcock JW, Penington AJ, Morrison WA, Thompson EW, Mitchell GM. Endothelial precursor cells home to a vascularized tissue engineering chamber by application of the angiogenic chemokine CXCL12. *Tissue Eng Part A In Press.*

53. Lokmic Z, Mitchell GM. Engineering the microcirculation. *Tissue Eng Part B Rev* **14** (2008) 87.

54. Thomas GP, Hemmrich K, Abberton KM, McCombe D, Penington AJ, Thompson EW, Morrison WA. Zymosan-induced inflammation stimulates neo-adipogenesis. *Int J Obes (Lond)* **32** (2008) 239.

55. Hemmrich K, Thomas GP, Abberton KM, Thompson EW, Rophael JA, Penington AJ, Morrison WA. Monocyte chemoattractant protein-1 and nitric oxide promote adipogenesis in a model that mimics obesity. *Obesity (Silver Spring)* **15** (2007) 2951.

56. Messina A, Bortolotto SK, Cassell OC, Kelly J, Abberton KM, Morrison WA. Generation of a vascularized organoid using skeletal muscle as the inductive source. *FASEB J* **19** (2005) 1570.

57. Fassnacht D, Portner R. Experimental and theoretical considerations on oxygen supply for animal cell growth in fixed-bed reactors. *J Biotechnol* **72** (1999) 169.

58. Miller Q, Bird E, Bird K, Meschter C, Moulton MJ. Effect of subatmospheric pressure on the acute healing wound. *Curr Surg* **61** (2004) 205.

59. Khouri RK, Schlenz I, Murphy BJ, Baker TJ. Nonsurgical breast enlargement using an external soft-tissue expansion system. *Plast Reconstr Surg* **105** (2000) 2500.

60. Liu W, Chen B, Deng D, Xu F, Cui L, Cao Y. Repair of tendon defect with dermal fibroblast engineered tendon in a porcine model. *Tissue Eng* **12** (2006) 775.

61. Hankemeier S, van Griensven M, Ezechieli M, Barkhausen T, Austin M, Jagodzinski M, Meller R, Bosch U, Krettek C, Zeichen J. Tissue engineering of tendons and ligaments by human bone marrow stromal cells in a liquid fibrin matrix in immunodeficient rats: results of a histologic study. *Arch Orthop Trauma Surg* **127** (2007) 815.

62. Ouyang HW, Goh JC, Mo XM, Teoh SH, Lee EH. The efficacy of bone marrow stromal cell-seeded knitted PLGA fiber scaffold for Achilles tendon repair. *Ann N Y Acad Sci* **961** (2002) 126.

63. Butler DL, Goldstein SA, Guilak F. Functional tissue engineering: the role of biomechanics. *J Biomech Eng* **122** (2000) 570.

64. Juncosa-Melvin N, Shearn JT, Boivin GP, Gooch C, Galloway MT, West JR, Nirmalanandhan VS, Bradica G, Butler DL. Effects of mechanical stimulation on the biomechanics and histology of stem cell-collagen sponge constructs for rabbit patellar tendon repair. *Tissue Eng* **12** (2006) 2291.

65. Morgan M, Kostyuk O, Brown RA, Mudera V. In situ monitoring of tendon structural changes by elastic scattering spectroscopy: correlation with changes in collagen fibril diameter and crimp. *Tissue Eng* **12** (2006) 1821.

66. Dennis RG, Kosnik PE, 2nd. Excitability and isometric contractile properties of mammalian skeletal muscle constructs engineered *in vitro*. *In Vitro Cell Dev Biol Anim* **36** (2000) 327.

67. Dhawan V, Lytle IF, Dow DE, Huang YC, Brown DL. Neurotization improves contractile forces of tissue-engineered skeletal muscle. *Tissue Eng* **13** (2007) 2813.

68. Beier JP, Stern-Straeter J, Foerster VT, Kneser U, Stark GB, Bach AD. Tissue engineering of injectable muscle: three-dimensional myoblast-fibrin injection in the syngeneic rat animal model. *Plast Reconstr Surg* **118** (2006) 1113.

69. Moon du G, Christ G, Stitzel JD, Atala A, Yoo JJ. Cyclic mechanical preconditioning improves engineered muscle contraction. *Tissue Eng Part A* **14** (2008) 473.

70. Lane JM, Tomin E, Bostrom MP. Biosynthetic bone grafting. *Clin Orthop Relat Res* **S107**, 1999.

71. Holt GE, Halpern JL, Dovan TT, Hamming D, Schwartz HS. Evolution of an *in vivo* bioreactor. *J Orthop Res* **23** (2005) 916.

72. Gill DR, Ireland DC, Hurley JV, Morrison WA. The prefabrication of a bone graft in a rat model. *J Hand Surg [Am]* **23** (1998) 312.

73. Chen WJ, Zhang F, Mustain WC, Tucci M, Hu EC, Lineaweaver WC. Prefabrication of vascularized bone flap by demineralized bone matrix. *J Craniofac Surg* **18** (2007) 43.

74. Kneser U, Polykandriotis E, Ohnolz J, Heidner K, Grabinger L, Euler S, Amann KU, Hess A, Brune K, Greil P, Sturzl M, Horch RE. Engineering of vascularized transplantable bone tissues: induction of axial vascularization in an osteoconductive matrix using an arteriovenous loop. *Tissue Eng* **12** (2006) 1721.

75. Stevens MM, Marini RP, Schaefer D, Aronson J, Langer R, Shastri VP. *In vivo* engineering of organs: the bone bioreactor. *Proc Natl Acad Sci USA* **102** (2005) 11450.

76. Thomson RC, Mikos AG, Beahm E, Lemon JC, Satterfield WC, Aufdemorte TB, Miller MJ. Guided tissue fabrication from periosteum using preformed biodegradable polymer scaffolds. *Biomaterials* **20** (1999) 2007.

77. Cheng MH, Brey EM, Allori A, Satterfield WC, Chang DW, Patrick CW, Jr, and Miller MJ. Ovine model for engineering bone segments. *Tissue Eng* **11** (2005) 214.

78. Brey EM, Cheng MH, Allori A, Satterfield W, Chang DW, Patrick CW, Jr, and Miller MJ. Comparison of guided bone formation from periosteum and muscle fascia. *Plast Reconstr Surg* **119** (2007) 1216.

79. Dennis JE, Esterly K, Awadallah A, Parrish CR, Poynter GM, Goltry KL. Clinical-scale expansion of a mixed population of bone-marrow-derived stem and progenitor cells for potential use in bone-tissue regeneration. *Stem Cells* **25** (2007) 2575.

80. Dragoo JL, Lieberman JR, Lee RS, Deugarte DA, Lee Y, Zuk PA, Hedrick MH, Benhaim P. Tissue-engineered bone from BMP-2-transduced stem cells derived from human fat. *Plast Reconstr Surg* **115** (2005) 1665.

81. Di Bella C, Farlie P, Penington AJ. Bone regeneration in a rabbit critical-sized skull defect using autologous adipose-derived cells. *Tissue Eng Part A* **14** (2008) 483.

82. Johnson EO, Soucacos PN. Nerve repair: experimental and clinical evaluation of biodegradable artificial nerve guides. *Injury* **39**(Suppl. 3) (2008) S30.

83. Chong AK, Chang J. Tissue engineering for the hand surgeon: a clinical perspective. *J Hand Surg [Am]* **31** (2006) 349.

84. Colangelo AM, Finotti N, Ceriani M, Alberghina L, Martegani E, Aloe L, Lenzi L, Levi-Montalcini R. Recombinant human nerve growth factor with a marked activity *in vitro* and *in vivo*. *Proc Natl Acad Sci USA* **102** (2005) 18658.

85. Dowsing BJ, Hayes A, Bennett TM, Morrison WA, Messina A. Effects of LIF dose and laminin plus fibronectin on axotomized sciatic nerves. *Muscle Nerve* **23** (2000) 1356.

86. Lindahl A, Brittberg M, Peterson L. Cartilage repair with chondrocytes: clinical and cellular aspects. *Novartis Found Symp* **249** (2003) 175.

87. Liu K, Zhou GD, Liu W, Zhang WJ, Cui L, Liu X, Liu TY, Cao Y. The dependence of *in vivo* stable ectopic chondrogenesis by human mesenchymal stem cells on chondrogenic differentiation *in vitro*. *Biomaterials* **29** (2008) 2183.

88. Swieszkowski W, Tuan BH, Kurzydlowski KJ, Hutmacher DW. Repair and regeneration of osteochondral defects in the articular joints. *Biomol Eng* **24** (2007) 489.

CHAPTER 9

Nerve Engineering

*Gregory H. Borschel**

Rationale for Nerve Engineering

An "off the shelf" nerve substitute with a regenerative capacity equivalent to an autograft would be very useful to hand and peripheral nerve surgeons. Nerve injury is a common and debilitating problem. Thousands of nerve repair procedures are performed annually in the United States alone. Surgeons are repeatedly faced with the reconstructive challenges presented by nerve gaps. Autologous nerve grafting remains the gold standard by which all nerve gap reconstructions are measured. However, alternatives to nerve grafting have been developed. For smaller gaps, especially in distal, sensory nerves, bioengineered alternatives are appealing. In the future, such strategies will likely provide increasingly powerful alternatives to the current practice of autologous nerve grafting.

There are several theoretical advantages to using nerve substitutes instead of autologous nerve grafts. The harvesting of autografts presents a number of disadvantages: first, harvesting donor autologous nerve grafts produces donor site morbidity in the form of decreased sensory function in the region previously innervated by the harvested sensory nerve. The resulting hypesthesia and dysesthesia can be very troublesome to patients. The usual surgical risks of hypertrophic scarring and wound infection apply to the donor site incisions. Conversely, use of a nerve substitute presents a number of advantages: first, their use can decrease operative time. Second, because nerve recovery is improved when nerve reconstruction is carried out promptly, traumatic cases in which soft tissue coverage is a concern may be reconstructed immediately, without fear of losing nerve autografts to infection (instead of the common practice of waiting three months to achieve "soft tissue equilibrium").

Ideal Characteristics of a Nerve Substitute

There are many physiologic, engineering, and economic constraints imposed on the design of nerve substitutes. They should be flexible yet retain their shape *in vivo*. They should be biodegradable, nontoxic, elicit a minimal inflammatory response, be able to be stored in an operating room using conventional storage methods, and they should be cost effective to manufacture and implant. Their use should cost less than the overall equivalent cost of harvesting an autologous nerve graft, including operating room time, personnel costs, and professional fees.

Nerve Regeneration and Special Challenges with Nerve Tissue Engineering

Unlike tissue engineering of other tissues, such as muscle, bone, tendon, blood vessels and skin, tissue engineering of nerve primarily involves the extension of cellular processes (axons) over comparatively extremely long distances rather than proliferation and

*Corresponding author.

differentiation of cells within a matrix. The challenge is more one of guiding cellular regeneration from proximal to distal rather than engineering a *de novo* tissue for direct implantation.

In order to understand strategies that have been used to facilitate nerve regeneration, it is important to understand the biology of nerve regeneration. Much has been learned about the mechanisms responsible in peripheral nerve regeneration in the last few decades, the highlights of which will be detailed here.

Types of nerve injury

The response to a nerve injury depends on the degree of injury. Seddon proposed a nerve injury classification that was further expanded by Sunderland. In a first degree (neurapraxic) injury, the axons are intact but a conduction block is produced, often via a compression mechanism that results in an interruption of depolarization of the axon. A first degree injury always recovers complete function. A second degree injury (axonotmetic injury) results in disruption of the axons, while leaving the Schwann cell tubes intact. Recovery of a second degree injury would be expected to be complete, since axons would be expected to travel down their original pathways into the end organ. A third degree injury produces an injury to the Schwann cell tubes, but the perineurium remains intact. Therefore, regenerating axons may cross pathways, producing aberrant regeneration. Regeneration is often incomplete in Sunderland third degree injuries, especially if the injury is very proximal. The results of third degree injuries can be highly variable. Fourth degree injuries produce a neuroma in continuity, in which a few nerve fibers may conduct through the dense scar, but the majority of fibers have been transected and no longer conduct. This type of injury requires surgery in order to gain improvement in end organ function. A fifth degree injury consists of complete transection of the nerve. In these cases, immediate or early surgery produces the best clinical results. It is also important to realize that multiple degrees of nerve injury can exist within a single nerve injury depending on the mechanism.

Nerve regeneration and the response to nerve injury

Following sharp transection, the distal nerve stump undergoes Wallerian degeneration. After nerve transection, a proportion of the neurons die, depending on how proximal the lesion is. The more proximal the lesion, the more likely a neuron is to die. The process of Wallerian degeneration begins 48 to 96 hours after transection. During the first two or three days post-injury it is possible to electrically stimulate the distal stump intraoperatively, facilitating surgical reconstruction. During Wallerian degeneration, Schwann cells in the distal stump proliferate. The myelin sheaths in the distal stump undergo lysis as macrophages and Schwann cells remove debris. Bands of Bungner are formed by the Schwann cell basal lamina of the endoneurial tubes. These tubes provide a mechanical and biological scaffold within which the axons normally regenerate.

Regeneration from the proximal stump generally begins within the first few days after injury. The advancing axon produces a growth cone, which consists of filopodia rich in actin. The growth cone moves along the basal lamina tubes ("contact guidance") toward chemoattractant growth factors and extracellular matrix molecules present in the pathway of the distal stump ("neurotropism"). The resulting advancing axons are subsequently pruned as the axons make connections in the target organ. Myelin is produced by the Schwann cells during the maturation of the fiber.

The success of a surgically repaired nerve injury depends on many factors. Sharp transections regenerate better than those in which a broad zone of injury is present. In general, the earlier an injury is repaired, the faster the recovery. Distal injuries regenerate better than more proximal ones. Sensory recovery usually occurs faster and more completely than motor recovery. Sensory end organs are less susceptible to denervation atrophy than muscle. It becomes progressively more difficult with time to reinnervate motor end plates, but sensory organs in the skin have been successfully reinnervated even after ten years of denervation. The technique chosen in reconstruction is a major determinant of success: in many cases nerve transfers are better than nerve grafts; in others grafts are better than direct repairs. In addition, patient factors, such as age, size, and level of cooperation can also affect outcomes. For example, using autologous nerve grafts to reconstruct the entire brachial plexus is a suitable approach in a child, but not in an adult.

Historical treatment of nerve gaps

Millesi and others[1,2] have shown that minimizing tension on the repair is essential to optimal nerve regeneration. Prior to this work, the standard of care for nerve transection injuries was to débride and coapt the stumps under tension. Outcomes improved with autologous nerve grafting, and that practice has become the accepted "gold standard" for treatment of significant nerve gaps for the last 40 years. Autologous nerve grafts may be harvested from any of a number of convenient cutaneous sensory donor sites. For hand surgery, the sural, medial antebrachial cutaneous, posterior interosseous branch to the wrist joint are all common donor sites for autologous harvest. The donor site incision for the sural nerve leaves a scar on the posterior leg and a sensory defect involving the dorsolateral foot. The medial antebrachial cutaneous nerve is harvested through an incision beginning anteromedial to the medial humeral epicondyle; the sensory defect involves the medial forearm. The distal posterior interosseous nerve is harvested through the dorsal wrist, and it leaves part of the wrist capsule insensate (usually this is not noticed by the patient). The nerve grafts are then reversed and coapted under the operating microscope to the proximal and distal stumps with nylon microsutures. The grafts are reversed in orientation such that the part of the graft that was distal in the donor site becomes proximal in the recipient site. The rationale for this practice is to prevent small side branches from allowing escape of axons as they regenerate proximal to distal. Regeneration in humans proceeds at a rate of less than 1 mm per day in autologous nerve grafts.

Nerve engineering strategies

Many authors have proposed alternatives to autologous nerve grafting. Some of these strategies have now been commercially translated into wide clinical use. Investigators have used non-biodegradable conduits, conduits made of biodegradable scaffolds, autologous conduits made of vein or arterial segments, components of skeletal muscle, and nerve allografts. Growth factors and cell transplants have been investigated experimentally in animal models but they have not yet been translated clinically.

Synthetic "Biodurable" Conduits

Silicone conduits

The use of conduits for nerve repair was reported in the 19th century.[3] In 1982, Lundborg *et al.* described the use of a polydimethylsiloxane (PDMS) tube for bridging a 10 mm gap in a rat sciatic nerve.[4] Silicone is readily available, relatively inert, commonly used in medical devices, and familiar to surgeons. Regeneration proceeds through conduits in a stepwise fashion. First, a fluid phase occurs in which the surrounding extracellular fluid enters the chamber of the conduit within hours following implantation. This accumulated fluid has been shown to contain neurotrophic factors.[5,6] Within days, the fluid is replaced by a fibrin matrix, which tapers at either end as the fibrin matrix matures.[4,7–9] Following formation of the fibrin matrix, cells invade the chamber, including endothelial cells, Schwann cells, and fibroblasts. After two weeks, axons begin to enter the chamber. Regeneration within a chamber has been shown to be much slower than in autologous nerve grafts; in rats, sciatic nerve regeneration proceeded through a silicone conduit at a rate of 0.25 to 1 mm/d compared with nerve auto- or isografts, in which regeneration has been shown to proceed at a rate of 1–3 mm/d in rat.[4] Nevertheless, silicone conduits offer the theoretical advantage of being able to provide a controlled environment that can be manipulated to alter nerve regeneration in experimental studies. Silicone conduits have been used clinically with moderate success for reconstructing short gaps in peripheral nerves. A randomized prospective study was conducted in which 18 patients underwent repair of gaps 4 mm or less with either silicone tubes or with direct repair.[10] No differences were seen at one year in terms of motor or sensory recovery. In a follow-up study, no differences were seen at five years, except for cold intolerance, which was improved in the silicone conduit repairs compared with direct repair.[11] Implants required removal in seven patients secondary to compression. In a 1999 report by Braga-Silva,[12] 26 patients underwent silicone tube reconstruction of upper extremity defects up to 5 cm. Seven patients required removal of the implants due to late compressive effects; recovery was judged "satisfactory" in most cases.

Salubria hydrogel conduits (SaluBridge; SaluMedica, Atlanta, GA) were approved for human use by the United States Food and Drug Administration. They were also approved for use in Canada and the European Union. The product is not currently commercially available (the producer is seeking a distributor). The conduits are biodurable, composed of a synthetic non-biodegradable polyvinyl alcohol hydrogel the manufacturer says is similar to silicone. The conduits have inner diameters of 2, 5 and 10 mm, in lengths of 6.35 cm. The manufacturer documents one clinical case in which a digital nerve gap of 1 mm was repaired with the conduit. At three months, sensation had returned (data not published; http://www.salumedica.com/salubridgeinfodoc.htm).

Polytetrafluoroethylene (PTFE, GoreTex) conduits

PTFE is chemically extremely inert, immunologically nonreactive, easy to manufacture, and is non-biodegradable. The material has been used widely for implants in vascular surgery, abdominal hernia repair, and other applications.[13] A PTFE conduit was used to reconstruct an ulnar nerve defect of 3 cm in a 22 year old woman. She recovered sensory and motor function.[14,15] In another study, a series of 43 patients with defects of 15–60 mm underwent PTFE conduit reconstruction. Patients with short defects (15–40 mm) had substantially better results than those with longer gaps (41–60 mm), in whom only 13% recovered useful function, as determined by grip strength and two point discrimination.[15]

Synthetic Biodegradable Conduits

Polyglycolide, Poly(DL-Lactide-ε-Caprolactone) and other synthetic aliphatic polyesters

Synthetic aliphatic polyesters have been used to manufacture biodegradable devices for implantation, including sutures, plates and screws, and other devices. Materials include poly(glycolide), poly(lactide), poly(L-lactide), poly(D,L-lactide), P(LLA-GA), P(DLLA-GA), P(LLA-CL), and P(DLLA-CL). These biomaterials degrade through a process of hydrolysis. The absorption profile, tensile strength, flexibility, and 'feel' of the materials are highly dependent on the relative ratios of the constituent components and additives used in processing.

PGA conduits

Polyglycolic acid (PGA) has been used in commercially applied biomaterials, including Vicryl sutures, and nerve conduits.[16,17] The Neurotube (distributed by Synovis Micro Companies Alliance, Inc., a subsidiary of Synovis Life Technologies, Inc., St. Paul. Minnesota) has been used extensively for peripheral nerve gaps. It is made of poly(glycolide), the fibers of which are woven into a corrugated porous tube. The corrugated structure of the device allows the walls to be flexible but rigid enough to prevent the wall from collapsing. The conduits degrade *in vivo* though hydrolysis. They begin to undergo fragmentation as early as one month.[18] After six to eight months, the tubes are completely resorbed.[19]

Seckel *et al.* documented regeneration through a 10 mm PGA conduit in a rat sciatic model in 1984.[20,21] Dellon and Mackinnon studied a PGA tube in a 3 cm gap in monkey ulnar nerve and demonstrated histological and electrophysiological evidence of regeneration.[22] A PGA device based on their work was commercialized as the Neurotube.

The Neurotube is available as a 2.3 mm diameter conduit 4 cm in length, and also in diameters of 4 and 8 mm (but 2 cm in length). The 2.3 mm conduit is approved by the US Food and Drug Administration for use in gaps up to 3 cm in length. The device is applied to the prepared proximal and distal stumps using microsutures, and the lumen of the conduit is filled with saline solution (Figure 1).[2]

Clinical studies have demonstrated good results in distal sensory nerves. In 1990, Mackinnon and Dellon presented a prospective series of secondary digital nerve reconstruction, in which excellent or good recovery was noted in 86% of nerves. Fifteen patients were included in the series. They documented excellent two point discrimination in five patients, and good two point discrimination in eight patients.[23]

In 2000, Weber *et al.*[24] reported a randomized prospective study in digital nerves in which the Neurotube performed favorably compared to standard care (which was defined as either autologous nerve grafting or direct repair as determined by the surgeon). In "long" gaps (greater than 8 mm), the Neurotube performed better than nerve grafting. The Neurotube also performed better than standard care in "short gaps" of 4 mm or less. Intermediate gap performance was statistically not different between Neurotube and standard care.

Other investigators have studied the use of this device for longer gaps. In 2000, Casanas *et al.*[25] reported an abstract describing repair of digital nerve gaps between 2–3.5 cm. They reported recovery based on clinical evaluation and electrophysiology. In 2005, a study of facial nerve laceration repair with the Neurotube was reported. Gaps ranged from 1–3 cm. Motion returned in all seven patients in the series (satisfactory in 5 and poor in 2).[26] A 2005 report documented return of function in an iatrogenic injury to the spinal accessory nerve in which either a Neurotube or autologous greater auricular nerve graft was used. The Neurotube reconstruction resulted in M5 function three months after surgery; the autologous reconstruction reached M4 function 6 months after reconstruction.[27]

Battiston *et al.*[28] reported in 2005 a case series of 19 digital nerve repairs using the Neurotube. Thirteen patients had two point discrimination of less than 15 mm, which they rated as "very good."

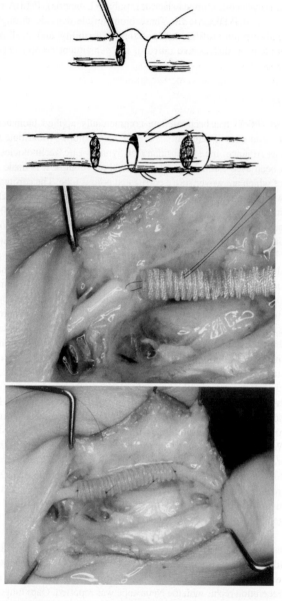

Figure 1. (Above) Line drawing of conduit-repair technique. The minimum distance between nerve stumps is 5 mm, even in those instances in which the ends could be coapted without tension. (Center) Photograph of a nerve being secured in the polyglycolic acid conduit. (Below) Photograph of the completed conduit repair of the nerve. From: Weber RA, Breidenbach WC, Brown RE, Jabaley ME, Mass DP: A randomized prospective study of polyglycolic acid conduits for digital nerve reconstruction in humans. *Plast Reconstr Surg* 2000, 106(5): 1036–1045; discussion 1046–1038.

In 2007 Donoghoe *et al.*[29] reported using the Neurotube to reconstruct two patients, each of whom had 3 cm defects of the median nerve at the wrist. They reported return of sensation and motor function in both patients. Two patients, aged 43 and 61, were reported with 5-year follow-up of median nerve reconstruction in the distal forearm in which multiple conduits (Neurotube) were used to bridge 3-cm nerve defects. Four separate 2.3-mm diameter, 4.0 cm long conduits were used in each patient. Each patient recovered two-point discrimination with good localization in the thumb, index, and middle finger by 2 years after the nerve reconstruction. Both patients recovered abductor pollicis brevis function. The authors subsequently published an expanded series of six motor nerve reconstructions in 2008, in which one patient presented with injury to the spinal accessory nerve, two with median nerve injuries in the distal forearm, one with an ulnar nerve injury 6 cm proximal to the wrist, one with median and ulnar nerve injuries in the mid-brachium, and another with ulnar nerve injuries in the mid-brachium. The length of the gaps averaged 2.8 cm (range 1.5–4 cm). All patients had some return of motor function rated as M3 or greater by physical exam.[30]

Poly(DL-lactide-epsilon-caprolactone) (PLCL) conduits

Poly(DL-lactide-epsilon-caprolactone) has been used as a biomaterial in nerve conduits, both preclinically and clinically. A commercial device is currently available (NeuroLac, Polyganics B.V., Groningen, the Netherlands). The conduits are biodegradable, with degradation beginning after several weeks. Den Dunnen *et al.* showed that the conduit swells *in vivo*, and tensile strength becomes negligible at two months post-implantation.[31] In a later study, Meek *et al.*[32] found that cytotoxicity was minimal using a cell proliferation inhibition index on mouse fibroblasts *in vitro*, and they produced a mild foreign body reaction *in vivo*. Tensile strength went from 13.0 MPa at one week *in vivo* to 3.0 MPa at four weeks; at eight weeks *in vivo* the tensile strength was 0.1 MPa, and beyond that point the tensile strength was not measurable because the conduit had degraded to the point of global mechanical instability.[33] Den Dunnen *et al* evaluated use of a conduit in which the composition was 50% DL-lactide and 50% caprolactone; the lactide contained 85% L-lactide (LLA) and 15% D-lactide (DLA). These conduits demonstrated improved regeneration compared to isograft in preclinical studies: the conduit groups experienced improved sciatic function index, greater numbers of regenerated fibers, and improved electrical conductivity in a 10 mm sciatic nerve gap (Den Dunnen 1998).[34] A subsequent study, again in a 10 mm sciatic rat model, showed that at 15 weeks, 70% of sciatic functional index had returned, and 90% of sensory nerve recovery had returned.[35]

The NeuroLac conduit is available in the United States and Europe. A clinical study was published in 2005 by Bertleff, Meek and Nicolai.[36] The authors conducted a multi-center, randomized prospective study, in which 30 patients either underwent primary repair of digital nerve injuries or were treated using the NeuroLac device. For control patients treated without the device, the choice of primary coaptation or autologous graft was left to the surgeon's discretion. All of the control patients were treated without nerve grafts. There were cases in the control group in which a gap of 20 mm was managed by flexing the digit to facilitate repair. The average gaps measured 6–8 mm in both the control and experimental groups. Removal of the conduit was necessary in one patient. The authors concluded that sensory function was indistinguishable between the control and the experimental groups, using 2 point discrimination at one year. The authors stratified their gap length into three groups: less than 4 mm, 4–8 mm, and greater than 8 mm. However, they did not report results for each individual group; rather, they were reported in aggregate.

Both control and experimental groups had average moving and static two point discrimination of 8–9 mm.

Collagen conduits

Type I collagen is a highly abundant molecule, present in skin, bones, and peripheral nerves. Collagen has been used experimentally in multiple animal studies. Collagen nerve conduits have been fabricated from bovine type I collagen more often than other types of collagen, because the purification is relatively straightforward, the material is abundant and relatively inexpensive compared to other collagen types, and it is usually well tolerated *in vivo* from an immunologic standpoint. Type I collagen has also been used previously for other biomaterial applications, including hemostatic agents, topical wound dressings, and dural grafts. These conduits vary in their mechanical properties, degradation profiles, and cost of manufacturing.

A number of collagen-based nerve guidance products are currently marketed for clinical use in North America, including NeuraGen and NeuraWrap, by Integra LifeSciences Corporation, Plainsboro, New Jersey, and NeuroMatrix and NeuroFlex, by Collagen Matrix, Inc., Franklin Lakes, New Jersey.

Several preclinical studies using collagen conduits have been performed. In 1988 Eppley reported a collagen tube based repair of the rat mandibular nerve.[37] In 1991 Archibald *et al.* reported repair of the sciatic nerve in rats and the median nerve in monkeys, comparing regeneration to autografts. The regeneration gap was 4 mm. They concluded that the collagen tubes were as effective as the autografts. In later work they showed in macaca that nerve regeneration after repair of a 5 mm median nerve gap with a collagen nerve guide was similar to that after graft repair, and the final level of physiological recovery for both repair procedures is comparable to direct suture repair of the median nerve.[38-40] One preclinical study concluded that collagen conduits are superior to PGA conduits (see below).[41]

According to the manufacturer, the NeuraGen conduit, introduced in 2001 (Integra Life Sciences) is semipermeable, with a porous outer layer and a semi-permeable inner membrane. The conduits are composed of bovine type I collagen. The conduits are available in lengths of 2 cm or 3 cm. Both lengths are available in inner diameters of 1.5, 2, 3, 4, 5, 6, and 7 mm. The conduits are designed to entubulate the proximal and distal stumps. Saline solution is placed within the lumen prior to wound closure. The NeuraWrap product, introduced in 2004, is said to protect compressed, scarred or partially inured nerves. It is also composed of type I collagen, and is available in lengths of either 2 or 4 cm, in diameters of 3, 5, 7 and 10 mm. The manufacturer states that the NeuraGen is flexible to accommodate joint motion, and that it retains its shape to resist occlusive forces of surrounding tissues (http://www.integrals.com/products/?product=88).

NeuraGen was used as a cuff around repairs of lingual and inferior alveolar nerve injuries in a series of 9 patients, 8 of whom experienced improvements in sensory function.[42] Taras published a review in which it was stated that 75 patients had undergone treatment with the NeuraGen conduit. However, outcome data were not presented.[43] Bushnell *et al.* presented a retrospective case series of nine patients that underwent digital nerve repair with the NeuraGen conduit for gaps of 20 mm or less. At minimum follow-up of one year, the patients had good or excellent two point discrimination in 8 out of 9 patients. Semmes-Weinstein monofilament testing demonstrated full return in 5/9 nerves, and diminished touch in 2, with diminished protective sensation in 1, and lack of protective sensation in 1. The authors concluded that the collagen conduit reconstruction performed well compared to the gold standard of autologous nerve grafting in digital nerve repair. However, there was no control group in the study, and therefore conclusions should be taken with caution.[44]

In a 2009 case series of fifteen digital nerve lacerations repaired with the NeuraGen conduit, Lohmeyer *et al.*[45] reported excellent return of sensory function in four of the twelve patients that had follow-up of one year or more. The mean regeneration gap was 12.5 +/− 3.7 mm. Excellent function (S4) was defined as static two point discrimination of less than 7 mm. Good function (S3+), defined as static two point discrimination of less than 15 mm, was obtained in five patients. Poor function, defined as static two point discrimination of more than 15 mm, was obtained one patient. Two patients obtained no protective sensation.

Moore *et al.*[46] presented four cases that presented following failed reconstructions with the NeuraGen conduit, including two obstetrical brachial plexus reconstructions with supraclavicular trunk level gaps. They concluded that the indications for reconstruction with the NeuraGen conduit should be restricted to distal sensory nerves with short gaps.

Both Neuroflex and NeuroMatrix are marketed by Stryker, Inc. (Kalamazoo, Michigan; http://www.stryker.com/enus/products/Trauma/UpperExtremityHand/Hand/Neuro MatrixNeuroflex/index.htm). The Neuroflex product is designed to be a flexible collagen based nerve cuff. It is manufactured from bovine Achilles tendon type I collagen. According to the manufacturer, the conduit is non-immunogenic, resists kinking or collapsing, and is implanted using an entubulation technique. The NeuroMatrix product is a collagen type I conduit designed without anti kinking properties. Both products are available in lengths of 2.5 cm, with internal diameters of 2, 2.5, 3, 4, 5, and 6 mm. To date there is no clinical data published on the use of either product.

Revolnerv is a porcine collagen type I and type III conduit produced and marketed by Orthomed (Côte d'Azur, France; http://www.orthomed.fr/products/zone/hand/product/ revolnerv_protection.en.htm). It is surgically implanted using an entubulation method. It is available in lengths of 20 or 30 mm, with diameters of 2, 4 or 6 mm. It was studied pre-clinically in a rat sciatic model by Alluin *et al.*[47] The authors found that in a 10 mm gap, the conduits performed favorably, but did not surpass isografts histomorphometrically or functionally.

Tissue-based Conduits

Vein and arterial conduits

Vessels have been used for nerve conduits for decades. Büngner reported nerve regeneration using an arterial conduit in 1891 (Bungner, O. Von (1891) *Beitrage zurpathologischen Anatomie und zur allgemeinen. Pathologie,* 10, 321). Other authors have reported use of arteries as nerve conduits in animal models with mixed success.[48,49] Veins have also been used, with the theoretical advantage that veins are more expendable than arteries. In 1909 Wrede described the use of vein conduits for nerve gap repair (Wrede L. Uberbrueckung eines nervendefektes mittels seidennaht und leben venenstueckes (Bridging nerve defects by means of stitches and vein segments). Deutsches Med Wochenschrift 1909; 35:1125) Numerous small animal model and clinical studies have shown that vein grafts can bridge short gaps as well as autologous nerve grafts.[50–67]

A clinical series by Lee published in 2008 in which three patients underwent secondary vein grafting of digital nerve defects of 8–25 mm showed that sensory recovery was excellent (S3) at follow-up of three to eleven years.[51] A clinical study of 50 patients was performed in which 25 patients with digital nerve injuries were treated with sural nerve grafting and 25 were treated with vein conduits with an interposed posterior interosseous nerve segment. The authors found that the complication rate was better in the vein group, and that sensory recovery was equal in both groups, with mean recovery of 6 mm moving two point discrimination.

Veins, having thinner walls than arteries, are more susceptible to mechanical collapse from pressure from surrounding tissues. Nevertheless, their use as a readily available, inexpensive source of conduit material continues, and their role as a container for cells or growth factors is of continued interest.

Muscle-based conduits

The basal lamina of endoneural tubes contains extracellular matrix molecules that are also found in the basal lamina of skeletal muscle. Because of its relative ease of harvest, low donor site morbidity, and availability, muscle has been used as a conduit for nerve regeneration in experimental models. Acellularized muscle has also been used as a nerve conduit, with the rationale that the relatively insoluble extracellular matrix would support regeneration. Freeze-thaw treated rectus abdominis muscles have been used in the rat to support regeneration in the sciatic nerve up to 50 mm.[68] Histomorphometrically they produced about 25% of fiber regeneration as isograft controls. These grafts may become more relevant in the future with addition of Schwann cells.

Processed and decellularized nerve allografts

Peripheral nerve allografts have been investigated and clinically used for decades. However, the morbidity of immunosuppression has limited their clinical use to cases requiring extremely large amounts of nerve graft.[69] Various methods, including chemical decellularization, and freeze thaw have been used to remove cells from peripheral nerves in an effort to make them suitable for allografting without immunosuppression.[70–101]

Treatment with chondroitinase has been advocated for improving axonal regeneration in allografts.[102–105] Chondroitin-6-sulfate proteoglycan has been shown to be inhibitory to axonal regeneration.[106] Its removal has been shown to improve regeneration, especially within the first few millimeters into the graft.[102] Other agents have been investigated for use. Other extracellular matrix molecules, including heparan sulfate, have been shown to inhibit regeneration, and their degradation has similarly been used to enhance axonal regeneration.

One processed nerve allograft has resulted in a commercially available product (Avance, AxoGen, Inc., Alachua, Florida). The allografts are harvested from cadaveric human donors, chemically decellularized, and treated with chondroitinase. The product is commercially available in Europe, the United States and Canada. A preclinical study by Whitlock et al (In Press, Muscle and Nerve 2009), was conducted in a rat sciatic model. The authors showed that histomorphometrically, the processed allograft supported regeneration better than the NeuraGen conduit, and half as well as isograft.

Growth factors and cell-based based approaches

Various methods have been used experimentally in an attempt to use growth factors in peripheral nerve repair. The rational for this approach is that in a normal nerve healing environment, growth factors are normally elaborated post injury. Therefore, it may be possible to obtain better results with an environment more conducive to nerve healing.

While there have not been any human trials using growth factor application to date, a number of experiments have been conducted in small animal models that suggest that this approach may have value in subsequent human trials. Small animal models have used growth factors applied in various ways.

Similarly, there have been no clinical cases using cell transplantation to augment peripheral nerve regeneration. However, studies have been performed in small and large animal models examining the effectiveness of using Schwann cells and other cells for nerve gap repair. Schwann cells have been used in conduits. The underlying rationale is that the Schwann cells will produce growth factors that support the growth and development of the neuron and its axon. It is known that Schwann cells produce glial cell-derived neurotrophic factor, brain derived neurotrophic factor, artemin, neurturin, persephin, pleiotrophin, and other growth factors that regulate axon support. In the future, it is likely that further efforts will be directed at using cells and/or growth factors to improve axonal regeneration.

Conclusions

Nerve engineering, or the use of nerve conduits to harness and direct the body's innate healing capacity, will become an increasingly important area within the realm of tissue engineering as materials and methods are continually refined. Within hand surgery, the reconstruction of nerves remains a critical aspect of restoring or maintaining function after injury. In addition, the use of engineering nerve grafts can decrease operative time, allow for immediate reconstruction, and reduce the risk of losing nerve autografts to infection. As discussed, numerous biomaterials have been developed and tested in conduits to augment the natural healing process. Unfortunately, surprisingly little work has been done comparing various biomaterials for nerve reconstruction *in vivo*. Waitayawinyu *et al.*[41] studied the use of the NeuraGen conduit compared to the NeuroTube and isograft in a 10 mm rat sciatic gap model. They found that the collagen based NeuraGen supported greater numbers of fibers than the PGA based NeuroTube conduit, and they also concluded that the collagen conduits produced results equivalent to isograft. However, different inner diameters were used for the conduits, and the study was limited in number of animals, and only a single time point (15 weeks) was used. Thus the challenges in nerve engineering in the hand include not only demonstrating efficacy when compared to nerve autografts, but developing an understanding of which conduit and nerve engineering technologies provide the best patient prognosis relative to the others.

References

1. Millesi H. Techniques for nerve grafting. *Hand Clin* **16**(1) (2000) 73–91, viii.
2. Millesi H. Nerve transplantation for reconstruction of peripheral nerves injured by the use of the microsurgical technic. *Minerva Chir* **22**(17) (1967) 950–951.
3. Ijpma FF, Van De Graaf RC, Meek MF. The early history of tubulation in nerve repair. *J Hand Surg Eur Vol* **33**(5) (2008) 581–586.
4. Lundborg G, Dahlin LB, Danielsen N, Gelberman RH, Longo FM, Powell HC, Varon S. Nerve regeneration in silicone chambers: influence of gap length and of distal stump components. *Exp Neurol* **76**(2) (1982) 361–375.
5. Longo FM, Manthorpe M, Skaper SD, Lundborg G, Varon S. Neuronotrophic activities accumulate *in vivo* within silicone nerve regeneration chambers. *Brain Res* **261**(1) (1983) 109–116.
6. Longo FM, Skaper SD, Manthorpe M, Williams LR, Lundborg G, Varon S. Temporal changes of neuronotrophic activities accumulating *in vivo* within nerve regeneration chambers. *Exp Neurol* **81**(3) (1983) 756–769.

7. Williams LR, Longo FM, Powell HC, Lundborg G, Varon S. Spatial-temporal progress of peripheral nerve regeneration within a silicone chamber: parameters for a bioassay. *J Comp Neurol* **218**(4) (1983) 460–470.

8. Lundborg G, Dahlin LB, Danielsen N, Hansson HA, Johannesson A, Longo FM, Varon S. Nerve regeneration across an extended gap: a neurobiological view of nerve repair and the possible involvement of neuronotrophic factors. *J Hand Surg [Am]* **7**(6) (1982) 580–587.

9. Lundborg G, Longo FM, Varon S. Nerve regeneration model and trophic factors *in vivo*. *Brain Res* **232**(1) (1982) 157–161.

10. Lundborg G, Rosen B, Dahlin L, Danielsen N, Holmberg J. Tubular versus conventional repair of median and ulnar nerves in the human forearm: early results from a prospective, random-ized, clinical study. *J Hand Surg [Am]* **22**(1) (1997) 99–106.

11. Lundborg G, Rosen B, Dahlin L, Holmberg J, Rosen I. Tubular repair of the median or ulnar nerve in the human forearm: a 5-year follow-up. *J Hand Surg [Br]* **29**(2) (2004) 100–107.

12. Braga-Silva J. The use of silicone tubing in the late repair of the median and ulnar nerves in the forearm. *J Hand Surg [Br]* **24**(6) (1999) 703–706.

13. Schlosshauer B, Dreesmann L, Schaller HE, Sinis N. Synthetic nerve guide implants in humans: a comprehensive survey. *Neurosurgery* **59**(4) (2006) 740–747; discussion 747–748.

14. Stanec S, Stanec Z. Ulnar nerve reconstruction with an expanded polytetrafluoroethylene conduit. *Br J Plast Surg* **51**(8) (1998) 637–639.

15. Stanec S, Stanec Z. Reconstruction of upper-extremity peripheral-nerve injuries with ePTFE conduits. *J Reconstr Microsurg* **14**(4) (1998) 227–232.

16. Meek MF, Coert JH. US Food and Drug Administration/Conformit Europe-approved absorbable nerve conduits for clinical repair of peripheral and cranial nerves. *Ann Plast Surg* **60**(1) (2008) 110–116.

17. Meek MF, Coert JH. Clinical use of nerve conduits in peripheral-nerve repair: review of the literature. *J Reconstr Microsurg* **18**(2) (2002) 97–109.

18. Keeley RD, Nguyen KD, Stephanides MJ, Padilla J, Rosen JM. The artificial nerve graft: a comparison of blended elastomer-hydrogel with polyglycolic acid conduits. *J Reconstr Microsurg* **7**(2) (1991) 93–100.

19. Merrell JC, Russell RC, Zook EG. Polyglycolic acid tubing as a conduit for nerve regeneration. *Ann Plast Surg* **17**(1) (1986) 49–58.

20. Seckel BR, Ryan SE, Gagne RG, Chiu TH, Watkins E, Jr, Target-specific nerve regeneration through a nerve guide in the rat. *Plast Reconstr Surg* **78**(6) (1986) 793–800.

21. Seckel BR, Chiu TH, Nyilas E, Sidman RL. Nerve regeneration through synthetic biodegrad-able nerve guides: regulation by the target organ. *Plast Reconstr Surg* **74**(2) (1984) 173–181.

22. Dellon AL, Mackinnon SE. An alternative to the classical nerve graft for the management of the short nerve gap. *Plast Reconstr Surg* **82**(5) (1988) 849–856.

23. Mackinnon SE, Dellon AL. Clinical nerve reconstruction with a bioabsorbable polyglycolic acid tube. *Plast Reconstr Surg* **85**(3) (1990) 419–424.

24. Weber RA, Breidenbach WC, Brown RE, Jabaley ME, Mass DP. A randomized prospective study of polyglycolic acid conduits for digital nerve reconstruction in humans. *Plast Reconstr Surg* **106**(5) (2000) 1036–1045; discussion 1046–1038.

25. Casanas J, Serra J, Orduna M, Garcia-Portabella M, Mir X. Repair of digital sensory nerves of the hand using polyglycolic acid conduits. *J Hand Surg [Br]* **25**, (2000).

26. Navissano M, Malan F, Carnino R, Battiston B. Neurotube for facial nerve repair. *Microsurgery* **25**(4) (2005) 268–271.

27. Ducic I, Maloney CT, Jr, Dellon AL. Reconstruction of the spinal accessory nerve with auto-graft or neurotube? Two case reports. *J Reconstr Microsurg* **21**(1) (2005) 29–33; discussion 34.

28. Battiston B, Geuna S, Ferrero M, Tos P. Nerve repair by means of tubulization: literature review and personal clinical experience comparing biological and synthetic conduits for sensory nerve repair. *Microsurgery* **25**(4) (2005) 258–267.

29. Donoghoe N, Rosson GD, Dellon AL. Reconstruction of the human median nerve in the forearm with the Neurotube. *Microsurgery* **27**(7) (2007) 595–600.

30. Rosson GD, Williams EH, Dellon AL. Motor nerve regeneration across a conduit. *Microsurgery* **29**(2) (2009) 107–114.

31. den Dunnen WF, Meek MF, Grijpma DW, Robinson PH, Schakenraad JM. *In vivo* and *in vitro* degradation of poly[(50)/(50) ((85)/(15)(L)/(D))LA/epsilon-CL], and the implications for the use in nerve reconstruction. *J Biomed Mater Res* **51**(4) (2000) 575–585.

32. Meek MF, van der Werff JF, Klok F, Robinson PH, Nicolai JP, Gramsbergen A. Functional nerve recovery after bridging a 15 mm gap in rat sciatic nerve with a biodegradable nerve guide. *Scand J Plast Reconstr Surg Hand Surg* **37**(5) (2003) 258–265.

33. Meek MF, Jansen K. Two years after *in vivo* implantation of poly(DL-lactide-epsilon-caprolactone) nerve guides: Has the material finally resorbed? *J Biomed Mater Res A* 2008.

34. Den Dunnen WF, Meek MF, Robinson PH, Schakernraad JM. Peripheral nerve regeneration through P(DLLA-epsilon-CL) nerve guides. *J Mater Sci Mater Med* **9**(12) (1998) 811–814.

35. Meek MF, Dijkstra JR, Den Dunnen WF, Ijkema-Paassen J, Schakenraad JM, Gramsbergen A, Robinson PH. Functional assessment of sciatic nerve reconstruction: biodegradable poly (DLLA-epsilon-CL) nerve guides versus autologous nerve grafts. *Microsurgery* **19**(8) (1999) 381–388.

36. Bertleff MJ, Meek MF, Nicolai JP. A prospective clinical evaluation of biodegradable neurolac nerve guides for sensory nerve repair in the hand. *J Hand Surg [Am]* **30**(3) (2005) 513–518.

37. Eppley BL, Delfino JJ. Collagen tube repair of the mandibular nerve: a preliminary investigation in the rat. *J Oral Maxillofac Surg* **46**(1) (1988) 41–47.

38. Archibald SJ, Shefner J, Krarup C, Madison RD. Monkey median nerve repaired by nerve graft or collagen nerve guide tube. *J Neurosci* **15**(5 Pt 2) (1995) 4109–4123.

39. Li ST, Archibald SJ, Krarup C, Madison RD. Peripheral nerve repair with collagen conduits. *Clin Mater* **9**(3–4) (1992) 195–200.

40. Archibald SJ, Krarup C, Shefner J, Li ST, Madison RD. A collagen-based nerve guide conduit for peripheral nerve repair: an electrophysiological study of nerve regeneration in rodents and nonhuman primates. *J Comp Neurol* **306**(4) (1991) 685–696.

41. Waitayawinyu T, Parisi DM, Miller B, Luria S, Morton HJ, Chin SH, Trumble TE. A comparison of polyglycolic acid versus type 1 collagen bioabsorbable nerve conduits in a rat model: an alternative to autografting. *J Hand Surg [Am]* **32**(10) (2007) 1521–1529.

42. Farole A, Jamal BT. A bioabsorbable collagen nerve cuff (NeuraGen) for repair of lingual and inferior alveolar nerve injuries: a case series. *J Oral Maxillofac Surg* **66**(10) (2008) 2058–2062.

43. Taras JS, Jacoby SM. Repair of lacerated peripheral nerves with nerve conduits. *Tech Hand Up Extrem Surg* **12**(2) (2008) 100–106.

44. Bushnell BD, McWilliams AD, Whitener GB, Messer TM. Early clinical experience with collagen nerve tubes in digital nerve repair. *J Hand Surg [Am]* **33**(7) (2008) 1081–1087.

45. Lohmeyer JA, Siemers F, Machens HG, Mailander P. The clinical use of artificial nerve conduits for digital nerve repair: a prospective cohort study and literature review. *J Reconstr Microsurg* **25**(1) (2009) 55–61.

46. Moore AM, Kasukurthi R, Magill CK, Farhadi HF, Borschel GH, Mackinnon SE. Limitations of Conduits in Peripheral Nerve Repairs. *Hand (NY)* 2009.

47. Alluin O, Wittmann C, Marqueste T, Chabas JF, Garcia S, Lavaut MN, Guinard D, Feron F, Decherchi P. Functional recovery after peripheral nerve injury and implantation of a collagen guide. *Biomaterials* **30**(3) (2009) 363–373.

48. Doolabh VB, Hertl MC, Mackinnon SE. The role of conduits in nerve repair: a review. *Rev Neurosci* **7**(1) (1996) 47–84.

49. Anderson PN, Turmaine M. Axonal regeneration through arterial grafts. *J Anat* **147** (1986) 73–82.

50. Tang J, Wang XM, Hu J, Luo E, Qi MC. Autogenous standard versus inside-out vein graft to repair facial nerve in rabbits. *Chin J Traumatol* **11**(2) (2008) 104–109.

51. Lee YH, Shieh SJ. Secondary nerve reconstruction using vein conduit grafts for neglected digital nerve injuries. *Microsurgery* **28**(6) (2008) 436–440.

52. Calcagnotto GN, Braga Silva J. [The treatment of digital nerve defects by the technique of vein conduit with nerve segment. A randomized prospective study]. *Chir Main* **25**(3–4) (2006) 126–130.

53. Lundborg G. Alternatives to autologous nerve grafts. *Handchir Mikrochir Plast Chir* **36**(1) (2004) 1–7.

54. Meek MF, Varejao AS, Geuna S. Use of skeletal muscle tissue in peripheral nerve repair: review of the literature. *Tissue Eng* **10**(7–8) (2004) 1027–1036.

55. Geuna S, Tos P, Battiston B, Giacobini-Robecchi MG. Bridging peripheral nerve defects with muscle-vein combined guides. *Neurol Res* **26**(2) (2004) 139–144.

56. Mersa B, Agir H, Aydin A, Sen C. Comparison of expanded polytetrafluoroethylene (ePTFE) with autogenous vein as a nerve conduit in rat sciatic nerve defects. *Kulak Burun Bogaz Ihtis Derg* **13**(5–6) (2004) 103–111.

57. Keskin M, Akbas H, Uysal OA, Canan S, Ayyldz M, Agar E, Kaplan S. Enhancement of nerve regeneration and orientation across a gap with a nerve graft within a vein conduit graft: a functional, stereological, and electrophysiological study. *Plast Reconstr Surg* **113**(5) (2004) 1372–1379.

58. Geuna S, Raimondo S, Nicolino S, Boux E, Fornaro M, Tos P, Battiston B, Perroteau I. Schwann-cell proliferation in muscle-vein combined conduits for bridging rat sciatic nerve defects. *J Reconstr Microsurg* **19**(2) (2003) 119–123; discussion 124.

59. Pogrel MA, Maghen A. The use of autogenous vein grafts for inferior alveolar and lingual nerve reconstruction. *J Oral Maxillofac Surg* **59**(9) (2001) 985–988; discussion 988–993.

60. Tos P, Battiston B, Geuna S, Giacobini-Robecchi MG, Hill MA, Lanzetta M, Owen ER. Tissue specificity in rat peripheral nerve regeneration through combined skeletal muscle and vein conduit grafts. *Microsurgery* **20**(2) (2000) 65–71.

61. Di Benedetto G, Zura G, Mazzucchelli R, Santinelli A, Scarpelli M, Bertani A. Nerve regeneration through a combined autologous conduit (vein plus acellular muscle grafts). *Biomaterials* **19**(1–3) (1998) 173–181.

62. Strauch B, Ferder M, Lovelle-Allen S, Moore K, Kim DJ, Llena J. Determining the maximal length of a vein conduit used as an interposition graft for nerve regeneration. *J Reconstr Microsurg* **12**(8) (1996) 521–527.

63. Giardino R, Nicoli Aldini N, Perego G, Cella G, Maltarello MC, Fini M, Rocca M, Giavaresi G. Biological and synthetic conduits in peripheral nerve repair: a comparative experimental study. *Int J Artif Organs* **18**(4) (1995) 225–230.

64. Benito-Ruiz J, Navarro-Monzonis A, Piqueras A, Baena-Montilla P. Invaginated vein graft as nerve conduit: an experimental study. *Microsurgery* **15**(2) (1994) 105–115.

65. Lu XS. [Autogenous vein graft as a conduit for repairing the laryngeal nerve deficit]. *Zhonghua Wai Ke Za Zhi* **31**(1) (1993) 40–42.

66. Chiu DT, Strauch B. A prospective clinical evaluation of autogenous vein grafts used as a nerve conduit for distal sensory nerve defects of 3 cm or less. *Plast Reconstr Surg* **86**(5) (1990) 928–934.

67. Chiu DT, Janecka I, Krizek TJ, Wolff M, Lovelace RE. Autogenous vein graft as a conduit for nerve regeneration. *Surgery* **91**(2) (1982) 226–233.

68. Keilhoff G, Pratsch F, Wolf G, Fansa H. Bridging extra large defects of peripheral nerves: possibilities and limitations of alternative biological grafts from acellular muscle and Schwann cells. *Tissue Eng* **11**(7–8) (2005) 1004–1014.

69. Mackinnon SE, Doolabh VB, Novak CB, Trulock EP. Clinical outcome following nerve allograft transplantation. *Plast Reconstr Surg* **107**(6) (2001) 1419–1429.

70. Gulati AK, Cole GP. Immunogenicity and regenerative potential of acellular nerve allografts to repair peripheral nerve in rats and rabbits. *Acta Neurochir (Wien)* **126**(2–4) (1994) 158–164.

71. Moradzadeh A, Borschel GH, Luciano JP, Whitlock EL, Hayashi A, Hunter DA, Mackinnon SE. The impact of motor and sensory nerve architecture on nerve regeneration. *Exp Neurol* **212**(2) (2008) 370–376.

72. Hayashi A, Moradzadeh A, Tong A, Wei C, Tuffaha SH, Hunter DA, Tung TH, Parsadanian A, Mackinnon SE, Myckatyn TM. Treatment modality affects allograft-derived Schwann cell phenotype and myelinating capacity. *Exp Neurol* **212**(2) (2008) 324–336.

73. Hess JR, Brenner MJ, Fox IK, Nichols CM, Myckatyn TM, Hunter DA, Rickman SR, Mackinnon SE. Use of cold-preserved allografts seeded with autologous Schwann cells in the treatment of a long-gap peripheral nerve injury. *Plast Reconstr Surg* **119**(1) (2007) 246–259.

74. Hontanilla B, Auba C, Arcocha J, Gorria O. Nerve regeneration through nerve autografts and cold preserved allografts using tacrolimus (FK506) in a facial paralysis model: a topographical and neurophysiological study in monkeys. *Neurosurgery* **58**(4) (2006) 768–779; discussion 768–779.

75. Hontanilla B, Yeste L, Auba C, Gorria O. Neuronal quantification in cold-preserved nerve allografts and treatment with FK-506 through osmotic pumps compared to nerve autografts. *J Reconstr Microsurg* **22**(5) (2006) 363–374.

76. Fox IK, Schwetye KE, Keune JD, Brenner MJ, Yu JW, Hunter DA, Wood PM, Mackinnon SE. Schwann-cell injection of cold-preserved nerve allografts. *Microsurgery* **25**(6) (2005) 502–507.

77. Fox IK, Jaramillo A, Hunter DA, Rickman SR, Mohanakumar T, Mackinnon SE. Prolonged cold-preservation of nerve allografts. *Muscle Nerve* **31**(1) (2005) 59–69.

78. Grand AG, Myckatyn TM, Mackinnon SE, Hunter DA. Axonal regeneration after cold preservation of nerve allografts and immunosuppression with tacrolimus in mice. *J Neurosurg* **96**(5) (2002) 924–932.

79. Atchabahian A, Mackinnon SE, Hunter DA. Cold preservation of nerve grafts decreases expression of ICAM-1 and class II MHC antigens. *J Reconstr Microsurg* **15**(4) (1999) 307–311.

80. Evans PJ, MacKinnon SE, Midha R, Wade JA, Hunter DA, Nakao Y, Hare GM. Regeneration across cold preserved peripheral nerve allografts. *Microsurgery* **19**(3) (1999) 115–127.

81. Evans PJ, Mackinnon SE, Levi AD, Wade JA, Hunter DA, Nakao Y, Midha R. Cold preserved nerve allografts: changes in basement membrane, viability, immunogenicity, and regeneration. *Muscle Nerve* **21**(11) (1998) 1507–1522.

82. Strasberg SR, Mackinnon SE, Genden EM, Bain JR, Purcell CM, Hunter DA, Hay JB. Long-segment nerve allograft regeneration in the sheep model: experimental study and review of the literature. *J Reconstr Microsurg* **12**(8) (1996) 529–537.

83. Evans PJ, Mackinnon SE, Best TJ, Wade JA, Awerbuck DC, Makino AP, Hunter DA, Midha R. Regeneration across preserved peripheral nerve grafts. *Muscle Nerve* **18**(10) (1995) 1128–1138.

84. Hare GM, Evans PJ, Mackinnon SE, Wade JA, Young AJ, Hay JB. Phenotypic analysis of migrant, efferent lymphocytes after implantation of cold preserved, peripheral nerve allografts. *J Neuroimmunol* **56**(1) (1995) 9–16.

85. Hare GM, Evans PJ, Mackinnon SE, Nakao Y, Midha R, Wade JA, Hunter DA, Hay JB. Effect of cold preservation on lymphocyte migration into peripheral nerve allografts in sheep. *Transplantation* **56**(1) (1993) 154–162.

86. Sun XH, Che YQ, Tong XJ, Zhang LX, Feng Y, Xu AH, Tong L, Jia H, Zhang X. Improving Nerve Regeneration of Acellular Nerve Allografts Seeded with SCs Bridging the Sciatic Nerve Defects of Rat. *Cell Mol Neurobiol* 2008.

87. Zhang LX, Tong XJ, Sun XH, Tong L, Gao J, Jia H, Li ZH. Experimental study of low dose ultrashortwave promoting nerve regeneration after acellular nerve allografts repairing the sciatic nerve gap of rats. *Cell Mol Neurobiol* **28**(4) (2008) 501–509.

88. Niu Y, Hu M, E LL, Liang J, Sun MX, Wan SX. [Preliminary studies on histological changes after repairing the facial nerve defect with acellular facial nerve]. *Zhonghua Kou Qiang Yi Xue Za Zhi* **42**(12) (2007) 723–725.

89. Du HD, Zhou L, Tian HB, Tian J. [Repair of facial nerve with bioactive artificial nerve conduit: experiment with rats]. *Zhonghua Yi Xue Za Zhi* **87**(46) (2007) 3302–3306.

90. Connolly SS, Yoo JJ, Abouheba M, Soker S, McDougal WS, Atala A. Cavernous nerve regeneration using acellular nerve grafts. *World J Urol* **26**(4) (2008) 333–339.

91. Kim BS, Yoo JJ, Atala A. Peripheral nerve regeneration using acellular nerve grafts. *J Biomed Mater Res A* **68**(2) (2004) 201–209.

92. Borschel GH, Kia KF, Kuzon WM, Jr, Dennis RG. Mechanical properties of acellular peripheral nerve. *J Surg Res* **114**(2) (2003) 133–139.

93. Haase SC, Rovak JM, Dennis RG, Kuzon WM, Jr, Cederna PS. Recovery of muscle contractile function following nerve gap repair with chemically acellularized peripheral nerve grafts. *J Reconstr Microsurg* **19**(4) (2003) 241–248.

94. Osawa T, Ide C, Tohyama K. Nerve regeneration through cryo-treated xenogeneic nerve grafts. *Arch Histol Jpn* **50**(2) (1987) 193–208.
95. Osawa T, Ide C, Tohyama K. Nerve regeneration through allogenic nerve grafts in mice. *Arch Histol Jpn* **49**(1) (1986) 69–81.
96. Hudson TW, Zawko S, Deister C, Lundy S, Hu CY, Lee K, Schmidt CE. Optimized acellular nerve graft is immunologically tolerated and supports regeneration. *Tissue Eng* **10**(11–12) (2004) 1641–1651.
97. Hudson TW, Liu SY, Schmidt CE. Engineering an improved acellular nerve graft via optimized chemical processing. *Tissue Eng* **10**(9–10) (2004) 1346–1358.
98. Liu XL, Arai T, Sondell M, Lundborg G, Kanje M, Dahlin LB. Use of chemically extracted muscle grafts to repair extended nerve defects in rats. *Scand J Plast Reconstr Surg Hand Surg* **35**(4) (2001) 337–345.
99. Arai T, Kanje M, Lundborg G, Sondell M, Liu XL, Dahlin LB. Axonal outgrowth in muscle grafts made acellular by chemical extraction. *Restor Neurol Neurosci* **17**(4) (2000) 165–174.
100. Sondell M, Lundborg G, Kanje M. Vascular endothelial growth factor stimulates Schwann cell invasion and neovascularization of acellular nerve grafts. *Brain Res* **846**(2) (1999) 219–228.
101. Sondell M, Lundborg G, Kanje M. Regeneration of the rat sciatic nerve into allografts made acellular through chemical extraction. *Brain Res* **795**(1–2) (1998) 44–54.
102. Graham JB, Neubauer D, Xue QS, Muir D. Chondroitinase applied to peripheral nerve repair averts retrograde axonal regeneration. *Exp Neurol* **203**(1) (2007) 185–195.
103. Neubauer D, Graham JB, Muir D. Chondroitinase treatment increases the effective length of acellular nerve grafts. *Exp Neurol* **207**(1) (2007) 163–170.
104. Krekoski CA, Neubauer D, Graham JB, Muir D. Metalloproteinase-dependent predegeneration *in vitro* enhances axonal regeneration within acellular peripheral nerve grafts. *J Neurosci* **22**(23) (2002) 10408–10415.
105. Zuo J, Neubauer D, Graham J, Krekoski CA, Ferguson TA, Muir D. Regeneration of axons after nerve transection repair is enhanced by degradation of chondroitin sulfate proteoglycan. *Exp Neurol* **176**(1) (2002) 221–228.
106. Pizzi MA, Crowe MJ. Matrix metalloproteinases and proteoglycans in axonal regeneration. *Exp Neurol* **204**(2) (2007) 496–511.

CHAPTER 10

Tendon Engineering

Johan Thorfinn,†,‡, Ioannis Angelidis*,†, Brian Pridgen*,†
and James Chang*,†*

Introduction

Tissue engineered tendon constructs would be of clinical utility in cases of extensive hand trauma or in certain clinical conditions, such as tumors, degenerative disease, and congenital malformations. Traumatic mutilating hand injuries often result in loss of tendons that must be replaced in the acute stage or at secondary reconstruction. Tumor resection in the hand and rheumatoid disease often leave the patient with a tendon deficit, either in regard to the actual tendons, their function, or both.

The only option in available practice today for the hand surgeon is to use autologous grafts from other parts of the body. This practice was described already in 1912 by Lexer, who presented a series of patients with free flexor tendon grafting. Four years later, Mayer published three articles describing the fundamentals of flexor tendon grafting.[1] Since then, others have modified and refined the techniques, but the main principles established by Mayer still apply.

The healing and outcome of a tendon repair is dependent on a multitude of factors. Examples of particular interest include the combination of multi-strand core sutures and an epitendinous repair, ensuring equal tension across all strands, and early post-operative mobilization to lessen the probability for scarring and adhesion and to improve the tensile strength of the repaired tendon.[2] These surgical principles are important when an autologous graft or tissue engineered construct is placed to restore a tendon defect.

In autologous transplantation, removing a healthy and functional tissue from one part of the body to repair another results in some donor site morbidity depending on the size and location of the harvested tissue. To minimize donor site morbidity, there are several tendons that are most commonly harvested for grafting. The most common donor tendon is the palmaris longus tendon when it is present (in 85% of the general population), but other options include the plantaris tendon and the extensor digitorum longus tendons in the lower extremity, or the extensor indicis proprius and the flexor digitorum superficialis tendons in the upper extremity.[3]

Ultrastructure of tendon

To understand the principles and the challenges of tissue engineering tendons of the hand, it is important to have an understanding of the ultrastructure of tendons. The main component of tendons is collagen, a protein comprising three interwoven strands. Collagen

---

*Corresponding authors.
†Department of Plastic & Reconstructive Surgery, Stanford University Medical Center, Stanford, CA, USA.
‡Department of Plastic Surgery, Hand Surgery, and Burns, University Hospital, Linköping, Sweden.

synthesis begins within the cell until it is secreted by exocytosis as pro-collagen. Extracellular pro-collagen peptidases then cleave the pro-collagen, and the resulting tropocollagen is arranged to form a coil that is linked by covalent bonds into fibrils that in turn form larger fibers. Tendons consist of mainly type I collagen, but types III and V are also present.[4-7]

Collagen is synthesized by several cell types, including fibroblasts and tenocytes. In the normal tendon, tenocytes are found within the tendon (endotenon). Tenocytes are located between the collagen fibers with focal attachments by gap junctions. They secrete collagen, glycosaminoglycans, cytokines, and enzymes, and they are responsible for the normal turnover of collagen and ground substance.[8] The collagen fibrils secreted by endotenon tenocytes organize into fibers and ultimately into larger fascicles. Several fascicles coalesce and are bound by surrounding loose connective tissue termed the epitenon. The epitenon contains a few fibroblast-like cells and sparse blood and lymphatic vessels.

Tendons can be either intrasynovial or extrasynovial. Intrasynovial tendons are surrounded by a sheath of synovial cells that provides a gliding surface for the tendon, unlike extrasynovial tendons, which lack this feature.[9,10] Extrasynovial grafts are inferior to intrasynovial grafts because they are more susceptible to forming adhesions and scars.[11] However, only a few of the available tendon grafts in humans are intrasynovial, an example of which is the flexor digitorum longus of the foot. These tendons are theoretically more attractive as a donor tendon than extrasynovial tendons, though the intrasynovial portion of this tendon is shorter than what might be needed to restore a flexor tendon deficit in the hand.[12] Thus, one of the aims of tissue engineered tendon grafts is to culture a pseudo-membrane consisting of sheath-like fibroblasts around the graft that mimics the properties of the actual synovial layer present in intrasynovial flexor tendons.[13]

Goals of tissue engineered tendon constructs

The ultimate goal of tissue engineering tendons is to provide easy access to tissue that resembles fresh autologous tendon tissue histologically and morphologically, closely matches the biomechanical properties of native tendon, and provides a clinical outcome comparable or superior to repairs done with autologous tendon grafts. Examples of current research addressing these goals include understanding the mechanisms of tendon healing and scarring, understanding the effect of different materials and cell types on biomechanical properties, and optimizing culture conditions for more rapid *in vitro* expansion and maturation of cells seeded onto tendon scaffolds.[14]

Most research in tendon engineering has been focused on flexor tendons, which is probably due to the fact that lack of these tendons (as opposed to extensor tendons) is more likely to constitute a greater challenge in clinical practice when it comes to reconstruction. Also, due to the anatomy of the native flexor tendons, it is easier to mimic these structures than the extensor tendons that are more complicated in its design in zones III and IV.

Despite this, several attempts have been made to engineer extensor tendons. Wang *et al.* successfully seeded human fetal tenocytes onto a polyglycolic acid scaffold that was shaped to imitate the central portion and the lateral bands of the finger (Figure 1). These constructs were subjected to *in vivo* loading by suturing them subcutaneously to muscle fascia in nude mice. Compared to unloaded controls placed subcutaneously, the loaded tendon constructs were significantly stronger and exhibited a significantly greater stiffness.[15]

Cell based strategies for tissue engineering tendons of the hand include *de novo* manufacturing and guided regeneration.[2,16] *De novo* manufacturing would allow clinicians

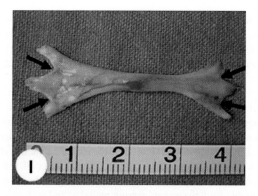

Figure 1. Tissue engineered extensor tendon construct. From Wang *et al.* "Engineering of extensor tendon complex by an *ex vivo* approach", *Biomaterials*, July 2008.

to control the production of a new tendon, dictating features such as its size, material properties, chemical composition, and embedded signaling molecules. This process would include using elements of native tendons, such as decellularized collagen material or its cells, to build a cell-cultured three dimensional construct. However, using such a tissue engineered tendon routinely will require the clinician and researcher to take several issues in consideration that are not the case with autologous grafts. For example, harvesting cells from the patient and culturing them *in vitro* prior to transplantation requires at least a two-step procedure, whereas the autologous graft is harvested at the time of reconstruction. This could be overcome by harvesting cells or tissue components from a donor ahead of reconstruction, but in turn the risk of disease transmission between individuals cannot be ruled out.

Guided regeneration describes manipulation of the *in vivo* environment to provide a more favorable milieu for host cell proliferation. This might be achieved by implanting a cell-free scaffold that promotes chemotaxis of cells from the recipient. This three-dimensional scaffold promoting neo-tendogenesis should facilitate cell-adhesion and growth, while providing sufficient mechanical support.[17] Another example of guided regeneration is placing cells that facilitate tendon healing at the site of injury. For example, Pacini *et al.* injected mesenchymal stromal cells from bone-marrow at a partial superficial tendon defect of the superficial digital flexor tendon in race horses. Horses receiving the cell-suspension injection healed faster upon examination with ultrasound compared to controls that received standard therapy for these injuries. The treated horses had not relapsed at two-year follow up in contrast to the control group, in which all horses were diagnosed again with the same condition after a median time of 7 months.[16]

Principles of Tendon Tissue Engineering

Successful design of a functional biologic substitute for human tendons will require selection of a pre-defined cell type and provision of optimal conditions for these cells to proliferate on a three-dimensional scaffold. This requires an understanding of the complex pathways that regulate proliferation, differentiation, maturation, remodeling, and interaction of tenocytes and sheath cells during normal conditions, healing, and remodeling. Several approaches to identify cells such as dermal fibroblasts or adipose-derived stem cells that are easy to harvest and culture have been attempted and have shown promising results. These cells must then be seeded onto a scaffold. A scaffold must be mechanically stable in order to withstand *in vivo* forces before complete healing has occurred, have low immunogenicity, and ultimately result in a construct that closely emulates native tendon.

Animal models

Although there have been no *in vivo* tendon tissue engineering experiments in humans, a number of animal models have been used. These include rabbits,[18,19] hens,[20] pigs,[21] dogs[11] and race horses.[16] These animals all exhibit a flexor tendon apparatus that resembles that of humans, with a deep and superficial tendon. The diameter of the tendons of these species differs from that of humans and varies with the size of the animal, but it does not make them less suitable for a tissue engineering model.

In our experience, rabbits are suitable as a tissue engineering model (Figure 2). Rabbits heal well and rarely have problems with infections, are easy to handle, and have a moderate inflammatory response.[22] Even when using allografts, the scarring and gross inflammatory response is low. Both the forepaw and rear paw deep and superficial flexor tendons can be accessed under anesthesia without difficulty. At the metacarpal and metatarsal level, these tendons are approximately 2–3 millimeters wide, which make them technically uncomplicated to suture under loupe magnification. The flexor tendon system on the digits of the forepaw is anatomically similar to humans. Though there are also similarities in the rear paw, the superficial flexor digitorum tendons merge at the level of the proximal metatarsals to form a single tendon that runs on top of the Achilles tendon, connecting to it with loose fibrous bands where the Achilles tendon inserts in the bony portion of the calcaneous. The deep flexor digitorum tendons also merge to form a dense, tendinous band at the same level as the superficial tendon, but runs medially to the calcaneous next to the neurovascular bundle that supplies the rear paw. The Achilles tendon of the rabbit has also been used as a tendon model by some investigators.[23,24]

Figure 2.   The rear paw of the New Zealand white rabbit with tissue engineered flexor tendon grafts in zone I-III. The grafts are sutured proximal to the metacarpophalangeal joints, and at the distal insertion (arrows).

Cell Types Used for Tendon Tissue Engineering

Two main cell based approaches can be identified: 1) the use of a stem cell or otherwise pluripotent cell type that may differentiate further into the desired target cell, or 2) a mature target cell. Several cell types have been successfully employed to engineer tendon constructs. Kryger *et al.* showed that there was no difference in collagen expression between rabbit epitenon tenocytes, tendon sheath fibroblasts, bone marrow-derived multipotent stromal cells and adipose derived multipotent stromal cells in culture. When each of these cell types was seeded onto acellularized flexor tendons and grafted into rabbits, the four cell types were shown to be similarly viable within the constructs for up to 8 weeks.[22]

Stem cells are pluripotent cells, which are defined by their self-renewal capacity and multilineage differentiation, and they are widely considered in tissue engineering.[25-27] The pluripotent stem cell is capable of forming new cells in all three germ lineages, which makes it ideal for tissue engineering purposes. These cells can be harvested from fetal tissues, but more mature fetal cells have also been used in tendon tissue engineering. For example, human fetal extensor tenocytes have been effectively cultured on a polyglycolic acid scaffold to create a construct with macroscopic and microscopic resemblance of the extensor tendon apparatus of a finger.[15] However, the use of fetal tissues is controversial regardless of whether the cells are pluripotent or are more mature. To avoid these ethical debates, the current research is aimed at finding stem cells in adult tissues or finding a way to reverse the differentiation process in already differentiated adult cells.[28] Several groups have used adult stem cell based strategies for tendon tissue engineering in animals. These adult stem cells have been harvested from various sources, such as bone-marrow, periosteum, and adipose tissue.[29] It has been suggested that most tissues contain stem cells that are capable of differentiating into other more mature tissues.[29,30] In the literature, the term *mesenchymal stem cell,* or MSC, is often used to refer to a cell type that originates from an adult individual and has the capability to differentiate into different mature cell types.

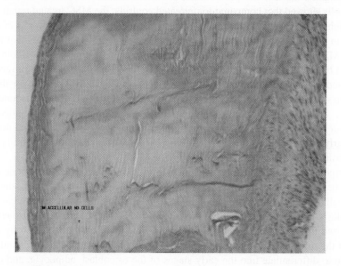

Figure 3. Acellularized rabbit flexor tendon explanted after 3 weeks *in vivo*. Note the lack of cells in the endotenon, but cells with small, rounded nuclei have populated the epitenon. These cells represent a mixture of inflammatory response of the host, and tenocyte-like fibroblasts.

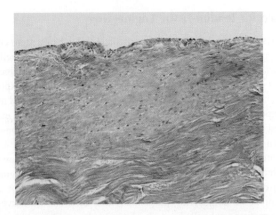

Figure 4.   Tenocyte re-seeded rabbit flexor tendon explanted after 10 weeks *in vivo*. The endotenon as well as the epitenon exhibits cells that are tenocyte-like.

Mesenchyme is by definition tissue from the developing embryo that later differentiates into numerous other types of tissues or organs, including cartilage, bone, blood cells, and epithelium. However, MSCs are found only within adult connective, or *stromal*, tissue and so the term *multipotent stromal cell* would be a better alternative. In tendon tissue engineering, it is suggested that MSCs: 1) differentiate into tenocytes that take part in remodeling and tissue repair, or 2) act as a plentiful source of growth factors that facilitate migration, differentiation and maturation of other cells into cell types that in turn facilitate healing.

MSCs are commonly used for tissue engineering purposes. Chong *et al.* isolated MSCs from iliac crest bone marrow of adult rabbits (bone marrow stromal cell, or BMSC) and investigated whether the presence of these cells enhanced Achilles tendon healing. Their results showed that compared to untreated controls, the MSC treated repair site had a 32% higher elastic modulus when tested biomechanically after three weeks of healing, but later in the healing process, this finding was not evident.[23] This implies that the presence of MSCs at the repair site facilitated more rapid healing. These findings were supported by Pacani *et al.*, who acquired MSCs from sternal bone marrow in race horses and applied a MSC cell suspension to partial lesions of the superficial flexor digitorum tendon. In nine out of ten animals the tendon had a greater density on ultrasound examination as early as one month following treatment. These horses continued racing throughout the two year follow-up period, unlike untreated controls, which did not heal or had an early relapse.[16] In addition, MSCs have been shown to have the capability to grow well on collagen scaffolds in a gel or sponge form.[14,31,32]

Pluripotent stem cells can be obtained from adult tissues other than bone marrow. Adipose tissue has been shown to contain undifferentiated cells (adipose tissue-derived stromal cells, or ATSC) with the capability to develop into other more mature tissues such as fibroblasts.[33] Compared to BMSCs, the advantage of ATSCs, sometimes also referred to as processed lipoaspirate (PLA) cells, is that they are more easily harvested by a simple biopsy of the subcutaneous tissue.[33] Lee *et al.* showed that both human ATSCs and BMSCs were able to differentiate into the early stages of the osteoblast, adipocyte, and chondrocyte lineages. ATSCs and the BMSCs showed similar characteristics on flow cytometric analyses. Interestingly, the ATSCs were able to maintain their proliferating capacity to a greater extent than the BMSCs.[34]

More mature cell types have also been widely used in tissue engineering. Liu *et al.* used pigs to compare dermal fibroblasts with tenocytes that were seeded onto a polygly-colic acid fiber scaffold and sutured to a defect in the flexor digitorum superficialis tendon. The constructs were very similar in strength (approximately 75% to that of natu-ral tendon) and histological appearance. The seeded fibroblasts, which typically express collagen III, began expressing collagen I, suggesting a possible change in phenotype; this is the same collagen type expressed by tenocytes. This study suggests that dermal fibrob-lasts may be a suitable substitute for tenocytes on seeded tendon constructs.[21] These results are supported by He *et al.*, who seeded fetal dermal fibroblasts onto a construct based on human amnion tissue encapsulating a core-sutured tendon defect of 1 cm in the rabbit Achilles tendon. They found that these fetal dermal fibroblasts grew well, and that the tensile strength of these constructs was 81% of that of normal tendon. Also, their overall biomechanical properties were better compared to controls with unseeded amnion encapsulation, as well as when compared to a standard polydioxanone (PDS) suture alone.[24]

Synovial cells have also been cultured *in vitro* in order to fabricate a construct that can be used as a biomembrane, which helps prevent adhesions and scarring in tendon repair. The rationale for this is that an intact tendon sheath provides better gliding proper-ties and that the synovial fluid produced by the synovial membrane supplies an extrinsic source of nutrients for the healing tendon. Ozturk *et al.* have started to explore this approach by harvesting and expanding synovial cells from the rodent Achilles tendon and knee joint. These cells were later seeded onto a collagen I matrix for 2 weeks. The results show that the synovial cells produced hyaluronic acid, providing evidence of functional synovial cells.[13] Despite the fact that harvesting synovial cells is an invasive procedure and leaves the donor animal with a seemingly serious tendon and joint capsule defect, these are promising results that need to be elucidated further.

Scaffold Materials

Numerous scaffold types have been tried for tissue engineering of tendons, but no scaffold material to date has proven to be ideal. An ideal scaffold for tissue engineering of tendons would 1) be derived from cytocompatible material that allows remodeling after implanta-tion with a controllable rate of biodegradability, 2) consist of a micro-architecture that allows the seeded cells as well as the appropriate migrating host cells to attach, expand and remodel after transfer, 3) have biomechanical characteristics that withstand the *in vivo* post surgical forces until healing or remodeling have occurred, 4) for artificial scaffolds, be manufactured to the desired size and shape, 5) be devoid of cellular components from the donor individual (applies to biological scaffolds), whether allogenic or xenogenic, to avoid an excess inflammatory response and graft versus host-reactions.[17,35]

In tendon tissue engineering, two major groups of scaffold materials can be identi-fied: 1) biological or biomimetic scaffolds and 2) synthetic scaffolds. The former can be further divided into four different types: polysaccharides, polyesters, collagen derivates, and biological collagen scaffolds.[17,36]

Polysaccharides

Chitosan, alginate, and hyaluronan, are all polysaccharides that have been tried in tissue engineering, either alone or in combination. These substances mimic some components of the extra-cellular matrix, such as the glycosaminoglycans. Thus, they are also thought to exhibit similar biofunctionality in attracting growth factors, promoting cell adhesion, and interacting with receptor proteins.[17,37]

Chitosan is a cationic, linear polysaccharide. Due to its hydrophilic properties, it promotes better cell proliferation and adhesion than some of its polyester counterparts that are hydrophobic. It is also known to evoke a minimal foreign-body reaction, and it can be formed to porous structures that promote cellular ingrowth.[17] Majima *et al.* implanted fibroblast seeded chitosan constructs in a rotator cuff defect in rabbits. The constructs were seeded for 4 weeks prior to the surgery, and the animals were sacrificed at 4 and 12 weeks following implant of the seeded chitosan constructs. The results showed that the seeded constructs had more collagen I as measured with immunohistochemical stains, and a better tensile strength and tangent modulus than the non-seeded constructs at 12 weeks.[38,39] Bagnaninchi *et al.* seeded porcine tenocytes on a porous chitosan scaffold and managed to have the cells to attach and grow. Unfortunately these constructs were not tested biomechanically.[40]

Hyaluronan, which is an important constituent of the extra-cellular matrix, and alginate, are two anionic polysaccharides that can be combined with chitosan to further enhance the biomechanical and biological properties of the polysaccharide scaffold. There is an electrostatic interaction between these anionic molecules and the cationic chitosan. Funakoshi *et al.* demonstrated that hyaluronan coated chitosan scaffolds used in tendon tissue engineering, when compared to uncoated chitosan scaffolds, have increased cell adhesion, proliferation, collagen production, and tensile strength.[41]

Polysaccharides play an important role in cell signaling and the immune system. When these signaling mechanisms are better understood, polysaccharides may serve an important role as scaffolds in tissue engineering of tendons by promoting cell adhesion and proliferation.[17]

Polyesters

Polyester scaffolds, which are made of polymers such as polylactic acid and polyglycolic acid, have shown great promise; they provide good mechanical stability and are easily manufactured. Within the body, these materials are degraded into lactic acid and glycolic acid, both of which are naturally present. In a study by Lu *et al.* comparing scaffolds of polyglycolic acid, poly-L-lactic acid, and polylactic-co-glycolic acid, the polyglycolic acid scaffold had the highest biomechanical strength after one week of *in vitro* culture. However, following the first week, the polyglycolic acid scaffold rapidly lost its strength and failed. The authors concluded that poly-L-lactic acid had better overall performance. Additionally, they found that the seeded cells (rabbit anterior cruciate ligament cells) attached and grew better on this scaffold.[42] It is noteworthy to point out that degradation of the polyesters may result in metabolites (acids) that have systemic or local effects on the host.[17]

Polyesters have been tried in several animal experiments. In a study by Cao *et al.*, a polyglycolic acid fiber construct was seeded with avian tenocytes and put under constant tension during culture. This experiment resulted in a tendon construct with newly synthesized collagen fibers, although it was too narrow in diameter to be used to repair a tendon defect.[43] The same group also seeded a polyglycolic unwoven fiber construct with hen tenocytes to bridge a flexor digitorum profundus defect in the hen. Compared to unseeded constructs of the same material after 14 weeks, the tissue engineered tendon was stronger (83% of normal tendon) than the unseeded construct (9% of normal tendon), and they were nearly indistinguishable from normal tendon macroscopically and histologically.[20] Yokoya *et al.* used a polyglycolic acid scaffold to repair an infraspinatus tendon defect in the rabbit.

The contralateral rotator cuff was either repaired with a poly-L-lactate-ε-caprolactone absorbable material or left with a similar, but unrepaired defect. Although none of the scaffolds were seeded with cells, histology of the polyglycolic acid scaffold demonstrated a population of cells organized in a fibrocartilage layer. The other two groups had little or no histological resemblance to normal tendon-to-bone insertion. The polyglycolic acid repair also had a greater strength than the other groups. None of the scaffolds in this experiment were seeded with cells, suggesting that polyglycolic acid functions as an attractant for cells and promotes cell adhesion.[44]

Recently, electrospinning of nano- and microfibers (diameter 3 nm–6 μm, and up to several meters in length) has emerged as a new technique for polymer scaffold production. The electrospinning of nanofibers provides a material with a high surface area-volume ratio, variable pore size, and high porosity, that mimics the nanostructure of extracellular matrix of natural tissue, and such a surface is known to aid cell adhesion, enhance proliferation, and facilitate deposition of extra cellular matrix. Efforts have been made to coat the nanosurfaces with bioactive molecules, in order to further modulate cellular behavior.[45–47]

Sahoo and colleagues tested a polyester scaffold (D,L-lactide–co–glycolide) composed of a knitted base that provided mechanical support covered by an electrospun nanofiber surface of the same polyester. This construct was seeded with porcine BMSCs and compared to constructs of the same polyester without electrospun nanofibers coating the surface. The results showed that cells in the nanofiber group proliferated faster and that the cells of these constructs expressed more collagen I, decorin, and biglycan genes.[47]

Collagen

The native tendon is composed mainly of collagen I, and thus, some argue that it is desirable to use a collagen scaffold to make tissue engineered tendons. Collagen is known to have excellent biocompatibility and better support for cell proliferation and cell adhesion than other materials.[17] There are two major approaches to using collagen as a scaffold for tendon tissue engineering: 1) *de novo* manufacturing of collagen scaffold derivates, or 2) the use of biological collagen scaffolds that originate from native autologous or allogenic tendons.

Collagen derivates

The structure of collagen fibers is complex, which makes it difficult to manufacture collagen constructs with consistent quality and properties. Gel-based collagen scaffolds have been tried, but they have a poor mechanical strength, which make them less suitable in tendon tissue engineering than other materials. However, by introducing collagen fibers into the gel and freeze-drying the mixture, sponge-like collagen scaffolds have been produced, which have been shown to exhibit greater mechanical stability than gel-based collagen scaffolds alone.[18,31] Nirmalandhan *et al.* studied how the structure of a collagen scaffold is related to the response of the construct to mechanical stimulation. They showed that there was a significant increase in the linear modulus and linear stiffness of the collagen-sponge scaffold compared to the collagen-gel scaffold when these constructs were seeded with MSCs and subjected to cyclic strain for 2 weeks. This might be explained by the greater collagen content in the collagen-sponge constructs.[31,32]

Studies have demonstrated that the three-dimensional structural properties of the scaffold play an important role in determining the spatial growth of the seeded cells and their reaction to mechanical stimulation. In a series of experiments using MSC seeded constructs, longer collagen scaffolds (51 mm) subjected to cyclic loading had a significantly higher linear stiffness *in vitro* than shorter scaffolds (11 mm) after 14 days in a culture. This is independent of the collagen content in the scaffold.[31] Although it is not known why this phenomenon occurs, this finding suggests that it is important to make the scaffold longer than the defect.

Biological collagen scaffolds

Rather than manufacturing artificial tendon scaffolds, some investigators are using scaffolds made from acellularized tendons harvested from humans or animals. The collagen structure and supportive framework of a native tendon offers superior cell adhesion and structural integrity when compared to artificial tendon scaffolds. However, this approach presents unique challenges. Using harvested tendons would limit the supply to the number of willing donors as well as introduce variable quality between individuals, neither of which would be a concern with manufactured artificial tendon scaffolds. An additional problem is that, unlike an artificial scaffold, a native tendon is populated with host tenocytes and other cellular material, which might elicit an immunogenic reaction in an allograft recipient. Thus, the native tendon must first be acellularized before being re-seeded with cells as with other scaffold materials. Acellularization can be achieved by a number of means, including washes in hypotonic and hypertonic solutions, trypsin, and detergents as well as freeze-thawing. The acellularization process inevitably affects the structural components of collagen. This might render the tendon somewhat weaker compared to native, fresh tendon.[35,48,49,50]

Despite the initial weakening due to acellularization, acellularized tendons re-seeded with cells become stronger, unlike un-seeded acellularized tendons. In a study by Chong *et al.*, acellularized constructs re-seeded with tenocytes returned to normal elastic modulus and ultimate tensile strength.[50] This data was confirmed by Saber *et al.* in a similar study.[48]

Using an acellularized tendon as a scaffold is still appealing due to its structural composition, and because it is superior to manufactured scaffold materials. One drawback with this strategy, though, is that the acellularized tendon has to be re-seeded with the patient's own cells to improve in strength before transplantation. Harvesting, expanding *in vitro*, and re-seeding the acellularized scaffold would take 2–4 weeks. In major reconstructive trauma cases, this would require multi-stage surgical procedures.

Biological scaffolds other than acellularized tendons have been tried in tissue engineering of tendons. Submucosa from porcine small intestine has been tested because it contains a large percentage of collagen I, as well as a small amount of growth factors and chemotactic factors.[51,52,55] Liang and colleagues confirmed that the small intestine porcine submucosa works well in a rabbit medial collateral ligament injury model as a tendon substitute. They found that the scaffold expressed larger collagen fibrils after 6 weeks compared to a group of animals that had an uncovered tendon defect.[51]

Another example of a biological scaffold is human amnion tissue. He *et al.* seeded this construct with fetal fibroblasts in a rabbit model and draped the construct around an Achilles tendon defect. After 3 months, the fibroblasts had organized in a parallel or radiating fashion in the amnion matrix scaffold.[24]

In addition, fibrin, fibronectin and silk can be used as scaffold materials.[36,53] Silk is a protein synthesized by worms or spiders that is stronger than collagen and would therefore be a potential candidate for a tendon scaffold.[54] Takezawa *et al.* studied scaffolds composed of silk and collagen type I seeded with fibroblasts. *In vitro* testing has shown that the fibroblasts grow well on these scaffolds, though there have not yet been any *in vivo* studies.[56]

Synthetic scaffolds

Synthetic materials to date have not proven successful in tendon tissue engineering.[57,58] Synthetic scaffolds have several disadvantages over biological scaffolds. Firstly, biodegradation of synthetic scaffolds can result in production of cytotoxic and antimitotic break-down products.[59] Secondly, the characteristics of a biological scaffold or bioderivative material are difficult to emulate with a synthetic material. The challenge is finding a synthetic biocompatible material with tendon-like biomechanical properties that allows cellular in-growth and remodeling, as well as a minimal inflammatory reaction and scar formation. Despite numerous attempts, synthetic materials have not been widely adopted in tendon tissue engineering due to a lack of desired characteristics and the superiority of biological scaffold materials.[60–62]

Actions to Improve Results

Tissue engineering of tendons is a field that involves different aspects, such as choosing a cell type and appropriate scaffold material. Additional measures can also be taken in order to enhance the end result.

Loading

Uniaxial loading of engineered tendons has proven to be effective for enhancing the characteristics of a tendon construct.[48,63–65] *In vitro* loading can either be applied by subjecting cell culture dishes to stretching or by placing a seeded tendon construct in cell medium within a device that stretches the tendon at defined intervals.[48,66] *In vivo* loading can be done by implanting a tendon construct in a freely moving animal or by exercising the native tendon before removal and analysis.[15,67]

It has been shown that tenocytes respond to mechanical changes in their environment by altering their metabolism, which results in changes of the mechanical and structural properties of the tendon.[11,68] Collagen synthesis by tenocytes is controlled by a mechanosensitive process that is sensitive to the frequency, duration, and intensity of the force applied to the tendon. For example, twenty-four hours of *in vitro* loading of rodent tendon fascicles at 1 Hz resulted in significantly more collagen synthesis compared to 10 min of loading during the 24 h period.[69] Constant *in vitro* loading of a polyester scaffold seeded with tenocytes yielded a strong but narrow construct with highly compacted collagen fibers. Even though constant loading in this study enhanced tissue maturation, these results suggest that constant loading is less beneficial than cyclic loading.[43] Cyclic strain of reseeded acellularized tendons improves ultimate tensile stress and elastic modulus compared to unloaded constructs, resulting in tendons with similar strength to fresh tendons. This cyclic strain causes the tenocytes to become organized parallel to the collagen fibrils as opposed to unloaded tendons where the cells remain unaligned throughout the tendon.[48] This has also been shown by Riboh *et al.* for other cell types, such as BMSCs and ATSCs, seeded onto acellularized tendons that underwent cyclic strain.[70]

Nirmalanandhan *et al.* studied the effect of the scaffold material on the linear stiffness and linear modulus of tendon constructs after mechanical stimulation. A collagen-sponge construct seeded with MSCs was stretched every 5 minutes for 8 hours, and compared to a collagen-gel construct that was similarly seeded. Seeded, unstimulated sponge and gel constructs were used as controls. The results showed that the unstimulated sponge and gel constructs had the weakest linear stiffness and modulation, and that the collagen-sponge constructs had a three-fold stronger linear stiffness and modulus compared to the stimulated collagen-gel constructs.[32]

Tenocytes in native avian flexor tendons subjected to loading *in vivo* on a treadmill for 1 hour and 45 minutes have been shown to upregulate genes that code for certain proteins known to be important for flexor tendon biology. Genes coding for collagen I and cysteine protease inhibitors were upregulated, the latter thought to be important in tendon remodeling.[67] In another study by the same group, avian flexor tendon tenocytes in cell culture that were mechanically stimulated at a frequency of 1 Hz for 8 hours had a two to three-fold increase in DNA synthesis compared to unloaded dishes when PDFG-BB was added to the medium.[66] Interestingly, there was not such a big difference between loaded and unloaded dishes without the growth factor supplementation.

Loading of a tendon construct not only elongates the scaffold but also induces compression as the construct narrows. Spalazzi *et al.* have shown that a contracting nanofiber collar (15% for 24 hours) around bovine patellar tendon kept for up to 2 weeks decreases the number of tenocytes in the tendon, but on the other hand results in a more organized extra-cellular matrix with smaller fibers than uncompressed controls. The compressed tendons also expressed up-regulation of fibrocartilage markers.[71]

Perfusion bioreactors have been used as another means of adjusting the growth environment for cultured tendon constructs. Bagnaninchi *et al.* oriented a porous chitosan scaffold parallel to the direction of fluid flow through a perfusion bioreactor during cell culture. They found that this treatment increased the cell occupation ratio compared to identical scaffolds cultured in static medium.[40] This finding is likely due to the perfusion bioreactor providing enhanced delivery of nutrients and removal of waste products.

Molecular Enhancement

Another area of great promise in tendon regeneration and healing is the study of the complex interactions between growth factors, which influence pathways of growth, differentiation and proliferation.[72] With recent advances in the molecular sciences, it has become possible to identify these growth factors and their functions and to produce them for use in controlled experiments.

Growth factors refer to a diverse group of proteins that are produced by various cell types and are used as autocrine or paracrine signal molecules in response to external stimuli. Growth factors typically bind to a membrane bound receptor on the target cell that triggers an intra-cellular reaction by means of phosphorylation, which modulates the response of the target cell.

Understanding the healing process of normal tendons is the first step in understanding how the application of growth factors can be utilized in tissue engineering of tendons. Current research has focused on basic fibroblast growth factor (bFGF), transforming growth factor beta (TGF-β), platelet derived growth factor (PDGF), insulin-like growth factor one (IGF-1), and vascular endothelial growth factor (VEGF).[72–80]

bFGF

Basic fibroblast growth factor (bFGF) is a 146 amino acid peptide known to modulate embryonic development, angiogenesis, and wound healing. It typically binds to extra-cellular and cell membrane bound heparan and heparan-sulfate proteoglycans. Fibroblasts, endothelial cells, smooth muscle cells, mast cells, and chondrocytes, produce bFGF. It has a variety of effects on each of its target cells. In fibroblasts, bFGF stimulates the production of collagen and initiates the formation of granulation tissue in early healing. Duffy *et al.* showed that bFGF is present in normal intrasynovial flexor tendons. Chang *et al.* demonstrated that bFGF was increased in the healing flexor tendon and sheath up to 8 weeks after repair. When applied exogenously to tendons, bFGF induced wound closure *in vitro*, most likely by induction of fibroblast proliferation. An *in vitro* experiment by Chan *et al.* confirmed this finding, and the authors concluded that the effect is most likely due to modulation of proliferation rather than chemotaxis.[72,75,77,78]

TGF-β

Tranforming growth factor-β has systemic effects in humans. TGF-β exists in three different isoforms (β1-β3), and when bound to any of its three membrane receptors (RI-III), it starts an intracellular cascade that involves the activation of serine-threonine kinase activity. The presence of TGF-β results in chemotaxis, extracellular matrix deposition (collagen I), and angiogenesis. In wound healing, the main source of TGF-β is from platelet degranulation, although it is secreted by nearly all cell types involved in healing such as lymphocytes, macrophages, endothelial cells, smooth muscle cells, and fibroblasts. TGF-β1 mRNA has been found at increased levels in the healing tendon sheath as well as the tendon itself, especially in proximity to the epitenon.[73] It has been shown to promote healing but also scarring and fibrosis as a result of disorganized collagen excretion.[81] At these sites of increased mRNA expression, the corresponding TGF-β receptors on fibroblasts are also increased.[82] In a study by Zhang *et al.*, neutralizing TGF-β antibody was applied to rabbit flexor tendon cell cultures, which resulted in decreased deposition of collagen and TGF-β bioactivity.[74] In a study by Chang *et al.*, TGF-β neutralizing antibodies were applied to surgically repaired rabbit patellar tendons. Following repair the treated group had an increased range of motion compared to controls that did not receive the neutralizing antibody.[83] These studies demonstrate the important role of TGF-β in tendon healing, and a possible role in the process of scar and adhesion formation in and around healing tendons.

PDGF

In 1974, platelet derived growth factor (PDGF) was thought to be a protein that stimulated proliferation of smooth muscle cells and fibroblasts.[84] In recent years, however, there has been much focus of the effect of PDGF on wound healing. Topical PDGF is commercially available in the United States (becaplermin, Johnson & Johnson, Raritan, NJ) to improve healing of diabetic foot ulcers.[85] PDGF is an initiator of the events in early wound healing that later result in fibroblast proliferation, angiogenesis, and collagen deposition.[86] PDGF consists of two polypeptide chains (A and B) that make up the PDGF macromolecule, and the three combinations of these (AA, BB, and AB) form three different isoforms of PDGF. PDGF-BB binds to the α receptor, and all three forms, including the BB-form, bind to the α receptor. Both of these receptors are of the tyrosinase kinase type.[87] As the name suggests, PDGF is secreted by platelets, but other cell types such as keratinocytes, fibroblasts, endothelial cells, and macrophages also secrete PDGF.[88] Its primary effects on smooth muscle cells, fibroblasts, and endothelial cells, are that it induces mitosis and stimulates the formation of hyaluronic acid and fibronectin.[86] The BB isoform promotes fibroblast proliferation and differentiation, collagen deposition, and angiogenesis. PDGF has similar effects in tenocytes. It stimulates tenocyte collagen deposition, as well as DNA synthesis and proteoglycan formation.[72] Duffy *et al*. have shown that PDGF is increased in healing canine flexor tendons compared to uninjured flexor tendons.[74]

IGF-1

Insulin-like growth factor-1 plays a major role in fetal development and in wound healing. In regards to wound healing, evidence suggests that IGF-1 expression is dependent on PDGF. Some of the observed effects of PDGF in wound healing may even be IGF-1 mediated.[89] IGF-1 is a single chain polypeptide that shares approximately 50% amino acid homology with insulin and the protein IGF-2.[90] Growth hormone, sex steroids, and parathyroid hormone have all been shown to regulate the levels of IGF-1. It can have autocrine, paracrine, or endocrine effects. Circulating IGF-1 is secreted mainly by the liver and is primarily bound in serum to six different IGF binding proteins, while the locally acting IGF-1 exists in its free and unbound form.[91] Many different cell types are known to secrete IGF-1, including fibroblasts, osteoblasts, and keratinocytes.[92] Because of its mitogenic and anabolic properties on a large number of cell types, it is not surprising that it is involved in the stages of wound healing in general.[72] IGF-1 is present in both epitenon and endotenon of uninjured avian tendons. A vibration-induced injury model of the Achilles tendon showed the presence of IGF-1.[93,94] Dahlgren *et al*. showed in a study on equine flexor digitorum superficialis tendon that the levels of IGF-1 decreased to 40% of normal levels during the first two weeks following a collagenase-induced injury. It was not until after 4 weeks that the IGF-1 levels were increased, and it remained elevated until 8 weeks post-injury.[89] In an uninjured rabbit tendon model, exogenously applied IGF-1 resulted in cell proliferation and an increased synthesis of collagen, proteoglycans, and DNA.[95] These effects may one day be used to promote more rapid healing of injured tendons. In a tendinitis model in horses it has been shown that an intratendinous injection of IGF-1 results in smaller lesions compared to saline-injected controls evaluated with ultrasound 3 weeks after injury. Additionally, cell proliferation and collagen content were also higher in the treated group.[96]

VEGF

Vascular endothelial growth factor is known to be secreted by tumor cells, but it also plays an essential role in wound healing, retinal angiogenesis, and embryonic cardiovascular development. In wound healing VEGF is important for neoangiogenesis and for regulation of other growth factors such as bFGF, TGF-β, and PDGF.[72,97] This regulatory role is mediated by VEGF binding to heparan sulfate proteoglycans on the extra-cellular matrix and cell surface, resulting in the release of proteoglycan-bound growth factors.[98] The local effects mediated by VEGF occur after binding to one of its three receptors on the cell membrane. These effects include increased expression of α–integrins, induction of enzymes (collagenases and gelatinases) that break down the vascular basement membrane, increased permeability and vascular vasodilation, and endothelial cell proliferation.[72] All of these effects promote formation of new vessels. Neoangiogenesis is a vital part of wound healing, and consequently, VEGF has been shown to improve both ischemic wound healing when applied topically and re-endothelialization after vessel injury.[99,100] Similarly, VEGF is necessary for the formation of new vessels in tendons. In injured Achilles tendons the VEGFR-1 receptor was expressed in the microvessels of the tendon, and the level of VEGF was increased.[101] Bidder *et al*. have shown that VEGF mRNA levels are increased in the canine tendon after an injury, with the highest concentrations being at the repair site and the lowest concentrations in the epitenon at the periphery.[102] In another rodent study by Zhang *et al.*, repaired Achilles tendons injected with VEGF proved to have greater tensile strength compared with uninjected controls.[103] However, Wang *et al.* found little effect on collagen production in proliferating tenocytes *in vitro* when transfected with the VEGF gene. However, they did find that expression of the TGF-β gene increased, suggesting that VEGF might contribute to scarring and adhesion by increasing levels of TGF-β.[104]

Multi growth factor approach

Wound healing, including that of tendons, involves a complex interaction between different growth factors secreted during different stages of the repair process. The understanding of this cascade of events in tendon healing is not completely understood. Although many of the studies focus on examining one single growth factor and its role in tendon wound healing, there have been some studies using multiple growth factors. Costa *et al.* have investigated the effects of IGF-1, bFGF and PDFG-BB on rabbit flexor digitorum tendon cells alone and in combination. The results showed that each individual growth factor had a positive effect on proliferation, but an even greater effect in combination. The greatest effect (251% to that of unsupplemented controls) was seen on sheath cells with a combination of all three growth factors.[105]

Even with a study design like this, the conditions are not the same as during physiological healing or regeneration. Thus, there have been attempts to better reproduce the optimal combination of growth factors involved in governing the differentiation and proliferation of cells involved in tendon healing. It has been done by subjecting cultured cells *in vitro*, or healing tendons *in vivo*, to platelet-rich matrices that contain a mixture of the growth factors present in platelets. The hypothesis is that this better resembles the true composition of growth factors present during tendon wound healing. The drawback of this approach, however, is that it is more difficult to control and study the sequence of events and growth factors interactions. Moreover, the growth factor cocktail contained in the platelet-rich matrix corresponds to one composition at a certain time point when it is harvested, but during the dynamic process of wound healing, the composition of growth factors is most likely to change over the course of time. In a study by Sanchez *et al.*, athletes with a ruptured Achilles tendon who were undergoing open surgery received a treatment with a platelet-rich matrix containing IGF-1, TGF-β, PDGF, and VEGF, combined with epidermal growth factor and hepatocyte growth factor. They concluded that this treatment had a positive effect on range of motion, complication rate, and the time to resuming training activities compared to untreated controls.[106] However, it is not known by what mechanisms a platelet-rich matrix promotes healing, as demonstrated in a study by de Mos *et al.*, in which either platelet-rich clot releasate (PRCR) containing PDGF and VEGF or platelet-poor clot releasate (PPCR) was used to treat cultured human tenocytes. The results surprisingly showed that both PRCR and PPCR promoted cell proliferation and increased total collagen production. Metalloproteinases, enzymes that degrade extracellular matrix and promote healing by increasing matrix remodeling, were upregulated in the PRCR group but not in the PPRC group.[107]

Gene transfer

It would be more desirable to have the cells involved in the healing or regenerating process to secrete the growth factors, rather than administering them artificially. This might ensure a stable and long-term delivery of the growth factors, instead of a peak at the time of delivery followed by a decline in concentration. By modifying or adding specific genes, the sequence of remodeling or healing can be controlled and changed. Examples of potential targets include cell surface receptors, extra-cellular matrix components, enzymes, or growth factors. Although initial results have been promising, use of this technology remains far from clinical practice.

Transfer of a novel gene into the existing DNA of a cell can be done either by the help of viral vectors or by non-viral transfer. Viral vectors are viruses that are no longer able to replicate and which have the desired gene inserted into the viral DNA.[57] The virus infects the cell and delivers genetic material into the nucleus, and the protein for which the gene encodes is expressed by the cell. This is the most common mode of gene transfer. Examples of non-viral transfer include direct injection, microseeding, or liposome mediated transfer. Liposomes are small membrane surrounded particles that are taken up by the cells. Although they are less effective than viral vectors, liposomes do not contain pathogenic viral DNA in addition to the genetic material intended for gene transfer.[2]

Cells can be transfected *in vivo*, or they can be extracted from the target tissue, expanded, transfected *in vitro*, and re-implanted. The first method is simpler and less costly, but it is less effective and more difficult to control. With this method, not only are the target cells transfected but also other adjacent cell types, which might have unforeseen consequences. Also, the efficiency of gene transfer is not easy to control. On the other hand, extracting the target cells and transfecting them *in vitro* enables greater control of the process, but this involves additional steps and a higher level of sophistication.[2,57]

Adenovirus mediated transfer of genes has been tried successfully in a rabbit flexor tendon model. Mehta *et al.* injected rabbit forepaw and rear paw flexor digitorum tendons with a solution of adenoviruses that contained genes coding for green fluorescent protein and bone morphogenic protein 13. Twelve days following injection, the animals were terminated, and the tendons were harvested. The results showed expression of the transgenic genes in the tenocytes and that the expression was dose dependent to the viral titer.[108] Rat tendon fibroblasts have been successfully transfected by a retrovirus containing genes that encode for PDGF or IGF-1, in order to stimulate collagen deposition and DNA synthesis, and then seeded onto a polymer scaffold and cultured *in vitro*. This construct was later used as to repair a torn rotator cuff in the rat. Subsequent analysis showed that the repair was nearly complete with normalized histology exhibiting longitudinally aligned collagen bundles.[109]

In an attempt to lessen the probability for scarring, Basile *et al.* have developed a mouse flexor digitorum longus model in which freeze-dried tendon grafts have been loaded with an adenovirus that contains a gene that encodes for growth and differentiation factor 5 (Gdf5). Deficiency of Gdf5 has been shown to delay healing of mouse Achilles tendon injuries, and Gdf5 treatment of Achilles tendons in rats results in stronger tendons. The results showed that the Gdf5 gene could be transferred successfully to the seeded cells, and that the tendons had better gliding properties and flexion range at the metatarsophalangeal joint compared to control tendons that were transfected with the same virus but without the Gdf5 gene.[49]

Gene transfer might very well be a standard in tendon tissue engineering in the future, but many challenges remain, including identifying genes that are relevant in tendon wound healing and finding the safest and most effective way to distribute these genes to their designated targets with minimal side effects.

Future Perspectives

Although impressive progress and optimistic results have been achieved in the field of tendon tissue engineering and guided regeneration, many challenges await and much work remains to be done.[57] It is likely that a combination of different strategies must be embraced in order to optimize the tissue engineered tendon.[65] These strategies include growth factor enhancement, mechanical modulation of the scaffold, and gene delivery, among others.

Further advances in this field will be needed to fully understand the complex inter-actions, and several questions are yet to be answered. What are the optimal ratio and concentrations of growth factors? When during the regeneration process is it ideal to apply them? How can slow release of growth factors be controlled at the site of repair, and are there any systemic effects? When is the best timing for the delivery of growth factors or gene transfer that modulates growth and maturation of the involved cell types to promote healing and lessen the inflammatory response and scarring? Moreover, some growth factors have both upregulating and downregulating effects on wound healing, which makes it difficult to use them to promote healing. However, advances in technology during the last decade, such as the ability to produce custom recombinant proteins, promises to further advance the research on growth factors.

Regarding the choice of the most appropriate cell type for tendon tissue engineering, there is no conclusive data to favor either the stem-cell approach or the mature cell approach. An ideal candidate would be a multipotent cell that is easy to identify, readily accessible for minimally invasive harvesting, easily cultured, and capable of differentiat-ing into both a tenocyte and sheath-like type cell *in vitro* and *in vivo*.

As for scaffold materials, the development of nanotechniques will open new opportunities to improve scaffold strength even further. Rather than having to choose between manufactured or biological scaffolds as is done today, an acellularized tendon might be improved by using nano-coating techniques to apply custom made, extra-cellular matrix proteins and growth factors to enhance cellular growth and maturation.

Once an ideal tendon with acceptable biomechanical and healing properties is tissue engineered, additional challenges in the field will include integrating this tendon in a more complex biomechanical environment together with the skeletal framework and muscle and improving the interaction with the tendon sheaths and gliding surfaces. Moreover, it is necessary to show not only short-term clinical results, but also the safety in using the tissue engineered tendon in a longer perspective. This is important since the process most likely will involve gene transfer, growth factors and cultured cells before these techniques can be commercially available for human use in the clinical setting.

References

1. Mayer L. The physiological method of tendon transplantations III. Experimental and clinical perspectives. *Surg Gynecol Obstet* **22** (1916) 472.
2. Luo J, Mass DP, Phillips CS, He TC. The future of flexor tendon surgery. *Hand Clin* **21** (2005) 267–273.
3. White WL. Tendon Grafts: a consideration of their source, procurement and suitability. *Surg Clin North Am* **40** (1960) 403–413.
4. Kannus P. Structure of the tendon connective tissue. *Scand J Med Sci Sports* **10** (2000) 312–320.

5. Cen L, Liu W, Cui L, Zhang W, Cao Y. Collagen tissue engineering: development of novel biomaterials and applications. *Pediatr Res* **63**(5) (2008) 492–496.
6. Traub W, Yonath A, Segal DM. On the molecular structure of collagen. *Nature* **221**(5184) (1969) 914–917.
7. Wess TJ, Hammersley AP, Wess L, Miller A. Molecular packing of type I collagen in tendon. *J Mol Biol* **275**(2) (1998) 255–267.
8. McNeilly MC, Banes BA, Benjamin M, Ralphs JR. Tendon cells *in vivo* form a three dimensional network of cell process linked by gap junctions. *J Anat* **189** (1996) 593–600.
9. Lundborg G, Myrhage R. The vascularization and structure of the human digital tendon sheath as related to flexor tendon function. An angiographic and histological study. *Scand J Plast Reconstr Surg* **11**(3) (1977) 195–203.
10. Uchiyama S, Amadio PC, Coert JH, Berglund LJ, An KN. Gliding resistance of extrasynovial and intrasynovial tendons through the A2 pulley. *J Bone Joint Surg Am* **79**(2) (1997) 219–224.
11. Seiler JG 3rd, Chu CR, Amiel D, Woo SL, Gelberman RH. Autogenous flexor tendon grafts. Biologic mechanisms for incorporation. *Clin Orthop Relat Res* **354** (1997) 239–247.
12. Seiler JG 3rd, Reddy AS, Simpson LE, Williams CS, Hewan-Lowe K, Gelberman RH. The flexor digitorum longus: an anatomic and microscopic study for use as a tendon graft. *J Hand Surg [Am]* **20**(3) (1995) 492–495.
13. Ozturk AM, Yam A, Chin SI, Heong TS, Helvacioglu F, Tan A. Synovial cell culture and tissue engineering of a tendon synovial cell biomembrane. *J Biomed Mater Res A* **84**(4) (2008) 1120–1126.
14. Juncosa-Melvin N, Shearn JT, Boivin GP, Gooch C, Galloway MT, West JR, Nirmalanandhan VS, Bradica G, Butler DL. Effects of mechanical stimulation on the biomechanics and histology of stem cell-collagen sponge constructs for rabbit patellar tendon repair. *Tissue Eng* **12**(8) (2006) 2291–2300.
15. Wang B, Liu W, Zhang Y, Jiang Y, Zhang WJ, Zhou G, Cui L, Cao Y. Engineering of extensor tendon complex by an *ex vivo* approach. *Biomaterials* **29**(20) (2008) 2954–2961.
16. Pacini S, Spinabella S, Trombi L, Fazzi R, Galimberti S, Dini F, Carlucci F, Petrini M. Suspension of bone marrow-derived undifferentiated mesenchymal stromal cells for repair of superficial digital flexor tendon in race horses. *Tissue Eng* **13**(12) (2007) 2949–2955.
17. Liu Y, Ramanath HS, Wang DA. Tendon tissue engineering using scaffold enhancing strategies. *Trends Biotechnol* **26**(4) (2008) 201–209.
18. Juncosa-Melvin N, Boivin GP, Gooch C, Galloway MT, West JR, Dunn MG, Butler DL. The effect of autologous mesenchymal stem cells on the biomechanics and histology of gel-collagen sponge constructs used for rabbit patellar tendon repair. *Tissue Eng* **12**(2) (2006) 369–379.
19. Awad HA, Butler DL, Boivin GP, Smith FN, Malaviya P, Huibregtse B, Caplan AI. Autologous mesenchymal stem cell-mediated repair of tendon. *Tissue Eng* **5**(3) (1999) **5**(3) 267–277.
20. Cao Y, Liu Y, Liu W, Shan Q, Buonocore SD, Cui L. Bridging tendon defects using autologous tenocyte engineered tendon in a hen model. *Plast Reconstr Surg* **110**(5) (2002) 1280–1289.
21. Liu W, Chen B, Deng D, Xu F, Cui L, Cao Y. Repair of tendon defect with dermal fibroblast engineered tendon in a porcine model. *Tissue Eng* **12**(4) (2006) 775–788.
22. Kryger GS, Chong AK, Costa M, Pham H, Bates SJ, Chang J. A comparison of Tenocytes and Mesenchymal Stem Cells for Use in Flexor Tendon Tissue Engineering, *J Hand Surg* **32A** (2007) 597–605.
23. Chong AK, Ang AD, Goh JC, Hui JH, Lim AY, Lee EH, Lim BH. Bone marrow-derived mesenchymal stem cells influence early tendon-healing in a rabbit achilles tendon model. *J Bone Joint Surg Am* **89**(1) (2007) 74–81.
24. He Q, Li Q, Chen B, Wang Z. Repair of flexor tendon defects of rabbit with tissue engineering method. *Chin J Traumatol* **5**(4) (2002) 200–208.
25. Yamahara K, Nagaya N. Stem cell implantation for myocardial disorders. *Curr Drug Deliv* **5**(3) (2008) 224–229.
26. Bluteau G, Luder HU, De Bari C, Mitsiadis TA. Stem cells for tooth engineering. *Eur Cell Mater* **16** (2008) 1–9.

27. Zhang CP, Fu XB. Therapeutic potential of stem cells in skin repair and regeneration. *Chin J Traumatol* **11**(4) (2008) 209–221.

28. Ratajczak MZ, Zuba-Surma EK, Wysoczynski M, Wan W, Ratajczak J, Wojakowski W, Kucia M. Hunt for pluripotent stem cell — regenerative medicine search for almighty cell. *J Autoimmun* **30**(3) (2008) 151–162.

29. Richardson LE, Dudhia J, Clegg PD, Smith R. Stem cells in veterinary medicine — attempts at regenerating equine tendon after injury. *Trends Biotechnol* **25**(9) (2007) 409–416.

30. Pelagiadis I, Dimitriou H, Kalmanti M. Biologic characteristics of mesenchymal stromal cells and their clinical applications in pediatric patients. *J Pediatr Hematol Oncol* **30**(4) (2008) 301–309.

31. Nirmalanandhan VS, Dressler MR, Shearn JT, Juncosa-Melvin N, Rao M, Gooch C, Bradica G, Butler DL. Effect of scaffold material, construct length and mechanical stimulation on the *in vitro* stiffness of the engineered tendon construct. *J Biomech* **41**(4) (2008) 822–828.

32. Nirmalanandhan VS, Dressler MR, Shearn JT, Juncosa-Melvin N, Rao M, Gooch C, Bradica G, Butler DL. Mechanical stimulation of tissue engineered tendon constructs: effect of scaffold materials. *J Biomech Eng* **129**(6) (2007) 919–923.

33. Zuk PA, Zhu M, Mizuno H, Huang J, Futrell JW, Katz AJ, Benhaim P, Lorenz HP, Hedrick MH. Multilineage cells from human adipose tissue: implications for cell-based therapies. *Tissue Eng* **7**(2) (2001) 211–228.

34. Lee RH, Kim B, Choi I, Kim H, Choi HS, Suh K, Bae YC, Jung JS. Characterization and expression analysis of mesenchymal stem cells from human bone marrow and adipose tissue. *Cell Physiol Biochem* **14**(4–6) (2004) 311–324.

35. Whitlock P, Smith T, Poehling G, Shilt J, van Dyke M. A naturally derived, cytocompatible, and architecturally optimized scaffold for tendon and ligament regeneration. *Biomaterials* **28** (2007) 4321–4329.

36. Brown RA, Phillips JB. Cell responses to biomimetic protein scaffolds used in tissue repair and engineering. *Int Rev Cytol* **262** (2007) 75–150.

37. Suh JK, Matthew HW. Application of chitosan-based polysaccharide biomaterials in cartilage tissue engineering: a review. *Biomaterials* **21**(24) (2000) 2589–2598.

38. Majima T, Irie T, Sawaguchi N, Funakoshi T, Iwasaki N, Harada K, Minami A, Nishimura SI. Chitosan-based hyaluronan hybrid polymer fibre scaffold for ligament and tendon tissue engineering. *Proc Inst Mech Eng [H]* **221**(5) (2007) 537–546.

39. Funakoshi T, Majima T, Iwasaki N, Suenaga N, Sawaguchi N, Shimode K, Minami A, Harada K, Nishimura S. Application of tissue engineering techniques for rotator cuff regeneration using a chitosan-based hyaluronan hybrid fiber scaffold. *Am J Sports Med* **33**(8) (2005) 1193–1201.

40. Bagnaninchi PO, Yang Y, Zghoul N, Maffulli N, Wang RK, Haj AJ. Chitosan microchannel scaffolds for tendon tissue engineering characterized using optical coherence tomography. *Tissue Eng* **13**(2) (2007) 323–331.

41. Funakoshi T, Majima T, Iwasaki N, Yamane S, Masuko T, Minami A, Harada K, Tamura H, Tokura S, Nishimura S. Novel chitosan-based hyaluronan hybrid polymer fibers as a scaffold in ligament tissue engineering. *J Biomed Mater Res A* **74**(3) (2005) 338–346.

42. Lu HH, Cooper JA Jr, Manuel S, Freeman JW, Attawia MA, Ko FK, Laurencin CT. Anterior cruciate ligament regeneration using braided biodegradable scaffolds: *In vitro* optimization studies. *Biomaterials* **26**(23) (2005) 4805–4816.

43. Cao D, Liu W, Wei X, Xu F, Cui L, Cao Y. *In vitro* tendon engineering with avian tenocytes and polyglycolic acids: a preliminary report. *Tissue Eng* **12**(5) (2006) 1369–1377.

44. Yokoya S, Mochizuki Y, Nagata Y, Deie M, Ochi M. Tendon-bone insertion repair and regeneration using polyglycolic acid sheet in the rabbit rotator cuff injury model. *Am J Sports Med* **36**(7) (2008) 1298–1309.

45. Kumbar SG, James R, Nukavarapu SP, Laurencin CT. Electrospun nanofiber scaffolds: engineering soft tissues. *Biomed Mater* **3**(3) (2008) 034002.

46. Sahoo S, Cho-Hong JG, Siew-Lok T. Development of hybrid polymer scaffolds for potential applications in ligament and tendon tissue engineering. *Biomed Mater* **2**(3) (2007) 169–173.

47. Sahoo S, Ouyang H, Goh JC, Tay TE, Toh SL. Characterization of a novel polymeric scaffold for potential application in tendon/ligament tissue engineering. *Tissue Eng* **12**(1) (2006) 91–99.

48. Saber S, Zhang AY, Ki SH, Lindsey D, Smith RL, Riboh J, Pham HM, Chang J. Flexor tendon Tissue Engineering: bioreactor Cyclic Strain Increases Construct Strength. Submitted.

49. Basile P, Dadali T, Jacobson J, Hasslund S, Ulrich-Vinther M, Søballe K, Nishio Y, Drissi MH, Langstein HN, Mitten DJ, O'Keefe RJ, Schwarz EM, Awad HA. Freeze-dried tendon allografts as tissue-engineering scaffolds for Gdf5 gene delivery. *Mol Ther* **16**(3) (2008) 466–473.

50. Chong AK, Riboh J, Ford CN, Lindsey DP, Pham HM, Chang J. Biomechanical properties of acellular flexor tendons. Submitted.

51. Liang R, Woo SL, Nguyen TD, Liu PC, Almarza A. Effects of a bioscaffold on collagen fibrillogenesis in healing medial collateral ligament in rabbits. *J Orthop Res* **26**(8) (2008) 1098–1104.

52. Liang R, Woo SL, Takakura Y, Moon DK, Jia F, Abramowitch SD. Long-term effects of porcine small intestine submucosa on the healing of medial collateral ligament: a functional tissue engineering study. *J Orthop Res* **24**(4) (2006) 811–819.

53 Ahmed TA, Dare EV, Hincke M. Fibrin: a versatile scaffold for tissue engineering applications. *Tissue Eng Part B Rev* **14**(2) (2008) 199–215.

54. Vollrath F, Knight DP. Liquid crystalline spinning of spider silk. *Nature* **410**(6828) (2001) 541–548.

55. Musahl V, Abramowitch SD, Gilbert TW, Tsuda E, Wang JH, Badylak SF, Woo SL. The use of porcine small intestinal submucosa to enhance the healing of the medial collateral ligament — a functional tissue engineering study in rabbits. *J Orthop Res* **22**(1) (2004) 214–220.

56. Takezawa T, Ozaki K, Takabayashi C. Reconstruction of a hard connective tissue utilizing a pressed silk sheet and type-I collagen as the scaffold for fibroblasts. *Tissue Eng* **13**(6) (2007) 1357–1366.

57. Huang D, Balian G, Chhabra AB. Tendon tissue engineering and gene transfer: the future of surgical treatment. *J Hand Surg [Am]* **31**(5) (2006) 693–704.

58. Ricci JL, Gona AG, Alexander H. *In vitro* tendon cell growth rates on a synthetic fiber scaffold material and on standard culture plates. *J Biomed Mater Res* **25**(5) (1991) 651–666.

59. Garvin J, Qi J, Maloney M, Banes AJ. Novel system for engineering bioartificial tendons and application of mechanical load. *Tissue Eng* **9**(5) (2003) 967–979.

60. Bader KF, Curton JW. A successful silicone tendon prosthesis. *Arch Surg* **97**(3) (1968) 406–411.

61. Mahoney JL, Farkas LG, Lindsay WK. Quality of tendon graft healing in silastic pseudosheaths: breaking-strength studies. *Surg Forum* **27**(62) (1976) 572–573.

62. Beris AE, Darlis NA, Korompilias AV, Vekris MD, Mitsionis GI, Soucacos PN. Two-stage flexor tendon reconstruction in zone II using a silicone rod and a pedicled intrasynovial graft. *J Hand Surg [Am]* **28**(4) (2003) 652–660.

63. Joshi SD, Webb K. Variation of cyclic strain parameters regulates development of elastic modulus in fibroblast/substrate constructs. *J Orthop Res* **26**(8) (2008) 1105–1113.

64. Shearn JT, Juncosa-Melvin N, Boivin GP, Galloway MT, Goodwin W, Gooch C, Dunn MG, Butler DL. Mechanical stimulation of tendon tissue engineered constructs: effects on construct stiffness, repair biomechanics, and their correlation. *J Biomech Eng* **129**(6) (2007) 848–854.

65. Butler DL, Juncosa-Melvin N, Boivin GP, Galloway MT, Shearn JT, Gooch C, Awad H. Functional tissue engineering for tendon repair: A multidisciplinary strategy using mesenchymal stem cells, bioscaffolds, and mechanical stimulation. *J Orthop Res* **26**(1) (2008) 1–9.

66. Banes AJ, Tsuzaki M, Hu P, Brigman B, Brown T, Almekinders L, Lawrence WT, Fisher T. PDFG-BB and IGF-I and mechanical load stimulates DNA synthesis in avian tendon fibroblast *in vitro*. *J Biomech* **28** (1995) 1505–1513.

67. Banes AJ, Horesovsky G, Larson C, Tsuzaki M, Judex S, Archambault J, Zernicke R, Herzog W, Kelley S, Miller L. Mechanical load stimulates expression of novel genes *in vivo* and *in vitro* avian flexor tendon cells. *Osteoarthr Cartil* **7** (1999) 141–153.

68. Wang JH. Mechanobiology of tendon. *J Biomech* **39** (2006) 1563–1582.

69. Maeda E, Shelton JC, Bader DL, Lee DA. Time dependence of cyclic tensile strain on collagen production in tendon fascicles. *Biochem Biophys Res Commun* **362**(2) (2007) 399–404.

70. Riboh J, Chong AK, Pham H, Longaker M, Jacobs C, Chang J. Optimization of flexor tendon tissue engineering with a cyclic strain bioreactor. *J Hand Surg [Am]* **33**(8) (2008) 1388–1396.

71. Spalazzi JP, Vyner MC, Jacobs MT, Moffat KL, Lu HH. Mechanoactive scaffold induces tendon remodeling and expression of fibrocartilage markers. *Clin Orthop Relat Res* **466**(8) (2008) 1938–1947.

72. Hsu C, Chang J. Clinical implications of growth factors in flexor tendon wound healing. *J Hand Surg [Am]* **29**(4) (2004) 551–563.

73. Chang J, Most D, Stelnicki E, Siebert JW, Longaker MT, Hui K, Lineaweaver WC. Gene expression of transforming growth factor beta-1 in rabbit zone II flexor tendon wound healing: evidence for dual mechanisms of repair. *Plast Reconstr Surg* **100**(4) (1997) 937–944.

74. Zhang AY, Pham H, Ho F, Teng K, Longaker MT, Chang J. Inhibition of TGF-beta-induced collagen production in rabbit flexor tendons. *J Hand Surg [Am]* **29**(2) (2004) 230–235.

75. Duffy FJ Jr, Seiler JG, Gelberman RH, Hergrueter CA. Growth factors and canine flexor tendon healing: initial studies in uninjured and repair models. *J Hand Surg [Am]* **20**(4) (1995) 645–649.

76. Nakamura N, Shino K, Natsuume T, Horibe S, Matsumoto N, Kaneda Y, Ochi T. Early biological effect of *in vivo* gene transfer of platelet-derived growth factor (PDGF)-B into healing patellar ligament. *Gene Ther* **5**(9) (1998) 1165–1170.

77. Chan BP, Chan KM, Maffulli N, Webb S, Lee KK. Effect of basic fibroblast growth factor. An *in vitro* study of tendon healing. *Clin Orthop Relat Res* **342** (1997) 239–247.

78. Chang J, Most D, Thunder R, Mehrara B, Longaker MT, Lineaweaver WC. Molecular studies in flexor tendon wound healing: the role of basic fibroblast growth factor gene expression. *J Hand Surg [Am]* **23**(6) (1998) 1052–1058.

79. Abrahamsson SO. Similar effects of recombinant human insulin-like growth factor-I and II on cellular activities in flexor tendons of young rabbits: experimental studies *in vitro*. *J Orthop Res* **15**(2) (1997) 256–262.

80. Abrahamsson SO, Lohmander S. Differential effects of insulin-like growth factor-I on matrix and DNA synthesis in various regions and types of rabbit tendons. *J Orthop Res* **14**(3) (1996) 370–376.

81. Border WA, Noble NA. Transforming growth factor beta in tissue fibrosis. *N Engl J Med* **331**(19) (1994) 1286–1292.

82. Ngo M, Pham H, Longaker MT, Chang J. Differential expression of transforming growth factor-beta receptors in a rabbit zone II flexor tendon wound healing model. *Plast Reconstr Surg* **108**(5) (2001) 1260–1267.

83. Chang J, Thunder R, Most D, Longaker MT, Lineaweaver WC. Studies in flexor tendon wound healing: neutralizing antibody to TGF-beta1 increases postoperative range of motion. *Plast Reconstr Surg* **105**(1) (2000) 148–155.

84. Ross R, Glomset J, Kariya B, Harker L. A platelet-dependent serum factor that stimulates the proliferation of arterial smooth muscle cells *in vitro*. *Proc Natl Acad Sci USA* **71**(4) (1974) 1207–1210.

85. Steed DL. Clinical evaluation of recombinant human platelet-derived growth factor for the treatment of lower extremity diabetic ulcers. Diabetic Ulcer Study Group. *J Vasc Surg* **21**(1) (1995) 71–78; discussion 79–81.

86. Wu L, Brucker M, Gruskin E, Roth SI, Mustoe TA. Differential effects of platelet-derived growth factor BB in accelerating wound healing in aged versus young animals: the impact of tissue hypoxia. *Plast Reconstr Surg* **99**(3) (1997) 815–822; discussion 823–824.

87. Pierce GF, Mustoe TA, Altrock BW, Deuel TF, Thomason A. Role of platelet-derived growth factor in wound healing. *J Cell Biochem* **45**(4) (1991) 319–326.

88. Lynch SE, Nixon JC, Colvin RB, Antoniades HN. Role of platelet-derived growth factor in wound healing: synergistic effects with other growth factors. *Proc Natl Acad Sci USA* **84**(21) (1987) 7696–7700.

89. Dahlgren LA, Mohammed HO, Nixon AJ. Temporal expression of growth factors and matrix molecules in healing tendon lesions. *J Orthop Res* **23**(1) (2005) 84–92.

90. Le Roith D. Seminars in medicine of the Beth Israel Deaconess Medical Center. Insulin-like growth factors. *N Engl J Med* **336**(9) (1997) 633–640.

91. Jones JI, Clemmons DR. Insulin-like growth factors and their binding proteins: biological actions. *Endocr Rev* **16**(1) (1995) 3–34.

92. Murphy LJ, Murphy LC, Friesen HG. Estrogen induces insulin-like growth factor-I expression in the rat uterus. *Mol Endocrinol* **1**(7) (1987) 445–450.

93. Tsuzaki M, Brigman BE, Yamamoto J, Lawrence WT, Simmons JG, Mohapatra NK, Lund PK, Van Wyk J, Hannafin JA, Bhargava MM, Banes AJ. Insulin-like growth factor-I is expressed by avian flexor tendon cells. *J Orthop Res* **18**(4) (2000) 546–556.

94. Hansson HA, Dahlin LB, Lundborg G, Löwenadler B, Paleus S, Skottner A. Transiently increased insulin-like growth factor I immunoreactivity in tendons after vibration trauma. An immunohistochemical study on rats. *Scand J Plast Reconstr Surg Hand Surg* **22**(1) (1988) 1–6.

95. Abrahamsson SO, Lundborg G, Lohmander LS. Recombinant human insulin-like growth factor-I stimulates *in vitro* matrix synthesis and cell proliferation in rabbit flexor tendon. *J Orthop Res* **9**(4) (1991) 495–502.

96. Dahlgren LA, van der Meulen MC, Bertram JE, Starrak GS, Nixon AJ. Insulin-like growth factor-I improves cellular and molecular aspects of healing in a collagenase-induced model of flexor tendinitis. *J Orthop Res* **20**(5) (2002) 910–919.

97. Pepper MS, Ferrara N, Orci L, Montesano R. Potent synergism between vascular endothelial growth factor and basic fibroblast growth factor in the induction of angiogenesis *in vitro*. *Biochem Biophys Res Commun* **189**(2) (1992) 824–831.

98. Park JE, Keller GA, Ferrara N. The vascular endothelial growth factor (VEGF) isoforms: differential deposition into the subepithelial extracellular matrix and bioactivity of extracellular matrix-bound VEGF. *Mol Biol Cell* **4**(12) (1993) 1317–1326.

99. Corral CJ, Siddiqui A, Wu L, Farrell CL, Lyons D, Mustoe TA. Vascular endothelial growth factor is more important than basic fibroblastic growth factor during ischemic wound healing. *Arch Surg* **134**(2) (1999) 200–205.

100. Tsurumi Y, Kearney M, Chen D, Silver M, Takeshita S, Yang J, Symes JF, Isner JM. Treatment of acute limb ischemia by intramuscular injection of vascular endothelial growth factor gene. *Circulation* **96**(9 Suppl.) (1997) II-382–388.

101. Petersen W, Pufe T, Unterhausen F, Zantop T, Mentlein R, Weiler A. The splice variants 120 and 164 of the angiogenic peptide vascular endothelial cell growth factor (VEGF) are expressed during Achilles tendon healing. *Arch Orthop Trauma Surg* **123**(9) (2003) 475–480.

102. Bidder M, Towler DA, Gelberman RH, Boyer MI. Expression of mRNA for vascular endothelial growth factor at the repair site of healing canine flexor tendon. *J Orthop Res* **18**(2) (2000) 247–252.

103. Zhang F, Liu H, Stile F, Lei MP, Pang Y, Oswald TM, Beck J, Dorsett-Martin W, Lineaweaver WC. Effect of vascular endothelial growth factor on rat Achilles tendon healing. *Plast Reconstr Surg* **112**(6) (2003) 1613–1619.

104. Wang XT, Liu PY, Tang JB. Tendon healing *in vitro*: modification of tenocytes with exogenous vascular endothelial growth factor gene increases expression of transforming growth factor beta but minimally affects expression of collagen genes. *J Hand Surg [Am]* **30**(2) (2005) 222–229.

105. Costa MA, Wu C, Pham BV, Chong AK, Pham HM, Chang J. Tissue engineering of flexor tendons: optimization of tenocyte proliferation using growth factor supplementation. *Tissue Eng* **12**(7) (2006) 1937–1943.

106. Sánchez M, Anitua E, Azofra J, Andía I, Padilla S, Mujika I. Comparison of surgically repaired Achilles tendon tears using platelet-rich fibrin matrices. *Am J Sports Med* **35**(2) (2007) 245–251.

107. de Mos M, van der Windt AE, Jahr H, van Schie HT, Weinans H, Verhaar JA, van Osch GJ. Can platelet-rich plasma enhance tendon repair? A cell culture study. *Am J Sports Med* **36**(6) (2008) 1171–1178.

108. Mehta V, Kang Q, Luo J, He TC, Haydon RC, Mass DP. Characterization of adenovirus-mediated gene transfer in rabbit flexor tendons. *J Hand Surg [Am]* **30**(1) (2005) 136–141.

109. Uggen JC, Dines J, Uggen CW, Mason JS, Razzano P, Dines D, Grande DA. Tendon gene therapy modulates the local repair environment in the shoulder. *J Am Osteopath Assoc* **105**(1) (2005) 20–21.

CHAPTER 11

Skin

Deepak M. Gupta[‡], Nicholas J. Panetta[‡], Geoffrey C. Gurtner[‡]
and Michael T. Longaker[,†]*

Introduction

Among all the human body's organ systems, the integument is the largest.[1] It is one of the first to develop during embryogenesis, and it continuously evolves and renews itself as one ages. It consists of epidermis, dermis, the skin's numerous adnexae, subcutis and underlying connective tissues. It is conferred the responsibility to act as a primary barrier between our internal organs and harmful microbes and pathogens in our environment. It also plays a significant role in mediating temperature and water homeostasis, tactile sensation, as well as surveillance for our immunological organs and even an endocrine role. Our skin also plays a significant psychological role in that it largely defines how we are perceived by others and may elicit prejudices and judgments. Finally, the physical architecture and physiological function of integument vary widely depending on anatomical location. As such, skin anatomy and physiology are extremely complicated, arguably more than any other organ system.

Skin

Skin is made up of two principal layers — the dermis and the epidermis. The dermis underlies the epidermis, is significantly thicker and comprises approximately 95% of the thickness of the skin. The dermis is comprised of two domains, which are distinct at the microanatomical level. The papillary dermis underlies the epidermis and contains small-diameter fibers, while the deeper reticular dermis is less cellular and is largely deposited with thicker diameter collagen and elastin fibers.[2] The microvasculature of these compartments follows a similar pattern. The dermis also provides mechanical strength, flexibility and elasticity to the skin, as it is heavily populated by fibroblasts that deposit a heavy extracellular matrix composed predominantly of collagen fibrils interspersed with elastic fibers, proteoglycans and glycoproteins. Specifically, the reticular dermis provides skin with most of its mechanical properties, and loss of this layer can often lead to excessive scarring and wound contraction. The subcutis (or hypodermis) lies deep in the dermis and also contributes to the mechanical and thermoregulatory properties of skin.

*Corresponding author.
[†]Michael T. Longaker, MD, MBA Stanford University School of Medicine, 257 Campus Drive, Stanford, CA 94305-5148, Phone: 650-736-1707, Fax: 650-736-1705, Email: longaker@stanford.edu
[‡]Hagey Laboratory for Pediatric Regenerative Medicine, Division of Plastic and Reconstructive Surgery, Department of Surgery, Stanford University School of Medicine, Stanford, CA

The epidermis is superficial to the dermis and relatively thin. It is nonvascular and nonsensitive. It consists of four (or five) layers that vary in thickness depending on the anatomical location. Deepest and adjacent to the dermis-epidermis junction, the stratum basalis (or basal layer) rests on the epidermal basement membrane and is home to the numerous basal cells which proliferate and differentiate as they ascend to renew the superficial layers.[3] The stratum spinosum (or spinous layer) is the first suprabasal layer where the cells begin to undergo keratinization. The stratum granulosum (or granular layer) is the second suprabasal layer, and cells in this layer begin to exhibit nucleic extravasation and water-proofing lipid granule production. The stratum lucidum (or lucid layer) is the third suprabasal layer and is only found in thick skin (palmoplantar). The stratum lucidum helps reduce friction and shear forces between the stratum corneum and stratum granulosum. The stratum corneum (or cornified layer) is the most superficial suprabasal layer and gives strength to the epidermis and maintains hydration by absorbing water and preventing evaporation. These five layers are dynamic and renew approximately every 30 days. The principal cell in the epidermis is the keratinocyte.[4] Melanocytes and immune cells are also interspersed and provide pigmentation and surveillance of pathogens, respectively. The epidermis is devoid of endothelial cells as it is avascular and is nourished by diffusion from the dermis.

The epidermis also contains specialized cells and structures that maintain the functions of normal skin. Briefly, sweat glands are important for skin thermoregulatory function and fluid homeostasis. Sebaceous glands provide oil to the surface of the skin to prevent it from drying out and cracking as well as form a waterproof barrier. Hair follicles are also important for thermoregulatory function and serve as conduits for oily secretions. In cases of wounds, hair follicles are also important sources of proliferating keratinocytes that can mediate re-epithelialization.[5] While these specialized skin cells and structures are important in normal skin, their loss is less clinically significant than the loss of dermis and epidermis needed to cover and protect the underlying tissues in a wound. As such, tissue engineering of skin and skin replacements largely aims to address restoration and/or regeneration of the dermis and epidermis at the fundamental level. The dermis and epidermis represent two simple moving pieces which are critical for effective wound healing. The dermis is constituted by fibroblasts and extracellular matrix and provides mechanical support for the skin, and the epidermis is made up of stratified keratinocytes and provides for fluid homeostasis and a barrier to pathogens. Skin tissue engineering efforts have not commonly focused on reconstitution of blood vessels (endothelial cells) or immune cells, for example, as these components can usually be relied upon to migrate from the wound bed and repopulate the area of the wound quickly.[6] Likewise, melanocytes and Merkel cells are not critical for wound healing and in some reports, have been shown to arise from the wound bed or edges.

Skin of the Hand

The human body is covered by skin from head to toe; however, the skin exhibits significant variability along the way. For example, scalp skin is characterized by being thick, richly vascularized, and heavily populated by hair follicles while the skin of the axillae and pubic region are exclusive for localization of apocrine sweat glands. On the human hand, there are at least five types of skin; these five types of skin are regionally distinct.[7]

The majority of the dorsa of the hands is covered by thin skin, and contains numerous hair follicles. The thin skin is highly elastic, not adherent to the underlying structures and serves to cover a complex network of tendons, blood vessels, lymphatics and nerves, while allowing dexterity of the numerous musculoskeletal elements without added bulk.

The nail beds, also on the dorsal surface, are not covered by typical epidermis; instead, a heavily keratinized fingernail is tightly adherent and overlies the nail bed and matrix. The fingernail serves to protect and structurally support the fingertip, which is important for manipulations of fine objects. The fingernail also facilitates self-defense, removal of nits, and self-image, in some individuals.

On the volar surface, the majority of the skin is histologically classified as thick skin, and is also rich in sweat glands without hair follicles (glabrous). The palmar surface necessitates the thick skin "armor" to heavily protect against accidental insults with the use of the hand. The palmar surface is well-suited to resist underlying musculotendinous and neurovascular damage in cases of mild-to-moderate insults from cuts, scrapes, foreign body puncture/penetration, blunt impact, heat, cold, acid, alkali, etc. The skin here is not very elastic, but has flexure creases to absorb movements. Furthermore, it is firmly stabilized with the underlying fasciae to resist shearing stress with grasp.

Finally, the fingertip pads are covered by thick skin, and contain an extremely dense network of tactile (light touch) sensors as well as increased density of temperature and vibratory sensory nerve endings. The incredible sensitivity and tactile discrimination mediate skilled manipulation of fine objects and withdrawal reflexes from noxious stimuli.

To add to the complexity already described, the skin of the hand changes significantly with age. For example, skin texture, thickness, and wound healing change significantly as one considers infants, older children, adolescents, adults and the elderly.

Wound Healing

As explained above, skin is extremely complex, and as such, the outcome after tissue injury depends on an equally complicated process known as wound healing. This broad term can have many different definitions, depending on the time that has elapsed since injury, the age of the organism when the injury occurred, the character of the organism and the nature of the injury. In general, wound healing can result in a variety of tissue outcomes, and they range from chronic non-healing wounds to regeneration to scar to keloid. Wound healing is a complicated process that begins almost immediately upon injury and lasts days to months to years. In general, humans heal wounds with a non-functioning mass of fibrotic tissue known as scar. There are exceptions to this pattern, and early gestation embryos have been shown to regenerate skin wounds whereby the *de novo* deposited tissue resembles the pre-injury architecture and function.[8] This ability to regenerate skin tissue is restricted to early gestation embryos in humans; however, many other organisms retain the ability to regenerate throughout their lifetime.

In wound healing, three classic phases are usually described in terms of cellular content, cytokine profile and wound architecture. It is worth briefly reviewing them here so that the tissue engineer may consider where and when opportunities exist to potentially diverge from the normal wound healing pathway towards a regenerative response.

In the first stage (first 2–3 days), inflammatory and immune pathways as well as circulating coagulation components are activated in response to injury.[9] Local vasoconstriction and platelet clotting prevents further blood loss, and infiltration of neutrophils debrides devitalized tissue and microorganisms that may have entered upon skin barrier violation. At this stage, platelets and neutrophils dominate and high concentrations of platelet derived growth factor (PDGF), transforming growth factor-α (TGF-α), TGF-β, epidermal growth factor (EGF) and fibroblast growth factors (FGFs) are present in the wound.[10] These molecules have been shown to be critical modulators of subsequent events in normal wound healing, and modulation of any one of these molecules has been shown to significantly change the phenotype of wound repair. As such, these molecules represent unmet opportunities to accelerate and improve wound healing and tissue regeneration. Late in inflammation, circulating monocytes differentiate to macrophages and orchestrate the subsequent stage, new tissue formation (days 2–7).

In this second stage, cell proliferation and migration over the deposited fibrin matrix leads to the formation of granulation tissue, a conglomeration of endothelial cell-mediated capillary buds (angiogenesis) and fibroblasts in loose extracellular matrix.[11] Upon this "bed," keratinocyte migration can be seen arising from the epithelial edges to quickly close the wound and restore barrier function and fluid homeostasis. Vascular endothelial growth factor (VEGF) and basic fibroblast growth factor (bFGF, also known as FGF-2) are present in high concentrations during this stage.[12] These molecules are potent mitogens for vascular and mesenchymal cells, respectively. This stage of new tissue formation is also characterized by population by myofibroblasts, which arise from fibroblasts that have migrated into the wound and responded to macrophage signals to form contractile units that will help close the wound and deposit collagenous scar.

The third stage of wound healing is remodeling. It is marked by apoptosis and emigration of cells from the active wound and collagen remodeling for strength along the stress lines within the wound.[13] Remodeling occurs via secreted proteases (i.e. matrix metalloproteinases and collagenases) from fibroblasts, macrophages and myofibroblasts. The result is a relatively organized, acellular, avascular scar, which can take a year or more to mature.

Scar formation at the end of the complicated wound healing process can also take many characters.[14] For example, it can be atrophic, hypertrophic or keloidal in nature. In the first type, a dearth of tissue deposition leaves a depression in the skin between the wound edges that have contracted to proximity. In hypertrophic scar, abundant collagen deposition causes a raised area above the edges of the wound. Keloids are even further along this continuum and deposit exorbitant extracellular matrix such that the scar will extend beyond the original area of the wound. All three types of scar are in sharp contrast to fetal wound repair, which has been shown in several model systems to result in normal and functional tissue architecture.[15–17]

Studies related to wound healing, fibrosis, inflammation, scarring, scarless wound healing and fetal tissue regeneration have shed light on a long list of genes that are potentially exploitable in future tissue engineering strategies for skin. With further consideration of these molecules in the future, perhaps it will be possible to harness our latent regenerative abilities and mimic early intrauterine development or organisms that can regenerate tissue instead of deposit scar. This will lead to improved clinical outcomes and reduced socioeconomic burden of fibrosis.

Scarless Fetal Wound Healing

Fetal wound healing is characterized by an absence of scarring and fibrosis and instead, recapitulation of the developmental programs that occur during normal ectodermal differentiation and skin development.[18–20] This process has been shown in a number of animal models including sheep, pig, mouse, rat, rabbit, monkey and others. Detailed study of skin regeneration in these models has revealed several significant differences between fetal wound healing and tissue repair that occurs as an adult. First, inflammation is minimal or absent during fetal life, perhaps due to the immaturity of the immune system early in gestation or the absence of fibrin clot and platelet degranulation or both. Furthermore, collagen deposition follows normal architecture and skin is interspersed with hair follicles, capillaries and sebaceous and sweat glands.[21] Many hypotheses have been proposed and explored to explain this difference and remarkable ability of early gestation fetuses, yet our understanding is still immature. Some differences that have been reported include lack of fibrin clots and platelet degranulation, reduced inflammation and elevated levels of skin morphogenetic molecules, such as TGF-β3 and FGFs.[10] Furthermore, reduced platelet degranulation (fetal) also results in lower local concentrations (or absence) of TGF-β1, TGF-β2 and PDGF, which are potent stimuli for inflammatory mediators and fibroblast biosynthesis of collagen. As such, experimental manipulation (adult) of these molecules by neutralization has resulted in reduced scarring, compared to controls. In both knockout animals and neutralization/competitive inhibition models, the loss of TGF-β1 and/or TGF-β2 resulted in reduced scarring in adult animals.[22] However, when TGF-β3 was also neutralized, scarring was not improved.[23,24] Many studies have suggested that TGF-β3 is likely to underlie scarless healing in the fetus, as the homozygous null mouse exhibits impaired healing of wounds. The exogenous addition of TGF-β3 to mimic the fetal wound has also resulted in similar findings in the adult wound. Further study of the TGF-β3's actions has revealed that this isoform is specific for guiding fibroblast migration into the wound. While TGF-β1, TGF-β2 and TGF-β3 can all induce fibroblast proliferation, normal dermal architecture was only restituted in the presence of TGF-β3.[25] Subsequent studies have also revealed that intervention was necessary within 48 hours of the wound to effect reduced scar formation and improved wound healing. Upon detailed study, early actions of TGF-β1 and TGF-β2 that are unopposed by TGF-β3 can result in unabated inflammatory recruitment, autoinductive regulatory cascades and irreversible macrophage programming.[26] In fetal wounds, when TGF-β1 was added, wounds healed with a scar. Additionally, TGF-β3 in the critical window of the early wound healing phase has been shown to alter cells that would normally be directed towards healing with a scar, as well as encourage early fibroblast migration and matrix patterning.[25] Absence of TGF-β3 during this window is underscored by its impotence after the first 48 hours, when fibrin clot prevents fibroblast migration and second messenger pathways are "locked in" subsequent to TGF-β1 and TGF-β2 signaling. The 48 hour window is opportune for the clinical tissue engineer as it allows for theoretical administration upon surgery or injury without necessitating further treatment or patient follow-up/compliance. Based on the TGF-beta-related findings in *in vitro* and *in vivo* animal studies, manipulation of these molecules promises a novel therapeutic approach in humans, whereby the *ratio* of TGF-beta isoforms and *timing* of expression may be altered within a wound. TGF-β3 has already passed safety and clinical trials and is being commercialized by Renovo Group, Inc. (UK).[24,27]

Studies of fetal wound healing and regeneration have also shed light on deposition of extracellular matrix molecules including hyaluronan, fibronectin, various types of collagen and tenascin-C.[20,28,29] These molecules represent attractive targets for future skin tissue engineering efforts as highlighted by Ferguson and O'Kane; that is, scarring is a morphogenetic problem (failure to regenerate structure) and not a biochemical problem (abnormal composition of scar tissue).[25] Specifically, fetal tissue has the ability to recapitulate the basketweave orientation of large collagen fibrils of normal skin as opposed to adult tissue which deposits small, parallel collagen bundles.

The sterility of the fetal wound has also been proposed to underlie scarless healing. However, two observations have refuted this hypothesis. First, marsupial pouch young observe scarless healing, despite regular contamination with maternal urine and feces.[30] Second, adult sheep skin, when grafted to the unborn fetus and subsequently wounded, exhibit scar formation.[31] These observations suggest that instead of the aqueous or sterile nature of amniotic fluid, perhaps the degree of skin differentiation determines its ability to heal a wound.

While there is sharp contrast between the regenerative ability of unborn fetuses and adults in many regards, some have suggested that adults also maintain the ability to regenerate to some extent. This hypothesis is predicated on the observation that skin tissue has a differential ability to heal depending, first, on the *location* of the injury. For example, a 1 cm incision on the human shoulder is more likely to scar than a 1 cm incision on the face. Both of these incisions are in contrast to an incision in the gingiva, which will heal without scarring.[25] The *nature* of the injury can also determine outcome in the same tissue. For example, a small injury such as pin-prick or needle insertion in skin will heal without a scar. In contrast, a 1 cm incision made in the same piece of skin may scar. Detailed study of this model has suggested that small injuries do not observe the same bleeding and platelet degranulation, do not elicit the same inflammatory response, are not subject to the same tensile forces as incisional wounds and do not disrupt (or to a minimal degree) the extracellular matrix and adnexae of normal skin. These small injuries may also mirror the TGF-β signaling profiles of fetal wounds, where TGF-beta1 and TGF-beta2 are low, and TGF-beta3 is high. These observations imply there are other factors which may moderate signaling, wound healing and regeneration.

Thus, while many factors have been identified as *different* between embryonic and adult skin wounds, few have been characterized as *causative* in underlying scarless healing. This status can only be assigned to factors that have been shown to result in scarless healing (or reduced scar) when manipulated and exacerbation of scar when manipulated in the opposite direction. Following this algorithm rigorously, it will be possible to elucidate potential mechanisms and therapeutic targets for scarless wound healing.

Burden of Scarring

Why is detailed study of scar and wound healing so important? Novel strategies to accelerate and enhance wound healing are important because wounds, scar and fibrosis impose a significant biomedical burden on healthcare economy that is estimated to be on the order of tens of billions of dollars.[8] Scarring occurs with trauma, injury or surgery to any tissue or organ in the body, and is the result of failing to regenerate the disrupted tissue and instead, replacing it with extracellular matrix consisting of fibronectin and collagen (namely types I and III). As scarring is a form of maladaptive tissue deposition, it is associated with medical sequelae such as loss of tissue function, impaired growth, limitation of movement (when joints are affected), poor aesthetics and adverse psychological effects. Scarring is also permanent and in some cases, results in lifelong disability.

Traditionally, reconstructive options for skin defects and scars have included autologous sources (split thickness, full thickness and myofasciocutaneous), soft tissue expansion and subsequent harvest, decellularized dermal matrices, natural and synthetic biomaterials, cultured cellular grafts, pressure garments, steroid injections, and commercially available skin substitutes. The wide range of materials available for reconstruction, however, reflects both the inadequacy of each technique and the dire need to reconstruct the skin across a variety of clinical scenarios. While each method may achieve restoration, they all possess inherent limitations. The ineffectiveness of existing strategies to fully mimic the ability of living tissues, to adapt and remodel in response to environmental cues, is perhaps the single most important factor contributing to overall unsatisfactory results. Therefore, the principal challenge is to devise novel approaches that better guide the regenerative response at the site of skin injury. What is fundamentally needed are more biologically active strategies that provide potent pro-regenerative signals and a readily-available population of progenitor cells capable of proliferation and differentiation *in vivo*.

The Past: Skin Grafts and Tissue Expanders

Autologous skin grafts represent the gold standard for skin coverage in cases of loss due to burns, trauma and congenital absence or anomaly.[32] In these cases, split thickness (STSG), or sometimes full thickness (FTSG), skin grafts are harvested from a donor site that is distant from the recipient bed and not injured or diseased. The site from which the skin is harvested is determined by consideration of the skin quality and quantity at that site, and also the morbidity that harvesting the skin will impart – often times, the scarring and cosmesis are significant considerations with regard to the latter. As such, skin from the buttocks or lower extremity is often used to cover sites where both dermis and epidermal coverage are needed. The "take" of autologous skin grafts to a granulated wound bed is usually excellent and concerns regarding graft rejection and disease transmission are nil.[33,34] Split thickness skin grafts, however, are prone to contraction, scarring and altered pigmentation within the recipient bed.[35] Full thickness skin grafts (FTSG) can sometimes be used instead of STSGs, and these exhibit improved cosmesis and decreased contraction, but "take" at consistently lower rates.[36] Furthermore, the donor sites from which FTSGs are harvested are not only denuded, but dermal harvest necessitates either primary closure or STSG of the site.

Autologous skin grafts are also limited in quantity, especially where a large percentage of the total body surface area may be affected by tissue disease or destruction, such as in the case of major burns or tissue resection (exactly the cases when replacement skin coverage is needed most). In some cases, tissue expanders can be used to "grow" additional skin for reconstructing almost any part of the body.[37,38] Tissue expansion was perhaps first formally described in 1957.[39] Since then, it has become a fundamental strategy for reconstructive surgeons to achieve coverage of underlying structures when tissue is deficient. Perhaps most often, this strategy is used in breast reconstruction where excision of breast tissue, skin and/or nipple and areola prevents coverage of an implant, and also in cases of scalp repair or replacement, where hair growth makes it difficult to replace lost tissue with skin from other areas of the body.[40,41] Tissue expansion has also been used in cases of hand surgery where coverage was deficient.[42] For example, Levin *et al.* have reported successful tissue expansion in a case of a 9-year-old patient who sustained a traction avulsion injury of the right arm in a motor vehicle accident that resulted in near-complete supracondylar amputation of the arm.[43] Tissue expansion was used in this case to replace contracted scar tissue that affected the elbow and wrist. In the upper extremity, tissue expansion has also been achieved by external devices that impart negative pressure, thereby avoiding the need for surgical placement of tissue expanders and the potential morbidities associated with the additional procedure.[44] Tissue expansion for skin reconstruction is an excellent option that is associated with several significant advantages. First, the tissue is less prone to scarring or demise. It is often expanded directly within or adjacent to the recipient site, and therefore, direct feeds from the local vascular and nervous networks are established. Second, the nature of the skin is often preferred to that of a STSG — often the color, texture, and hair pattern nearly match that of the intended recipient site. Third, the tissue is autologous, which obviates concerns for immunologic rejection or disease transmission. There are also some disadvantages to tissue expansion. In some cases, especially where thick skin is required (e.g. back, torso), it may be difficult to achieve appropriate skin growth. Tissue expansion is also a lengthy process, which requires frequent visits to the treating physician for saline injections and monitoring, which can be inconvenient. Tissue expanders are also potential vehicles for infection, in which case the devices must be surgically removed and replaced at a later date once the infection has been treated. Finally, the process of tissue expansion also causes development of a bulge in tissue, and often, this is unsightly and can be an important determinant of patient compliance. With realistic expectations, tissue expansion has achieved remarkable results in reconstructing nearly every part of the body.

Both skin grafting and tissue expansion represent forms of *endogenous* skin tissue engineering. Endogenous tissue engineering is a strategy that has been applied to attempt to reconstruct the integument and minimize the need for subsequent surgery, prevent common sequelae, and achieve optimal post-reconstructive function and cosmesis. Put simply, endogenous tissue engineering aims at recapitulating molecular programming that occurs during embryonic development and/or organogenesis during postnatal life in a spatiotemporally specific manner to "generate" tissue instead of reconstruct or replace it. This process, ideally, and in the future, would be cued by surgeon-specified biochemical and/or biomechanical cues delivered to the site of deficiency.

In the case of split-thickness skin grafts, the donor site is denuded and a thin layer of dermis is also harvested. This donor site will reepithelialize from the wound edges and epidermal adnexae.[45] In most cases, the donor site can be used again for STSG or FTSG once epidermal coverage has been restored; the number of harvests from any one donor site is determined by the thickness of the dermis, which does not regenerate after harvesting a STSG. At the recipient site, studies have shown that endogenous cells within the recipient site will ultimately populate the graft and maintain its biologic functions long term.[46] It is generally believed that the grafted cells are replaced by cells in the recipient bed, however, the origin of these cells has not specifically been determined. The collagen matrix within the graft is also turned over relatively quickly (3 to 4 fold faster than that of unwounded skin) during healing in the recipient site.[47] As such, the delivery of a STSG to a recipient site elicits specific endogenous responses that ultimately result in integration of the graft.

Split thickness skin grafts can also be modified by several means to enhance the endogenous engineering of new skin. These techniques include pinch grafts, relay transplantation, meshing, Meek island grafts, and microskin grafts.[48–52] In general, these techniques increase the exposed edge area from which epidermal coverage can be restored. For example, in the case of meshing, STSG can be expanded up to 12 times its original harvested area by the introduction of lattice-organized slits.[53,54] This technique can accelerate epidermal coverage of a large area and can fit a complex shape, but also leaves a larger area to heal secondarily, which can have significant functional and cosmetic concerns. Meshing also facilitates fluid drainage and isolates potential nidi of infection. Relay transplantation is also a technique to accelerate epidermal coverage. In this strategy, strips of STSG are placed on a wound to allow epithelial growth from the graft onto the wound bed. After several days, the strips are removed and transplanted to another area, leaving the regenerated epithelium intact.

In the case of full-thickness skin grafts, enhanced biologic activity compared to STSG or skin substitute is maintained. FTSGs have the advantage that they are capable of sweating, oil secretion (sebaceous gland function) and hair growth.[55,56] Split thickness skin grafts, on the other hand, are unable to endogenously engineer the adnexae necessary to maintain these functions; instead, STSGs typically appear dry and produce little or no hair. Both STSG and FTSG are able to support endogenous engineering of capillary ingrowth with the recipient wound bed.

Finally, tissue expansion also represents a form of endogenous tissue engineering. Tissue expansion takes advantage of mechanotransduction, a phenomenon studied in a number of model systems, to simultaneously activate all the molecular machinery necessary to "grow" skin and underlying soft tissues with a single physical cue.[57] Taking advantage of mechanotransduction is a powerful way to endogenously engineer new tissue. Another well-known form of endogenous tissue engineering that relies on mechanotransductive underpinnings is distraction osteogenesis, whereby constant and gradual force across/away from an osteotomy site separates two "osteogenic fronts" and results in the formation of new, intervening bone. These results have been underlain by activation of FAK and MAPK/ERK (mechanotransduction), VEGF (angiogenesis and wound healing), and HIF-1 and SDF-1 (tissue hypoxia), which also, arguably, are likely to underlie soft tissue expansion.[58–61]

In summary, autologous skin grafting and tissue expansion techniques have numerous advantages and several disadvantages; perhaps their greatest disadvantage is their non-utility in cases of widespread, severe burns and significant tissue loss or resection, the exact situations where skin and soft tissue coverage is most highly demanded and critical.

The Present: Skin Substitutes

A number of commercially available biological materials are or have been available in addressing defects of the integument. These products are logical extensions of early studies performed by Rheinwald *et al.*, Bell *et al.*, Yannas *et al.*, and Burke *et al.* These early reports outlined *ex vivo,* allogeneic strategies for tissue engineering skin replacements. In 1975, Rheinwald *et al.* published *in vitro* studies demonstrating that cultured epidermal cells could form colonies and be separated from viable fibroblasts.[62] Several years later, Bell *et al.* performed *in vitro* experiments with fibroblasts embedded in a collagen lattice could support the growth of epidermal cells and be integrated into the skins of experimental animals.[63] In 1980, Yannas *et al.* published the first report of basic design principles for an artificial skin, including physicochemical, biochemical, and mechanical considerations.[64] Then, in 1981, Burke *et al.* documented perhaps the first case series of patients with major burns (50–95%) to receive bilayered, artificial skin.[65] They noted degradation of the implanted dermis as it was populated by fibroblasts and blood vessels from the wound bed, and ability to take epidermal autografts once the implant was integrated. These studies established the underpinnings upon which future epidermal, dermal, and composite skin substitutes were developed.

These materials have the advantage of being ready-to-use and off-the-shelf solutions to full thickness tissue loss in cases of trauma, burns, and chronic ulcers and are alternative strategies to autologous skin grafts in cases of insufficient or diseased donor tissue.[66] These skin replacements vary in their composition and design, and as such, require careful consideration of their unique advantages and disadvantages in each case. In general, these skin substitutes have met with a great deal of success in clinical applications; however, none has provided an outcome consistently comparable to an autologous skin graft.

Epidermal substitutes were perhaps the earliest seen skin substitutes used clinically, given the early studies by Rheinwald *et al.* and others, the early demonstration of the ability of keratinocytes to rapidly proliferate *in vitro* and the ability to form multiple cell layer-thick sheets of cells that could be transplanted to treat burn victims and restore barrier function and stave off bacterial sepsis. These constructs came to be known as cultured epithelial autografts (CEAs). The strategy to rapidly expand keratinocytes *in vitro* (from biopsied skin) and prepare CEAs was further supported by the ability of the cells, once transplanted, to assimilate back into the native host skin environment upon a wound bed and self-renew. Genzyme commercialized this technology in 1989 as the product, Epicel, which became the first cell-based product for tissue repair available in the United States.[67] In the 1980s, patients who suffered burn wounds received CEAs without necessitating further treatment, even long term.[68] The utility of cultured epithelial autografts has also been reported in cases involving surgery of the hand.[69] Clearly, the ability to cover wound areas with permanent biologically-active tissue, moderate donor site morbidity, and little risk of rejection seemed very attractive; however, these constructs fell out of favor given the long (and often impractical) period required to expand the cell sheets (three weeks or more), their fragility and proclivity to inconsistently take, form blisters and slough, and the exorbitant costs associated with their preparation. Some estimates have placed the cost of preparing an amount of CEA to cover 1% of total body surface area (TBSA) to range between $6,000 and $10,000.[69] Furthermore, the need for CEAs is perhaps greatest in cases where a large TBSA was affected by burn and autologous skin was insufficient, but it was in these cases that CEAs were least effective and most expensive. Given the promise of CEAs, other forms of epidermal constructs have been explored and are briefly reviewed:

- Laserskin, developed by Fidia Advanced Biopolymers, perhaps recognized the advantages of CEAs, but addressed the fragility that plagued its use. This construct also consisted of expanded keratinocytes *in vitro* from biopsied skin, but differed from earlier products in that the cells were cultured in sheets reinforced by a laser-perforated hyaluronic acid membrane, which improved mechanical stability. The construct has been used in animal models as well as human cases to replace lost skin tissue, treat vitiligo and resurface Integra.[70–72]

- Modex Therapeutiques' product EpiDex consisted of cultured autologous outer root sheath hair follicle cells, which demonstrated increased proliferative capacity and could be cryopreserved for repeated applications. EpiDex has been used clinically, and in one study, a multi-center, randomized study reported that it was as effective as split thickness autografts in the promotion of healing and complete closure of chronic vascular leg ulcers.[73] This technology has been further pioneered by Macneil *et al.* and is being improved to promise future applications to chronic wounds.[74]

- Myskin, developed by CellTran, consists of cultured autologous keratinocytes on a polyvinyl-chloride (PVC) polymer coated with a plasma-polymerized surface.[10] The purpose of incorporating the biomaterial was reported to encourage cell attachment and proliferation and provide a stable delivery vehicle. Myskin has been used in clinical cases to reconstruct nonhealing diabetic ulcers.[75,76] In this series, ulcers that received the construct seeded with cells healed or reduced their size by at least 50%; however, many recurred at a later date. Myskin eventually fell out of favor as it did not offer significant improvements over constructs already described and required a median of 10 (up to 17) repeated applications to enhance healing, after which wounds often recurred.

- CellSpray, developed by Clinical Cell Culture (C3), required the application of a keratinocyte cell suspension to a wound bed and demonstrated accelerated epidermal coverage in porcine models.[77] This strategy could be combined with fibrin or other carrier dressings (*e.g.* Myskin) in approximately half the time (and in some cases, as little as 2 days after biopsy) that was necessary to produce cultured epithelial sheets.[78] In patients, CellSpray has been used to treat major burns, accelerated wound healing and improved scar quality.[78,79]

As described above, epidermal substitutes have met with significant successes in clinical cases of generalized tissue loss such as burns and trauma. However, in general, they require a significant period of time to expand *in vitro*, are very expensive to fabricate and are fragile and prone to infection and blistering/sloughing/loss. Furthermore, in each case, a well-vascularized wound bed is necessary to allow the take of the epidermal graft and regeneration of biologically-active, renewable barrier. Numerous reports have documented the significant role the wound bed plays in the engraftment of the epidermal compartment, where without it, keratinocyte grafts have been of limited value.[80] As such, strategies to address the deficiency of a vascularized wound bed/dermis have also been devised, each with its own advantages and disadvantages.[81] There are several common themes that are seen with these dermal substitutes. First, the fibroblast, and not the keratinocyte, is the central player in these constructs.[82] Fibroblasts are the principal cell in connective tissue and secrete extracellular matrix, namely collagen.[83] Second, the matrix should allow for cell attachment, proliferation, differentiation to the fibroblast lineage and either passive or active compatibility for endogenous tissue engineering of a vascular bed to sustain the tissue long-term.[84,85] Third, the biologic implants that are used in these strategies have been widely commercialized and have come to the forefront of modern reconstruction for the integument in a number of models.

- Alloderm, developed by Life Cell Corporation, is one such biologic implant, which has perhaps received more attention and met with more success than any other direct competitor.[86] Its advent in 1996 heralded successes in wound bed, cleft palate, abdominal wall, and other sites of reconstruction and is used liberally and widely in surgery today.[87,88] Alloderm was developed from processed cadaver (allograft) skin and is available as a decellularized dermal matrix. The use of Alloderm is made possible by its processing which involves three steps – epidermal removal, cell solubilization, and dry preservation. This processing reduces antigenic components to the extent that it is comparable to autologous skin grafts with respect to inflammatory response and allows maintenance of the basement membrane, a critical component of reconstructed skin and epidermal coverage. Furthermore, Alloderm is available as a convenient off-the-shelf product that is readily available, but without the donor site morbidity and limitations on quantity that are associated with autologous skin grafts. The use of decellularized dermal matrix as a "biologic implant" to foster vascular ingrowth has relied on concurrent developments in sister fields such as bioengineering and transplant biology; how exactly this process occurs is still poorly understood, but the functional outcome is arguably still worth consideration in cases where wound bed coverage is deficient and scar tissue formation or other complication would likely predominate. Given the numerous advantages of Alloderm, investigators have widely explored its utility in the clinical setting. While Alloderm's track record is promising for long-term, uncomplicated wound healing and innovation in reconstructive surgery, issues of graft rejection and theoretical disease transmission should be considered.
- Integra is another dermal substitute that was commercialized by Johnson & Johnson.[89] It is comprised of bovine type I collagen and shark-derived chondroitin sulfate and was designed as a dressing for major burns.[90] It has been documented to support vascular and fibroblast ingrowth, which after removal of its synthetic polysiloxane polymer barrier, can sustain take of cultured epidermal grafts or split thickness skin autografts.[91] Integra has the disadvantage of using xenogeneic material, and as such, carries potential antigenic and disease transmission risks. Regardless, Integra has met with significant clinical success.

- Dermagraft, developed by Advanced Biohealing, Inc., is another dermal substitute which is composed of allogeneic neonatal fibroblasts on a three-dimensional bioabsorbable polyglactin (same material as Vicryl suture) scaffold.[10,92,93] Dermagraft is FDA-approved and has had notable success with reconstructive procedures to address nonhealing ulcers (i.e. diabetic foot).[80] The success is perhaps underlain by its potent proangiogenic activity and hospitable microenvironment for take of cultured epidermal grafts and autografts (the latter to restore barrier function of skin). These qualities of the "ideal skin substitute" are mediated by growth factors expressed within the construct including vascular endothelial growth factor (VEGF), platelet-derived growth factor (PDGF), insulin-like growth factor-I (IGF-I), colony stimulating factors (CSFs), interleukins (ILs), tumor necrosis factor (TNF), and transforming growth factor-β (TGF-β).[94] Dermagraft has been used in reconstructive surgery. Despite the clinical successes of Dermagraft in improving long-term outcomes and preventing complications in these cases and others, the product is not yet financially sensible.[66] Dermagraft also carries theoretical risks of disease transmission and graft rejection from fibroblast transplantation, although none have been reported in the literature.
- Transcyte, also developed by Advanced Biohealing, Inc., is related to the Dermagraft construct as the dermal component also contains allogeneic neonatal fibroblasts, but they are seeded on a collagen-coated nylon mesh.[10] The Transcyte construct also has an additional thin, silicone layer (Biobrane) as a temporary epidermal component to provide barrier function. Transcyte has namely been used clinically to dress wounds after burns.[95–97] Transcyte is also FDA-approved and promises improved skin reconstruction in cases of severe burns and other full-thickness injuries, but like Dermagraft, is not yet commercially viable. Transcyte also has the disadvantages where the nylon mesh is not absorbable and theoretical risks of disease transmission and graft rejection via fibroblast transplant exist.
- Permacol was developed by Tissue Sciences Laboratories and is another dermal substitute.[10] It is xenogeneic and is derived from porcine dermal matrix that has been decellularized and processed to reduce antigenicity. Permacol has met with success in some cases,[98] yet has not gained increased clinical popularity due to its decreased ability to support efficient revascularization. As such, it is not the ideal dermal matrix substitute to support an overlying epidermis.

The examples above outline several epidermal and dermal substitutes that have been used in animal models, clinically and/or in surgery of the hand. They promise improved clinical outcomes in cases of tissue loss and reduced complications. However, their use is not without disadvantages; in addition to the inherent problems that are discussed above, perhaps the greatest single disadvantage is the inability to reconstruct a full thickness wound with any one of these epidermal and dermal constructs alone. What is ultimately needed is an alternative biologic dressing to reconstruct the dermal wound bed that can provide mechanical strength, elasticity, cues for integration into the host including those for fibroblast proliferation and vascular ingrowth, along with a self-renewing, non-scarring, non-contracting, epidermal covering for barrier function and water homeostasis. To address this major concern, several products that contain both epidermal and dermal components have been commercialized and have been available for use clinically to reconstruct full thickness wounds after burn, trauma and other cases of tissue loss or deficiency.

- One such product is Apligraf, which was developed by Organogenesis.[99] Apligraf consists of both epidermal and dermal components in one readily available, off-the-shelf construct. The epidermal component consists of human allogeneic neonatal keratinocytes, and the dermal component consists of human allogeneic neonatal skin fibroblasts in bovine type I collagen. The construct has met with great success in several clinical scenarios, and while the risk of graft rejection and disease transmission from allogeneic keratinocytes and fibroblasts is potentially worrisome, no cases have been reported; some have hypothesized that the absence of endothelial cells in the transplanted construct make it less likely to be rejected. Apligraf "take" has also been observed to be comparable to autografts with good cosmetic results. Apligraf is a living skin equivalent that is FDA-approved for reconstruction of venous ulcers and diabetic lower extremity ulcers.[100]

- OrCel, developed by Ortec International, is another composite skin substitute with both epidermal and dermal components.[10] Like Apligraf, the epidermal domain is populated by human allogeneic neonatal keratinocytes and the dermal domain by human allogeneic neonatal foreskin fibroblasts in a bovine collagen sponge. OrCel provides a favorable environment for wound healing by delivering extracellular matrix components and growth factors, and was not intended as a permanent skin substitute. OrCel was originally developed as a treatment for epidermolysis bullosa.[101] OrCel also carries the potential risk for disease transmission and graft rejection.

- Permaderm, developed by Cutanogen and subsequently acquired by Cambrex, is another composite skin substitute with both epidermal and dermal domains; however, unlike Apligraf and OrCel, both domains are populated by autologous cells – keratinocytes in the epidermal domain, and fibroblasts in the dermal component.[102] Permaderm has the advantage of being autologous in nature, and as such, it will avoid any concerns for graft rejection; however, the xenobiotic nature of the cultures needed to prepare the materials and the bovine collagen construct upon which the cells are seeded still carry the potential risk for disease transmission. The autologous nature of the cells within the construct also necessarily implies a long period of *in vitro* culture necessary to expand the cells into confluent sheets, which is four weeks or more.

In cases where large areas of skin coverage are lost, skin substitutes have provided some relief and variable ability to restore coverage and barrier. The wide range of materials available for reconstruction reflects both the inadequacy of each skin substitute and the dire need to reconstruct the skin across a variety of clinical scenarios. While each method may achieve some degree of restoration, they all possess inherent limitations. The ineffectiveness of existing strategies to mimic the ability of living tissues, to adapt and remodel in response to environmental cues, and to maintain water and temperature homeostasis, is perhaps the single most important factor contributing to overall unsatisfactory results. Therefore, the principal challenge for the tissue engineer is to consider multiple characteristics of the ideal skin substitute and devise a novel approach that better guides the regenerative response at the site of injury and/or maintains the function of normal skin that has been lost.

The Future: Molecular and Gene Therapy, and Stem Cell Biology

As mentioned above, other cytokines have been supported by numerous studies as potentially promising in novel wound healing applications. Platelet-derived growth factor (PDGF) is one such molecule that can stimulate mesenchymal cells, monocytes, macrophages and fibroblasts.[103] It is derived from degranulating platelets in adult wounds, and when applied exogenously, PDGF can accelerate extracellular matrix deposition and stimulate healing in an otherwise refractory wound in a dose-dependent manner. It can also accelerate wound healing in normally healing rat wounds and increase final breaking strength. Specifically, PDGF can stimulate significant increases in neoangiogenesis and granulation tissue elaboration without an immunological reaction. Of interest, PDGF is present in reduced amounts in fetal wounds; as such, its presence is strongly associated with scar formation. Because PDGF has met with such promise in experimental reports, it moved quickly into clinical trials. In the human model, PDGF was effective in accelerating healing in chronic ulcers due to diabetes.[104] As such, PDGF is now FDA-approved for clinical use and is available in the product, Regranex. PDGF has also been trialed in cases of venous and pressure ulcers, however results have been equivocal.[105–107]

Basic FGF (bFGF, also known as FGF-2) is also found in healing wounds and is widely accepted as critical for wound healing. FGF-2 is first released from damaged endothelium and then secreted by activated macrophages. When FGF-2 is applied exogenously, it has been shown to enhance wound healing, accelerate neoangiogenesis and increase connective and granulation tissue growth.[108,109] Despite its successes in animal models, FGF-2 did not accelerate wound healing in a clinical trial for nonhealing ulcers in diabetic patients.[110,111] FGF-7 and FGF-10 are also FGF family members that have been shown to improve keratinocyte proliferation and wound closure after injury; their actions may be mediated by FGFR2-IIIb.[112,113]

Epidermal growth factor (EGF) has also received significant attention in the wound healing arena as it arises from platelet degranulation and macrophages.[114] EGF can stimulate keratinocyte and fibroblast proliferation, and may also guide keratinocyte migration *in vitro* and *in vivo*.[115,116] Epidermal growth factor has already been topically applied in the human setting.[117] In these cases, wound healing was enhanced. On a somewhat tangential note, the high concentration of EGF in saliva is also thought to mediate accelerated wound healing in animals that lick their wounds. With advents such as matrikine or matricryptin development, EGFR ligands will be expected to have an impact on novel approaches to wound healing.[118]

VEGF has broad effects in wounds such as inflammatory chemotaxis, recruitment of endothelial progenitors from bone marrow, and control of capillary permeability.[119] Most importantly, VEGF encourages development of new capillaries and granulation tissue formation in wound healing. These observations are supported by some models, such as diabetic mice, which fail to produce VEGF at the wound site, that demonstrate impaired healing; however, when VEGF was applied exogenously, healing was improved.[120–122] VEGF is elaborated in hypoxic environments and a report by Benest *et al.* suggests its cooperation with Ang-1 to generate significantly more mature microvessels compared to those grown under the influence of either growth factor alone.[123] VEGF has also been trialed in humans for limb ischemia and coronary artery revascularization.[124–126]

Stem Cell Biology

As explained above, the extracellular matrix and growth factors within a wound are important determinants of the character of the "soil" within which cells may mediate wound healing. These stem or progenitor cells play a significant role in wound healing and tissue regeneration and have the potential to be further exploited in tissue engineering strategies for skin. This may be especially true when one considers that current tissue engineering strategies that have made it to market are not ideal; these techniques offer improved wound healing in some cases, yet are still plagued by equivocal results, inconsistent vascularization and scarring. Furthermore, engineered constructs that rely on autologous biopsy as a source of cells are likely to prove less attractive when there is extensive tissue loss or disease of the host skin. As such, various sources from which stem (or progenitor) cells can be collected have been identified including the circulation, epidermis, dermis, hair follicles, bone marrow, adipose and embryonic tissue. These cells have the potential to promise novel therapies for wound healing and skin tissue engineering.

Epidermal cells have received a significant amount of attention in the recent literature. Investigators and clinicians have noted for years that hear-bearing areas tend to heal more quickly than areas of skin that lack hair follicles.[127-129] The exact relationship between the hair follicle and the epidermis, however, has long remained elusive and is still incompletely understood. However, several seminal reports since 2001 have begun to expand our understanding as to how the hair follicle may contribute to wound healing. These insights may offer improved perspectives when considering regenerative strategies for clinical wound healing. In general, investigators have consistently documented the hair follicle's lack of participation in normal epidermal homeostasis;[5] however, when skin is wounded, genetic lineage analyses indicate that local cell depletion within the epidermis resulted in activation and recruitment of progenitor cells from the hair follicle.[5,130-132] These cells then provide progeny, which migrate throughout the wound to mediate re-epithelialization and restore epidermal integrity. These observations generally establish that stem cells within hair follicles are not necessary for normal epidermal homeostasis, but rather are only utilized when the system is tipped to a stressful level. To further support this observation, when incisional wounds were made on tail skin of Edaradd mutant mice which lack hair follicles in this region, wound healing was delayed compared to control animals.[133]

Further study of the contribution of epidermal stem cells to wound healing has revealed that hair follicles may contain several distinct populations of progenitor keratinocytes.[134] Bulge cells have perhaps received the most attention and have become a household buzz word in epidermal stem cell-mediated wound healing.[5] However, the infundibulum and isthmus areas of the hair follicle are also continuous with the outer root sheath and interfollicular epidermis, and have also been reported to contribute to epidermal wound healing. Specifically, lineage analysis indicates that cells from these areas may also be activated in response to injury, and that perhaps these cells remain in the epidermis upon reepithelialization and participate in epidermal homeostasis as permanent residents.[134] This is in contrast to cells that originate from the bulge area, which largely do not persist in the newly formed epidermis. These studies hold exciting potential considering strategies to utilize progenitor cells.

Accumulating evidence suggests that cells within the dermis orchestrate the process of programming and deprogramming adjacent epithelial cells to guide the balance between differentiation, transdifferentiation, and apoptosis.[135] This is largely appreciated considering the regional variability of skin and the fibroblasts that populate these areas.[83] Furthermore, dermal cells have been reported to underlie hair follicle formation and give significant molecular character to the biology of the overlying epidermis.[136] For example, when epidermis that lacks hair (footpad) is transplanted over dermis in a hair-bearing area, follicular neogenesis was observed to be the result of BMP signaling from dermal papilla cells.[137] Furthermore, when volar and trunk keratinocytes were transplanted, they altered their keratin gene expression to match that of the original wound bed fibroblasts.[138]

The significant role that dermal cells may play in wound closure and tissue regeneration can also be seen in the results of urodele limb regeneration studies. The dermis of the urodele limb is composed primarily of fibroblast cells that form a loose connective tissue layer underlying the epidermis, and several studies have reported that small molecules arising from these cells largely establish the blueprint of limb patterning for tissue regeneration.[139] Furthermore, work by Endo *et al.* suggests that migration and de-differentiation of fibroblasts are central processes that may mediate regeneration of complex tissues in amphibians.[140] Despite being different from the human setting, unraveling the biological principles underlying this profound regenerative capacity in amphibians promises exciting implications for augmenting human wound healing.

Embryonic stem cells are different from the endogenous skin cells that are described above and are derived from the inner cell mass of the embryonic blastocyst.[141] These cells have been shown to proliferate *in vitro* indefinitely without losing their potential to differentiate. Most of what has been learned regarding embryonic stem cells has been derived from the mouse model, as human cell lines have only recently become available. Perhaps the greatest lessons of biology and medicine we have learned from embryonic stem cells in the mouse model comes from the generation of hundreds of genetic knockout models.[142–144] Nevertheless, there are still numerous ethical and technical problems that need to be overcome before the full potential of this type of cell can be realized. Alternatively, induced-pluripotent stem cells (iPS) are thought to be embryonic stem cell phenocopies without the destruction of an embryo.[145] These cells are derived from fibroblasts or other mature lineage cell from an adult animal (or human) and avoid the ethical hurdles that retard enthusiasm for ES cells. Notwithstanding iPS cells being relatively new in the literature, it will be exciting to follow the progress and potential of this technology in tissue engineering applications and beyond.

Bone marrow-derived mesenchymal (stromal) cells have also been investigated for their potential to enhance wound healing.[146] Several studies have reported on the high content of growth factors in bone marrow-derived mesenchymal cells that may accelerate wound healing. These molecules include FGF, HGF, IL-6, VEGF and GM-CSF.[146–148] Furthermore, bone marrow-derived mesenchymal cells are mobilized in response to injury and migrate to distant sites where they participate in new blood vessel formation. In some models, such as diabetes, endothelial progenitors and/or fibrocytes are not effectively mobilized; however, when bone marrow-derived endothelial progenitors were exogenously administered, cells trafficked to the wound and healing was improved. In diabetes, the few progenitor cells that do enter the circulation demonstrate defective incorporation mechanisms.[149] To circumvent this defect, some studies have directly applied these cells to wound sites and demonstrated improved wound healing. Bone marrow-derived cells have also been postulated to directly contribute to both normal skin and wound healing.[146,150,151] In unwounded skin, bone marrow-derived cells likely directly contribute to the active dermal fibroblast population, epidermal keratinocytes and hair follicles.[152–154] In the wound environment, the cells can take a fibrocyte phenotype from the peripheral circulation and contribute to wound healing;[146,155] however, this is a topic of significant debate largely stemming from a rudimentary understanding of their cell surface markers by which one may track these cells. To address this point of discussion, Yang *et al.* have reported on the utility of leukocyte-specific protein 1 (LSP-1) to track fibrocytes as they modulate dermal cells in wounds.[156] This is in contrast to the use of CD34, which has widely been shown to decrease as wound healing progresses, implying "transdifferentiation" of hematopoietic progenitor towards mature mesenchymal lineages, a phenomenon (plasticity) that was refuted in the myocardial injury model.[157,158] In general, fibrocytes have been postulated to participate in wound healing by one or more of several mechanisms including extracellular matrix production, antigen presentation, cell fusion and cytokine elaboration.[159,160]

While the direct mechanisms by which bone marrow-derived cells participate in normal skin renewal and wound healing are not completely understood, these cells certainly have the potential to regenerate epidermis and skin appendages under appropriate conditions.

Conclusion

The management of patients necessitating skin reconstruction has improved significantly over the last several decades. Important surgical advances have accompanied a multidisciplinary approach to patient care. Furthermore, technological innovations have become focused on improving functional results. However, research and clinical reports of innovation in skin reconstruction promise even greater improvements and further "minimalization" of surgery for the future.[161] For example, traditional reconstructive techniques are plagued with skin graft donor site morbidity, scar, immunologic rejection and poor aesthetic outcomes. These complications are unpleasant and have serious implications especially when aberrant wound healing and scarring lead to loss of function of an underlying joint. In contrast, research of tissue engineering applications for skin reconstruction may represent an alternative paradigm for surgical correction in these cases in the future. This chapter discusses several aspects of tissue engineering research that promise innovation in skin repair and improved clinical outcomes in the future.

References

1. Ehrenreich M, Ruszczak Z. Update on tissue-engineered biological dressings. *Tissue Eng* **12** (2006) 2407–2424.
2. Andreadis ST. Gene-modified tissue-engineered skin: the next generation of skin substitutes. *Advances in Biochemical Egineering/Biotechnology* **103** (2007) 241–274.
3. Marks JG, Miller JJ, Lookingbill DP, Lookingbill DP. *Lookingbill and Marks' Principles of Dermatology*, 331 p. (Saunders Elsevier, Philadelphia, PA, 2006).
4. Rook A, Burns T. *Rook's Textbook of Dermatology*, 4 v. (paged continuously) (Blackwell Science, Malden, Mass., 2004).
5. Ito M, *et al.* Stem cells in the hair follicle bulge contribute to wound repair but not to homeostasis of the epidermis. *Nat Med* **11** (2005) 1351–1354.
6. Lanza RP, Langer RS, Vacanti J. *Principles of Tissue Engineering*, xxvii, 1307 p. (Elsevier/Academic Press, Amsterdam; Boston, 2007).
7. Barron JN. The structure and function of the skin of the hand. *Hand* **2** (1970) 93–96.
8. Gurtner GC, Werner S, Barrandon Y, Longaker MT. Wound repair and regeneration. *Nature* **453** (2008) 314–321.
9. Singer AJ, Clark RA. Cutaneous wound healing. *N Engl J Med* **341** (1999) 738–746.
10. Metcalfe AD, Ferguson MW. Tissue engineering of replacement skin: the crossroads of biomaterials, wound healing, embryonic development, stem cells and regeneration. *J R Soc Interface* **4** (2007) 413–437.
11. Schaffer CJ, Nanney LB. Cell biology of wound healing. *Int Rev Cytol* **169** (1996) 151–181.
12. Kiritsy CP, Lynch AB, Lynch SE. Role of growth factors in cutaneous wound healing: a review. *Critical Reviews in Oral Biology and Medicine* **4** (1993) 729–760.
13. Lorenz HP, Longaker MT, (ed.) *Wound Healing: Repair Biology and Wound and Scar Treatment*, 209–234 (Elsevier, New York, 2006).
14. Bayat A, McGrouther DA, Ferguson MW. Skin scarring. *Bmj* **326** (2003) 88–92.
15. Adzick NS, Longaker MT. Scarless fetal healing. Therapeutic implications. *Ann Surg* **215** (1992) 3–7.
16. Longaker MT, Adzick NS. The biology of fetal wound healing: a review. *Plast Reconstr Surg* **87** (1991) 788–798.

17. Lorenz HP, *et al*. Scarless wound repair: a human fetal skin model. *Development* **114** (1992) 253–259.

18. Bullard KM, Longaker MT, Lorenz HP. Fetal wound healing: current biology. *World J Surg* **27** (2003) 54–61.

19. Ferguson MW, *et al*. Scar formation: the spectral nature of fetal and adult wound repair. *Plast Reconstr Surg* **97** (1996) 854–860.

20. Colwell AS, Longaker MT, Lorenz HP. Fetal wound healing. *Front Biosci* **8** (2003) s1240–1248.

21. Whitby DJ, Ferguson MW. The extracellular matrix of lip wounds in fetal, neonatal and adult mice. *Development* **112** (1991) 651–668.

22. Shah M, Foreman D, Ferguson M. Neutralizing antibody to TGF-B1, and 2 reduces cutaneous scarring in adult rodents. *J Cell Sci* **107** (1994) 1137–1157.

23. Shah M, Foreman DM, Ferguson MW. Neutralisation of TGF-beta 1 and TGF-beta 2 or exogenous addition of TGF-beta 3 to cutaneous rat wounds reduces scarring. *J Cell Sci* **108**(Pt 3) (1995) 985–1002.

24. Occleston NL, Laverty HG, O'Kane S, Ferguson MW. Prevention and reduction of scarring in the skin by Transforming Growth Factor beta 3 (TGFbeta3): from laboratory discovery to clinical pharmaceutical. *J Biomater Sci Polym Ed* **19** (2008) 1047–1063.

25. Ferguson MW, O'Kane S. Scar-free healing: from embryonic mechanisms to adult therapeutic intervention. *Philos Trans R Soc Lond B Biol Sci* **359** (2004) 839–850.

26. Erwig LP, Kluth DC, Walsh GM, Rees AJ. Initial cytokine exposure determines function of macrophages and renders them unresponsive to other cytokines. *J Immunol* **161** (1998) 1983–1988.

27. Occleston NL, O'Kane S, Goldspink N, Ferguson MW. New therapeutics for the prevention and reduction of scarring. *Drug Discov Today* **13** (2008) 973–981.

28. Colwell AS, Longaker MT, Lorenz HP. Mammalian fetal organ regeneration. *Adv Biochem Eng Biotechnol* **93** (2005) 83–100.

29. Longaker MT, *et al*. Studies in fetal wound healing, VII. Fetal wound healing may be modulated by hyaluronic acid stimulating activity in amniotic fluid. *J Pediatr Surg* **25** (1990) 430–433.

30. Armstrong JR, Ferguson MW. Ontogeny of the skin and the transition from scar-free to scarring phenotype during wound healing in the pouch young of a marsupial, Monodelphis domestica. *Dev Biol* **169** (1995) 242–260.

31. Longaker MT, *et al*. Adult skin wounds in the fetal environment heal with scar formation. *Ann Surg* **219** (1994) 65–72.

32. Skouge JW. *Skin grafting*, x, 81 p. (Churchill Livingstone, New York, 1991).

33. Rudolph, R, Fisher JC, Ninnemann JL. *Skin grafting*, xi, 205 p. (Little, Brown, Boston, 1979).

34. Klasen HJ. *History of free skin grafting: knowledge or empiricism?*, xii, 190 p. (Springer-Verlag, Berlin; New York, 1981).

35. Ratner D. Skin grafting. *Semin Cutan Med Surg* **22** (2003) 295–305.

36. Branham GH, Thomas JR. Skin grafts. *The Otolaryngologic clinics of North America* **23** (1990) 889–897.

37. Lari AA. Tissue Expansion. *Journal of the Royal College of Surgeons of Edinburgh* **37** (1992) 149–154.

38. Malata CM, Williams NW, Sharpe DT. Tissue Expansion: an overview. *Journal of Wound Care* **4** (1995) 37–44.

39. Neumann CG. The expansion of an area of skin by progressive distention of a subcutaneous balloon; use of the method for securing skin for subtotal reconstruction of the ear. *Plastic & reconstructive surgery* **19** (1957) 124–130.

40. Radovan C. Breast reconstruction after mastectomy using the temporary expander. *Plast Reconstr Surg* **69** (1982) 195–208.

41. Adson MH, Anderson RD, Argenta LC. Scalp expansion in the treatment of male pattern baldness. *Plast Reconstr Surg* **79** (1987) 906–914.

42. Meland NB, Smith AA, Johnson CH. Tissue expansion in the upper extremities. *Hand Clinics* **13** (1997) 303–314.

43. Heitmann C, Levin LS. Tissue expansion. *Techniques in Hand & Upper Extremity Surgery* **7** (2003) 7–11.

44. Lasheen AE. External tissue expansion using negative pressure in upper-extremity reconstruction. *The Journal of Hand Surgery* **31** (2006) 1694–1696.

45. Johnson TM, Ratner D, Nelson BR. Soft tissue reconstruction with skin grafting. *Journal of the American Academy of Dermatology* **27** (1992) 151–165.

46. Capla JM, *et al.* Skin graft vascularization involves precisely regulated regression and replacement of endothelial cells through both angiogenesis and vasculogenesis. *Plast Reconstr Surg* **117** (2006) 836–844.

47. Ohuchi K, Tsurufuji S. Degradation and turnover of collagen in the mouse skin and the effect of whole body x-irradiation. *Biochim Biophys Acta* **208** (1970) 475–481.

48. Zhang, M, Zhou G, Zhang P. [Microskin grafting in recent 15 years]. *Zhonghua Wai Ke Za Zhi* **39** (2001) 708–710.

49. Zhang ML [Retrospection and future of microskin grafting]. *Zhonghua Shao Shang Za Zhi* **24** (2008) 343–345.

50. Wheeland RG. The technique and current status of pinch grafting. *J Dermatol Surg Oncol* **13** (1987) 873–880.

51. Smahel J, Ganzoni N. Relay transplantation: a new method of expanding a free skin graft. *Br J Plast Surg* **28** (1975) 49–53.

52. Kreis RW, Mackie DP, Hermans RR, Vloemans AR. Expansion techniques for skin grafts: comparison between mesh and Meek island (sandwich) grafts. *Burns* **20**(Suppl. 1) (1994) S39–S42.

53. Vandeput J, Nelissen M, Tanner JC, Boswick J. A review of skin meshers. *Burns* **21** (1995) 364–370.

54. Richard R, Miller SF, Steinlage R, Finley RK, Jr. A comparison of the Tanner and Bioplasty skin mesher systems for maximal skin graft expansion. *J Burn Care Rehabil* **14** (1993) 690–695.

55. Rudolph R, Klein L. Healing processes in skin grafts. *Surg Gynecol Obstet* **136** (1973) 641–654.

56. Medawar PB. A second study of the behaviour and fate of skin homografts in rabbits: A Report to the War Wounds Committee of the Medical Research Council. *J Anat* **79**(4) (1945) 157–176.

57. Johnson TM, Lowe L, Brown MD, Sullivan MJ, Nelson BR. Histology and physiology of tissue expansion. *Journal of Dermatologic Surgery & Oncology* **19** (1993) 1074–1078.

58. Byun JH, Park BW, Kim JR, Lee JH. Expression of vascular endothelial growth factor and its receptors after mandibular distraction osteogenesis. *Int J Oral Maxillofac Surg* **36** (2007) 338–344.

59. Rhee ST, El-Bassiony L, Buchman SR. Extracellular signal-related kinase and bone morphogenetic protein expression during distraction osteogenesis of the mandible: *in vivo* evidence of a mechanotransduction mechanism for differentiation and osteogenesis by mesenchymal precursor cells. *Plast Reconstr Surg* **117** (2006) 2243–2249.

60. Loboa EG, *et al.* Mechanobiology of mandibular distraction osteogenesis: experimental analyses with a rat model. *Bone* **34** (2004) 336–343.

61. Bouletreau P, Longaker MT [The molecular biology of distraction osteogenesis]. *Rev Stomatol Chir Maxillofac* **105** (2004) 23–25.

62. Rheinwald JG, Green H. Serial cultivation of strains of human epidermal keratinocytes: the formation of keratinizing colonies from single cells. *Cell* **6** (1975) 331–343.

63. Bell E, Ivarsson B, Merrill C. Production of a tissue-like structure by contraction of collagen lattices by human fibroblasts of different proliferative potential *in vitro. Proceedings of the National Academy of Sciences of the United States of America* **76** (1979) 1274–1278.

64. Yannas IV, Burke JF. Design of an artificial skin. I. Basic design principles. *Journal of Biomedical Materials Research* **14** (1980) 65–81.

65. Burke JF, Yannas IV, Quinby WC, Jr., Bondoc CC, Jung WK. Successful use of a physiologically acceptable artificial skin in the treatment of extensive burn injury. *Ann Surg* **194** (1981) 413–428.
66. MacNeil S. Progress and opportunities for tissue-engineered skin. *Nature* **445** (2007) 874–880.
67. De Bie C. Genzyme: 15 years of cell and gene therapy research. *Regenerative Medicine* **2** (2007) 95–97.
68. Gallico GG, O'Connor NE, Compton CC, Kehinde O, Green H. Permanent coverage of large burn wounds with autologous cultured human epithelium. *New England Journal of Medicine, The* **311** (1984) 448–451.
69. Hansen SL, Voigt DW, Wiebelhaus P, Paul CN. Using skin replacement products to treat burns and wounds. *Advances in Skin & Wound Care* **14** (2001) 37–44; quiz 45.
70. Lam PK, *et al.* Development and evaluation of a new composite Laserskin graft. *The Journal of Trauma* **47** (1999) 918–922.
71. Andreassi L, Pianigiani E, Andreassi A, Taddeucci P, Biagioli M. A new model of epidermal culture for the surgical treatment of vitiligo. *International Journal of Dermatology* **37** (1998) 595–598.
72. Chan ES, *et al.* A new technique to resurface wounds with composite biocompatible epidermal graft and artificial skin. *The Journal of Trauma* **50** (2001) 358–362.
73. Tausche A-K, *et al.* An autologous epidermal equivalent tissue-engineered from follicular outer root sheath keratinocytes is as effective as split-thickness skin autograft in recalcitrant vascular leg ulcers. *Wound Repair and Regeneration* **11** (2003) 248–252.
74. Moustafa M, *et al.* A new autologous keratinocyte dressing treatment for non-healing diabetic neuropathic foot ulcers. *Diabet Med* **21** (2004) 786–789.
75. Haddow DB, Steele DA, Short RD, Dawson RA, Macneil S. Plasma-polymerized surfaces for culture of human keratinocytes and transfer of cells to an *in vitro* wound-bed model. *Journal of Biomedical Materials Research Part A* **64** (2003) 80–87.
76. Moustafa M, *et al.* Randomized, controlled, single-blind study on use of autologous keratinocytes on a transfer dressing to treat nonhealing diabetic ulcers. *Regenerative Medicine* **2** (2007) 887–902.
77. Navarro FA, *et al.* Sprayed keratinocyte suspensions accelerate epidermal coverage in a porcine microwound model. *Journal of Burn Care & Rehabilitation* **21** (2000) 513–518.
78. Zweifel CJ, *et al.* Initial experiences using non-cultured autologous keratinocyte suspension for burn wound closure. *Journal of Plastic, Reconstructive & Aesthetic Surgery* **61** e1–4 (2008).
79. Wood FM, Kolybaba ML, Allen P. The use of cultured epithelial autograft in the treatment of major burn wounds: eleven years of clinical experience. *Burns* **32** (2006) 538–544.
80. Ehrenreich M, Ruszczak Z. Update on dermal substitutes. *Acta Dermatovenerol Croat* **14** (2006) 172–187.
81. Lee KH. Tissue-engineered human living skin substitutes: development and clinical application. *Yonsei Med J* **41** (2000) 774–779.
82. Hata K. Current issues regarding skin substitutes using living cells as industrial materials. *J Artif Organs* **10** (2007) 129–132.
83. Nolte SV, Xu W, Rennekampff H-O, Rodemann HP. Diversity of fibroblasts — a review on implications for skin tissue engineering. *Cells, Tissues, Organs* **187** (2008) 165–176.
84. Brown RA, Phillips JB. Cell responses to biomimetic protein scaffolds used in tissue repair and engineering. *Int Rev Cytol* **262** (2007) 75–150.
85. Kolacna L, *et al.* Biochemical and biophysical aspects of collagen nanostructure in the extracellular matrix. *Physiol Res* **56**(Suppl. 1) (2007) S51–S60.
86. Supp DM, Boyce ST. Engineered skin substitutes: practices and potentials. *Clinics in Dermatology* **23** (2005) 403–412.
87. Candage R, *et al.* Use of human acellular dermal matrix for hernia repair: friend or foe? *Surgery* **144** (2008) 703–709 discussion 709.
88. Bello YM, Falabella AF, Eaglstein WH. Tissue-engineered skin. Current status in wound healing. *American Journal of Clinical Dermatology* **2** (2001) 305–313.

89. Winfrey ME, Cochran M, Hegarty MT. A new technology in burn therapy: INTEGRA artificial skin. *Dimens Crit Care Nurs* **18** (1999) 14–20.

90. Nguyen DQ, Dickson WA. A review of the use of a dermal skin substitute in burns care. *J Wound Care* **15** (2006) 373–376.

91. Myers SR, Partha VN, Soranzo C, Price RD, Navsaria HA. Hyalomatrix: a temporary epidermal barrier, hyaluronan delivery, and neodermis induction system for keratinocyte stem cell therapy. *Tissue Eng* **13** (2007) 2733–2741.

92. Pham C, Greenwood J, Cleland H, Woodruff P, Maddern G. Bioengineered skin substitutes for the management of burns: a systematic review. *Burns* **33** (2007) 946–957.

93. Waugh HV, Sherratt JA. Modeling the effects of treating diabetic wounds with engineered skin substitutes. *Wound repair and regeneration* **15** (2007) 556–565.

94. Jimenez PA, Jimenez SE. Tissue and cellular approaches to wound repair. *The American journal of surgery* **187** (2004) 56–64S.

95. Amani H, Dougherty WR, Blome-Eberwein S. Use of Transcyte and dermabrasion to treat burns reduces length of stay in burns of all size and etiology. *Burns* **32** (2006) 828–832.

96. Kumar RJ, Kimble RM, Boots R, Pegg SP. Treatment of partial-thickness burns: a prospective, randomized trial using Transcyte. *ANZ J Surg* **74** (2004) 622–626.

97. Tenenhaus M, Bhavsar D, Rennekampff HO. Treatment of deep partial thickness and indeterminate depth facial burn wounds with water-jet debridement and a biosynthetic dressing. *Injury* **38**(Suppl. 5) (2007) S39–S45.

98. Jehle KS, Rohatgi A. Use of porcine dermal collagen graft and topical negative pressure on infected open abdominal wounds. *Journal of Wound Care* **16** (2007) 36–37.

99. Zaulyanov L, Kirsner RS. A review of a bi-layered living cell treatment (Apligraf) in the treatment of venous leg ulcers and diabetic foot ulcers. *Clinical Interventions in Aging* **2** (2007) 93–98.

100. Cavorsi J, *et al.* Best-practice algorithms for the use of a bilayered living cell therapy (Apligraf) in the treatment of lower-extremity ulcers. *Wound Repair and Regeneration* **14** (2006) 102–109.

101. Eisenberg M, Llewelyn D. Surgical management of hands in children with recessive dystrophic epidermolysis bullosa: use of allogeneic composite cultured skin grafts. *British Journal of Plastic Surgery* **51** (1998) 608–613.

102. Boyce ST, *et al.* Cultured skin substitutes reduce requirements for harvesting of skin autograft for closure of excised, full-thickness burns. *The Journal of Trauma* **60** (2006) 821–829.

103. Hollinger JO, Hart CE, Hirsch SN, Lynch S, Friedlaender GE. Recombinant human platelet-derived growth factor: biology and clinical applications. *Journal of Bone and Joint Surgery* **90**(Suppl. 1) (2008) 48–54.

104. Steed DL. Clinical evaluation of recombinant human platelet-derived growth factor for the treatment of lower extremity ulcers. *Plastic and Reconstructive Surgery* **117** (2006) 143–149S; discussion 150S.

105. Margolis DJ, Lewis VL. A literature assessment of the use of miscellaneous topical agents, growth factors, and skin equivalents for the treatment of pressure ulcers. *Dermatologic Surgery* **21** (1995) 145–148.

106. Levi B, Rees R. Diagnosis and management of pressure ulcers. *Clin Plast Surg* **34** (2007) 735–748.

107. Kallianinen LK, Hirshberg J, Marchant B, Rees RS. Role of platelet-derived growth factor as an adjunct to surgery in the management of pressure ulcers. *Plastic and Reconstructive Surgery* **106** (2000) 1243–1248.

108. Broadley KN, *et al.* The diabetic rat as an impaired wound healing model: stimulatory effects of transforming growth factor-beta and basic fibroblast growth factor. *Biotechnol Ther* **1** (1989) 55–68.

109. Broadley KN, *et al.* Growth factors bFGF and TGB beta accelerate the rate of wound repair in normal and in diabetic rats. *Int J Tissue React* **10** (1988) 345–353.

110. Grazul-Bilska AT, *et al.* Effects of basic fibroblast growth factor (FGF-2) on proliferation of human skin fibroblasts in type II diabetes mellitus. *Exp Clin Endocrinol Diabetes* **110** (2002) 176–181.

111. Richard JL, *et al.* Effect of topical basic fibroblast growth factor on the healing of chronic diabetic neuropathic ulcer of the foot. A pilot, randomized, double-blind, placebo-controlled study. *Diabetes Care* **18** (1995) 64–69.

112. Tagashira S, Harada H, Katsumata T, Itoh N, Nakatsuka M. Cloning of mouse FGF10 and up-regulation of its gene expression during wound healing. *Gene* **197** (1997) 399–404.

113. Braun S, auf dem Keller U, Steiling H, Werner S. Fibroblast growth factors in epithelial repair and cytoprotection. *Philos Trans R Soc Lond B Biol Sci* **359** (2004) 753–757.

114. Gallin JI, Snyderman R. *Inflammation: Basic Principles and Clinical Correlates*, xxiii, 1335 p. (Lippincott Williams & Wilkins, Philadelphia, PA, 1999).

115. Hardwicke J, Schmaljohann D, Boyce D, Thomas D. Epidermal growth factor therapy and wound healing — past, present and future perspectives. *The Surgeon* **6** (2008) 172–177.

116. Pastore S, Mascia F, Mariani V, Girolomoni G. The epidermal growth factor receptor system in skin repair and inflammation. *The Journal of Investigative Dermatology* **128** (2008) 1365–1374.

117. Falanga V, *et al.* Topical use of human recombinant epidermal growth factor (h-EGF) in venous ulcers. *The Journal of Dermatologic Surgery and Oncology* **18** (1992) 604–606.

118. Tran KT, Lamb P, Deng JS. Matrikines and matricryptins: implications for cutaneous cancers and skin repair. *J Dermatol Sci* **40** (2005) 11–20.

119. Breen EC. VEGF in biological control. *Journal of Cellular Biochemistry* **102** (2007) 1358–1367.

120. Galiano RD, *et al.* Topical vascular endothelial growth factor accelerates diabetic wound healing through increased angiogenesis and by mobilizing and recruiting bone marrow-derived cells. *The American Journal of Pathology* **164** (2004) 1935–1947.

121. Frank S, *et al.* Regulation of vascular endothelial growth factor expression in cultured keratinocytes. Implications for normal and impaired wound healing. *J Biol Chem* **270** (1995) 12607–12613.

122. Lerman OZ, Galiano RD, Armour M, Levine JP, Gurtner GC. Cellular dysfunction in the diabetic fibroblast: impairment in migration, vascular endothelial growth factor production, and response to hypoxia. *The American Journal of Pathology* **162** (2003) 303–312.

123. Benest AV, *et al.* VEGF and angiopoietin-1 stimulate different angiogenic phenotypes that combine to enhance functional neovascularization in adult tissue. *Microcirculation* **13** (2006) 423–437.

124. Atluri P, Woo YJ. Pro-angiogenic cytokines as cardiovascular therapeutics: assessing the potential. *BioDrugs* **22** (2008) 209–222.

125. Sivakumar B, Harry LE, Paleolog EM. Modulating angiogenesis: more vs less. *JAMA* **292** (2004) 972–977.

126. Ferrara N. *Angiogenesis: From Basic Science to Clinical Applications*, 280 p. (CRC/Taylor & Francis, Boca Raton, 2007).

127. Bishop G. Regeneration after experimental removal of skin in man. *Am J Anat* **76** (1945) 153–181.

128. Brown JB, McDowell F. Epithelial Healing and the Transplantation of Skin. *Ann Surg* **115** (1942) 1166–1181.

129. Martinot V, Mitchell V, Fevrier P, Duhamel A, Pellerin P. Comparative study of split thickness skin grafts taken from the scalp and thigh in children. *Burns* **20** (1994) 146–150.

130. Jahoda CA, Reynolds AJ. Hair follicle dermal sheath cells: unsung participants in wound healing. *Lancet* **358** (2001) 1445–1448.

131. Claudinot S, Nicolas M, Oshima H, Rochat A, Barrandon Y. Long-term renewal of hair follicles from clonogenic multipotent stem cells. *Proc Natl Acad Sci USA* **102** (2005) 14677–14682.

132. Levy V, Lindon C, Harfe BD, Morgan BA. Distinct stem cell populations regenerate the follicle and interfollicular epidermis. *Dev Cell* **9** (2005) 855–861.

133. Langton AK, Herrick SE, Headon DJ. An extended epidermal response heals cutaneous wounds in the absence of a hair follicle stem cell contribution. *J Invest Dermatol* **128** (2008) 1311–1318.

134. Levy V, Lindon C, Zheng Y, Harfe BD, Morgan BA. Epidermal stem cells arise from the hair follicle after wounding. *Faseb J* **21** (2007) 1358–1366.

135. Fuchs E. Skin stem cells: rising to the surface. *J Cell Biol* **180** (2008) 273–284.
136. Tumbar T, *et al*. Defining the epithelial stem cell niche in skin. *Science* **303** (2004) 359–363.
137. Horne KA, Jahoda CA. Restoration of hair growth by surgical implantation of follicular dermal sheath. *Development* **116** (1992) 563–571.
138. Yamaguchi Y, *et al*. Regulation of keratin 9 in nonpalmoplantar keratinocytes by palmoplantar fibroblasts through epithelial-mesenchymal interactions. *J Invest Dermatol* **112** (1999) 483–488.
139. Gardiner DM, Endo T, Bryant SV. The molecular basis of amphibian limb regeneration: integrating the old with the new. *Semin Cell Dev Biol* **13** (2002) 345–352.
140. Endo T, Bryant SV, Gardiner DM. A stepwise model system for limb regeneration. *Dev Biol* **270** (2004) 135–145.
141. Thomson JA, *et al*. Embryonic stem cell lines derived from human blastocysts. *Science* **282**ÿ(1998) 1145–1147.
142. Hogan B. A shared vision. *Dev Cell* **13** (2007) 769–771.
143. Mak TW. Gene targeting in embryonic stem cells scores a knockout in Stockholm. *Cell* **131** (2007) 1027–1031.
144. Vogel G. Nobel Prizes. A knockout award in medicine. *Science* **318** (2007) 178–179.
145. Takahashi K, Yamanaka S. Induction of pluripotent stem cells from mouse embryonic and adult fibroblast cultures by defined factors. *Cell* **126** (2006) 663–676.
146. Wu Y, Wang J, Scott PG, Tredget EE. Bone marrow-derived stem cells in wound healing: a review. *Wound repair and regeneration* **15**(Suppl. 1) (2007) S18–S26.
147. D'Ippolito G, *et al*. Cooperative actions of hepatocyte growth factor and 1,25-dihydroxyvitamin D3 in osteoblastic differentiation of human vertebral bone marrow stromal cells. *Bone* **31** (2002) 269–275.
148. Metcalfe AD, Ferguson MW. Bioengineering skin using mechanisms of regeneration and repair. *Biomaterials* **28** (2007) 5100–5113.
149. Tepper OM, *et al*. Human endothelial progenitor cells from type II diabetics exhibit impaired proliferation, adhesion, and incorporation into vascular structures. *Circulation* **106** (2002) 2781–2786.
150. Bucala R, Spiegel LA, Chesney J, Hogan M, Cerami A. Circulating fibrocytes define a new leukocyte subpopulation that mediates tissue repair. *Mol Med* **1** (1994) 71–81.
151. Bellini A, Mattoli S. The role of the fibrocyte, a bone marrow-derived mesenchymal progenitor, in reactive and reparative fibroses. *Lab Invest* **87** (2007) 858–870.
152. Brittan M, *et al*. Bone marrow cells engraft within the epidermis and proliferate *in vivo* with no evidence of cell fusion. *J Pathol* **205** (2005) 1–13.
153. Fathke C, *et al*. Contribution of bone marrow-derived cells to skin: collagen deposition and wound repair. *Stem Cells* **22** (2004) 812–822.
154. Deng W, *et al*. Engrafted bone marrow-derived flk-(1+) mesenchymal stem cells regenerate skin tissue. *Tissue Eng* **11** (2005) 110–119.
155. Badiavas EV, Abedi M, Butmarc J, Falanga V, Quesenberry P. Participation of bone marrow derived cells in cutaneous wound healing. *J Cell Physiol* **196** (2003) 245–250.
156. Yang L, *et al*. Identification of fibrocytes in postburn hypertrophic scar. *Wound Repair Regen* **13** (2005) 398–404.
157. Balsam LB, *et al*. Haematopoietic stem cells adopt mature haematopoietic fates in ischaemic myocardium. *Nature* **428** (2004) 668–673.
158. Nygren JM, *et al*. Bone marrow-derived hematopoietic cells generate cardiomyocytes at a low frequency through cell fusion, but not transdifferentiation. *Nat Med* **10** (2004) 494–501.
159. Chesney J, Bacher M, Bender A, Bucala R. The peripheral blood fibrocyte is a potent antigen-presenting cell capable of priming naive T cells in situ. *Proc Natl Acad Sci USA* **94** (1997) 6307–6312.
160. Hartlapp I, *et al*. Fibrocytes induce an angiogenic phenotype in cultured endothelial cells and promote angiogenesis *in vivo*. *Faseb J* **15** (2001) 2215–2224.
161. Corcoran J, Hubli EH, Salyer KE. Distraction osteogenesis of costochondral neomandibles: a clinical experience. *Plast Reconstr Surg* **100** (1997) 311–315; discussion 316–317.

CHAPTER 12

Bone and Cartilage

Ashley Rothenberg* and Jennifer Elisseeff*

Cartilage and bone are intricately linked throughout life. The cells of both cartilage and bone develop from a common mesenchymal precursor. Furthermore, cartilage often serves as an intermediate precursor to bone in both embryonic development and adult wound repair. As a result of the intimate relationship between cartilage and bone, disease and damage to one tissue often results in phenotypic changes in the other. This relationship speaks of a constant communication between the cartilaginous and osseous tissues which extends beyond the developmental stage. As a result, it is important to consider both tissues when devising repair strategies.

Articular Cartilage Structure and Function

In the adult, hyaline cartilage covers the subchondral bone to provide a lubricating surface for joint movement as well as resistance to compressive and tensile forces. Chondrocytes, the cells in cartilage, and their extracellular matrix are organized in distinct zones related to their proximity to the subchondral bone. Each zone has a characteristic matrix composition and organization in addition to a unique cell density and morphology. Directly adjacent to the bone is a layer of calcified cartilage which consists of hypertrophic chondrocytes. These hypertropic cells express a variety of factors including Type X collagen and serve as an intermediate between the layers of cartilage and bone.[1] Moving away from the subchondral bone and calcified cartilage layers are the deep, middle, and superficial zones of cartilage, respectively. The deep zone is organized of alternating columns of cells and matrix which run perpendicular to the articular surface.[2] The deep zone consists of the highest proportion of aggrecan to collagen Type II and a low density of large round cells.[1–3] This limited cell density is paralleled by a limited fluid flow in the cartilage zones closest to the subchondral bone.[4] This makes sense given that the nutrient and oxygen supplies to cartilage are obtained via diffusion.[2] The middle zone of cartilage consists of a thick layer of a randomly organized matrix composed of aggrecan, Type II and VI collagen, and decorin.[1,2] The middle zone of cartilage consists of an intermediate density of rounded chondrocytes. Finally, chondrocytes of the superficial zone of cartilage are flattened in a matrix oriented parallel to the surface. Chondrocytes near the articular surface experience the highest levels of fluid flow and compressive strains which inevitably influences the matrix composition and organization.[4] These chondrocytes secrete lubricin, or superficial zone protein, which is believed to play a role in reducing friction on the articular surface.[1]

Bone Structure and Function

Bone tissues play a structural role in the skeleton. These tissues must both resist deformation without being too dense to permit body movement, while being flexible enough to absorb some forces and not fracture easily.[5] Generally, bone organization involves an inner

---

*Corresponding authors.

mass of spongy cancellous bone interwoven with bone marrow which is surrounded by a compact, dense, cortical bone. Bone is comprised of a Type I collagen scaffold which is then mineralized with calcium phosphate crystals or hydroxyapatite and non-collagenous proteins.[6,7] The non-collagenous components in bone include proteoglycans such as decorin, biglycan, osteoadherin, and osteoglycin/mimecan hyaluronan, and heparan sulfate.[8] In the cortical bone, collagen fibers run in cylindrical layers around canals to form osteons.[6] The canals provide an area in which blood vessels run to provide nutrients and waste removal from the component cells.[9] This micro architecture and mineral density in bone help define its ability to resist and or absorb mechanical deformations given the inverse relationship between mechanical flexibility and mineral content.[10,5] The variation in proteoglycan expression through different regions of bone also modulates the expression pattern of cytokines and growth factors.[8] In addition, organic components of bone may help regulate its growth. For example, osteocalcin is believed to play a role in controlling hydroxyapatite crystal formation through its calcium binding activity.[6]

Bone tissues undergo remodeling in response to mechanical loading.[11] Osteoclasts, the cells in bone responsible for resorption, and osteoblasts, the cells responsible for new bone growth, continue to play an active role throughout life since bone tissues are continually remodeling. However, the relative rate of bone formation to bone resorption has been shown to generally decrease with advancing age.[5] Interestingly, this decrease in remodeling corresponds with both an increase in the fatty content and decrease in hematopoietic content of the bone marrow.[12]

Natural Repair Capacity of Bone and Cartilage

Articular cartilage exhibits an inadequate capacity for self repair, possibly as a result of its avascular nature which causes it to rely on diffusion for nutrients and waste removal. Injuries restricted to the cartilage exhibit almost no repair aptitude whereas those involving the subchondral bone exhibit some, though limited, wound healing.[13] This suggests an important relationship between bone and cartilage health. Extensive research is therefore underway to develop enhanced repair strategies for cartilage damage and subsequent pain, many which involve both cartilage and bone.

Whereas cartilage has limited to no capacity for self repair, bone is often capable of repairing non-critical sized defects and fractures. Calcified cartilage and bone microfractures have been shown to heal through either calcification of the fracture region or resorption in the damaged area followed by deposition of new bone.[14] If however, the bone defect exceeds a certain size, then the defect is often filled with a fibrous tissue or otherwise fails to spontaneously heal.[15]

Given the lack of sufficient inherent repair capacity in cartilage and large bone defects, there exists a need for development of new methods for repair and replacement of diseased and damaged skeletal tissue. As such, research into the natural development and disease progression in these tissues is an important area of research.

Cartilage and Bone Development

During the 3rd to 8th week of embryonic development, the mesodermal layer of the embryo forms the beginnings of the skeletal system. Generally this entails two pathways of bone formation: endochondral ossification and intramembranous ossification.

Endochondral ossification involves creation of an intermediate cartilaginous structure which is later replaced by bone. This form of ossification is most pertinent to the current discussion as it is the pathway utilized for development of limb structures including the arms, legs, hands, and feet. During endochondral ossification, mesenchymal cells are recruited to the area and proliferate to form mesenchymal condensations. These mesenchymal cells produce a matrix composed of hyaluronan and Type I collagen before beginning differentiation towards a chondrocyte lineage. As chondrogenesis begins, the mesenchymal condensates express factors such as Gli 3, Sox 9, Hox A & D, as well as matrix components such as Type I and III collagen and fibronectin.[16,17] As differentiation continues, early chondroprogenitors are formed which express markers such as Sox 5,6,9, Runx2 (a.k.a. Cbfa1), bone morphogenic proteins (BMPs), and Type IIA collagen.[16,17] Finally, a proliferating chondrocyte is formed which continues to express Sox 5,6 and 9 while expressing other cartilage specific markers such as Type II B collagen, COMP, and aggrecan.[16,17] In skeletal regions that are to remain cartilaginous, Indian Hedgehog (Ihh) helps inhibit hypertrophy.[18] For skeletal regions that are to be replaced by bone, the chondrocytes become hypertropic and begin to express VEGF which stimulates vascularization. These cells degrade their matrix to allow room for the enlarged cell size in addition to the invasion of blood vessels and calcium deposition which ultimately allow for ossification.[19] At this stage, the hypertrophic chondrocytes express Type X collagen and Runx2.[16,17] Finally the mineralized cartilage tissue is invaded by a fresh population of mesenchymal cells which differentiate into osteoblasts and produce a matrix rich in Type I collagen. These cells express markers such as osteopontin, osteocalcin, and bone sialoprotein after reaching a mature osteocyte stage.[16,17,20]

Throughout this process, matrix metalloproteinase expression (MMP) and a disintegrin and metalloprotease with Type 1 thrombospondin motifs (ADAMTS) are vital. For example, expansion of precartilagenous mesenchymal condensates appears to be reliant on ECM remodeling.[21] In addition, both the invading osteoblasts and hypertrophic chondrocytes express proteinases, such as MMP-13, which degrades Type II collagen, to degrade their extra cellular matrix.[22] ADAMTS-1, -4 and -5, which cleave aggrecan and versican, are also important in the process of bone development in the limbs.[23,24]

In adolescents, a similar series of chondrogenic and osteogenic lineages may be observed simultaneously in consecutive zones of the growth plate. At the edge of the growth plate, a zone of resting chondrocytes sits adjacent to a region of proliferating chondrocytes.[18,25] This proliferative zone is then followed by a columnar zone of chondrocytes, a region of prehypertropic chondrocytes, and a zone of hypertrophy.[18,19] Finally, there exists a region of apoptotic chondrocytes which are bordered by the perichondrium and finally, the periosteum, blood vessels, and bone.[18,19] This also requires the function of MMP-9,-13, and -14 to properly control processes such as cell migration and death.[26]

Joint formation involves a sequential progression of elongation of the initial cartilage tissue via chondrocyte proliferation, branching to form a continuous cartilage element, and finally segmentation of the cartilage to form distinct skeletal elements.[27–29] The process of segmentation involves apoptosis of chondrocytes in areas between skeletal elements, such as the metacarpals and proximal phalanges in the hand.[29]

The other pathway for bone formation, the intramembranous pathway, involves differentiation of osteoblasts directly from mesenchymal condensations. This pathway is utilized primarily for formation of the membranous skeletal structures, such as the skull and thus will not be discussed in detail here. However, it should be noted that this pathway does play a role in the natural process of wound repair in the periosteum thus certain details of the process are of interest and will be discussed briefly later as it pertains to fracture healing.

Signaling in Development

The process of skeletal formation is controlled by a complex cascade of differentiation factors including the transforming growth factor B (TGF-B) superfamily, which includes the BMPs, the Sox family of transcription factors, and the Wnt, Hedgehog, and fibroblast growth factor (FGF) families.[19,29] These families of differentiation factors are controlled spatially along three axes by the apical ectodermal ridge, which controls the proximal-distal axis, the zone of polarizing activity, which controls patterning for the anterior-posterior axis, and finally, the surface ectoderm, which controls patterning along the dorsal-ventral axis.[30]

It is believed that the growth and differentiation factor (GDF) subfamily of the TGFB superfamily plays an important role in joint formation. For example, Gdf5 is believed to play a role in formation of the phalangeal skeletal elements during digit development.[27,28] In fact, absence of Gdf5 has been shown to induce apoptosis and failed development of phalanges.[31] This may be related to sonic hedgehog (Shh) signaling, since the forced expression of Shh has been shown to lead to fusion of digit elements and absence of Gdf5 and sFrp2.[32] Furthermore, Shh, which is mediated via Gli3, has been implicated in anterior-posterior patterning in the developing limb while FGF signaling has been implicated in proximal-distal patterning.[33,34]

Other members of the TGFB superfamily also play an important role in limb development and digit formation. Interplay between Activin and TGFB has been suggested to play a role in both initiation of digit formation and chondrogenesis.[35] Furthermore, BMPs have been shown to play an important role in segmentation via regulation of cell apoptosis and digit formation.[35] In addition, recent evidence from a human genetic study suggests that a frameshift mutation in BMP4 results in a poly/syndactyly phenotype.[36] The Hox genes have also been shown to play an important role in the specification of joints. Hox 13 has been shown to designate metacarpals and phalanges while Hox 12 specifies the carpals.[37] This gene family has also been implicated in specification of the proximal-distal patterning of the joints.[30]

Ultimately, a complex interplay amongst a variety of factors is necessary for successful joint formation. Evidence has been provided that Wnt-14 is important for initiation of joint formation.[28] In addition, Indian hedgehog (Ihh) has been implicated in digit elongation, while FGF8 expression prevents terminal phalange development.[34] Furthermore, Sulf1 expression has been shown to play a variety of roles including increasing aggregation and chondrogenesis of chondrocyte precursors, elimination of new phalange and joint formation, and stimulation of existing phalange enlargement.[38] Recent evidence has also suggested that the regulation of digit identity specification is related to a unique signature of SMAD1/5/8 activity in a particular region of the digital ray.[39]

Many of these pathways are still being fully elucidated, particularly in the context of joint formation. However, excellent current reviews on the potential interplay in these pathways in the growth plate and embryonic development exist and thus an extensive discussion of these pathways will not be included here. For example, an extensive review on BMP signaling in the growth plate was recently published by Pogue and Lyons.[25] In addition, regulation of the role of Sox, parathyroid hormone, and FGFs in endochondral ossification is outlined by both Crombrugghe et al. and Mackie et al.[18,19] and an overview of hedgehog signaling in development has been provided by Ehlen et al.[33] Finally, an overview of the process of synovial joint development is discussed by Khan et al.[37]

Human Cartilage Repair Following Injury

Following a partial thickness defect injury, an area of apoptosis initially surrounds the wound.[40] After the initial period of cell death, a period of enhanced cell proliferation and matrix synthesis occurs, however, this is insufficient to repair the defect.[40]

The optimal outcome which can currently be hoped for via the inherent repair capacity of cartilage is the replacement of the hyaline cartilage with fibrous tissue or fibrocartilage in response to a full thickness defect.[40,41] It is generally believed that the failure of partial thickness defects to demonstrate this same repair mechanism is related to the lack of access to the stem cell progenitors located in the bone marrow.[40,41,42] Since this repair mechanism relies on the occurrence of a full thickness defect, often cartilage excision and microfracture, a procedure to break the subchondral bone, will be performed in partial-thickness defects.[41,43] After such a break, blood entry into the wound initiates the clotting cascade which ultimately induces formation of fibrous tissue.[42]

Another tissue engineering repair strategy which is currently used in the clinic is autologous chondrocyte implantation (ACI). This method involves the harvest of a small biopsy (greater than or equal to 300 mg) of cartilage from a non weight bearing region of the patient.[42] The chondrocytes from this biopsy are then isolated, expanded, and implanted back into the patient into the cartilage defect under a piece of periosteum sutured over the defect and coated with fibrin glue.[40,42] Chondrocytes are commonly implanted at a minimum concentration of 30 million cells/ml.[42] This is related to research in which chondrocytes seeded in polylactide-polyglycolide scaffolds at densities below 10 million cells/ml were unable to adequately support formation of hyaline cartilage tissue, while an initial seeding density of 20–100 million cells/ml was found to be sufficient.[42] The autologous chondrocyte implantation procedure has been used, however, it has not been proven to be more effective than microfracture based on histological analysis or patient assessment at 2 and 5 year follow ups.[41]

Just as autologous tissues may be used in reconstruction of cartilage defects in the knee, use of autografts for finger joint reconstruction following trauma or tumor resection have also been examined. For example, osteochondral grafts have been used to treat articular cartilage defects in the hand.[44] Specifically, osteochondral autografts have been harvested from the rib[44] as well as the toe.[45]

Bone Tissue Repair Following Injury

Bone repair naturally occurs through a combination of endochondral ossification and intramembranous ossification. After injury, blood escapes from injured vasculature and a hematoma is formed through activation of the clotting cascade.[46,47] Concomitantly occurring is an inflammatory response during which chemotactic factors are released initiating a sequence of cellular invasions including an initial phase of leukocyte recruitment, followed by macrophages, and finally fibroblasts. This also instigates the recruitment and proliferation of mesenchymal stem cells and osteoprogenitor cells.[6,46] During the period of mesenchymal stem cell proliferation, the mesenchymal cells produce a fibrous tissue matrix rich in Type III collagen.[6,48] This is subsequently replaced by a fibrocartilage intermediate comprised primarily of Type II collagen.[6,46] The fibrocartilage intermediate is then replaced by bone tissue and a new matrix of Type I collagen.[6,46]

This process is accompanied by direct intramembranous ossification of new bone at the periosteum.[6,42] This area is invaded by mesenchymal progenitor cells which differentiate directly into osteoblasts.[6,49] Evidence suggests that tumor necrosis factor α (TNF-α)

plays a vital role in the recruitment of these osteoprogenitor cells.[50] These osteoblasts proceed to secrete a matrix primarily comprised of Type I collagen and to a lesser extent Type V collagen.[6]

Finally, remodeling of the bone through osteoclast resorption of necrotic areas and deposition of new bone by osteoblasts occurs in response to mechanical stimulation.[46] This allows for integration of the bone to form a functionally recovered tissue.

Repair of non-critical sized defects in bone are often accomplished simply by immobilization of the defect area (e.g. through casting) however, critical sized defects are accompanied by a lack of spontaneous healing. In critical sized defects, the defect area is too substantial and exceeds bone's inherent repair capacity and thus less than ten percent of the defect is regenerated.[15] Other defects may also experience nonunion when fibrous tissue forms in the place of bone, resulting in a non-fully functional recovery of the tissue.[15]

Finger joint reconstruction following trauma or tumor resection has employed both autograft and allograft techniques for repair. Autografting of the toe joint is one method used to reconstruct the joints in the hand.[45,51] Although autografting techniques are desirable because of their ability to preserve finger movement, they result in impaired function of the donor site. Furthermore, autografting results in a longer surgery, longer hospital stay, and in turn greater cost.[52] Finally, the supply and quality of autograft tissues is variable.[52] Osteocartilagenous allograft of cadaveric tissue is preferred because of the lack of donor site morbidity.[53] For example, following tumor resection an osteochondral allograft was utilized to reconstruct the middle phalanx and preserve a functional finger.[54] However, due to disease transmission and immunological concerns, tissues in which the cells have been removed prior to implantation are primarily used.[53] Furthermore, allograft availability remains a problem.[55]

Wound Healing and Limb Regeneration

Because the repair capacity of critically damaged cartilage and bone in its best case scenario results in functionally compromised fibrous tissue, many have turned to tissue engineering to provide replacements. Furthermore, an investigation of the process of wound healing in species which can inherently regenerate limbs, may provide important insight into the critical components required for regeneration.

Urodele amphibians (newts and salamanders) retain the capacity to regenerate severed limbs even as adults.[56–58] In these animals, limb amputation is followed by a progression of wound repair involving epidermal cell migration and formation of a wound epithelium. This is followed by a regenerative process involving formation of an apical epithelial cap (AEC), dedifferentiation and proliferation of cells to form a blastema of mesenchymal cells. Finally, these cells undergo redifferentiation and transdifferentiation of the dedifferentiated cells to form a fully functional limb.[56,57,59,60] In this process, the wound epithelium, and later the AEC, play a role similar to that which the AER plays in development.[59,60] In the newt, formation of the AEC relies on dedifferentiation.[60] This is in contrast to *Xenopus* (tadpoles) and *Planaria*. *Xenopus* rely on lineage specific progenitors and are only able to regenerate tissues at stage specific periods of development. *Planaria* rely on a stem cell population to regenerate tissues rather than utilizing dedifferentiation followed by subsequent proliferation and redifferentiation or transdifferentiation.[61,62]

Though the control of these processes is currently not well understood, many genes implicated in embryonic and growth plate development are also believed to play a role in regeneration. For example, FGF, Wnt/B-catenin, and BMPs/Msx have all been implicated in limb regeneration.[60,62] For example, BMP signaling is necessary for regenerative stages successive to wound epithelium formation.[63] Furthermore, Wnt and BMP signaling pathways have been shown to play a role in dedifferentiation.[64]

TGF-B signaling has also been implicated in several roles. For example, inhibition of TGF-B was found to switch the repair mechanism from regeneration to a healing process, suggesting that TGF-B may play a role in initiation of regeneration in *Xenopus.*[65] In addition, research suggests that TFG-B is required for epithelial wound repair in *Xenopus* tail regeneration.[65]

Matrix metalloproteinases (MMPs) play an important role in wound healing and regeneration because of the necessity of remodeling the extracellular matrix and therefore tissue breakdown.[62] In addition to their role in tissue remodeling, MMPs are required to help break down debris so it can be cleared after tissue damage.[60] In addition it has been suggested that MMP activity may play a role in the dedifferentiation of cells.[59] This is consistent with reports that local signals are important for maintenance of a state of differentiation, leading to the fully differentiated state in humans, and dedifferentiation in the newt.[57]

Apoptosis has also been found to play a key role in regeneration. Although, the exact reasons are unknown, some apoptosis has been shown to be required for regeneration.[56,62] It is important to note that too much apoptosis will also prevent regeneration from occurring.[62] It is possible that this requirement for apoptosis is related to the expression of heat shock proteins (Hsp). Hsp expression has been found to play a role in the regulation of apoptosis.[66,67] Hsp60 specifically has been found to have increased expression in the native regenerative *Xenopus* as compared to the nonregenerative transgenic *Xenopus.*[63] Hsp60 has been shown to be a proinflammatory molecule.[67] Thus, it makes sense that prolonged periods of Hsp60 expression might be inhibitory given that it would cause prolonged periods of inflammation and cell death. In contrast, initial periods of increased Hsp60 expression could be vital to mount the inflammatory response required to assemble progenitor cells with the proliferative and differentiation capabilities required for regeneration. This inflammation would also aid in the recruitment of macrophage and thus provide MMPs to allow for the tissue remodeling which is critical for successful regeneration. This is consistence with evidence that the liver, which is capable of regeneration over a period of several months in humans, undergoes regeneration in response to secretion of signals produced by macrophages as part of the innate immune response.[68]

Other factors have also been uncovered which are important for limb regeneration. For instance, retinoic acid (RA) has been shown to play an important role in limb patterning in both tadpoles and newts.[69,70] In addition, the newt anterior gradient (nAG) is a mitogen which has been shown to be permissive of digit regeneration in the absence of enervation, however it is not sufficient to restore nerve regrowth.[58,71] In humans, AG2 which has high consensus to nAG, is highly expressed in some metastatic tumors thereby supporting the suggested role of nAG in cell division.[58]

The ability of *Xenopus* to regenerate limbs during its tadpole stage, but not as an adult suggests a time dependent nature for regeneration in some species early in life. For example, mammals such as juvenile mice are capable of regenerating their digit tips.[60] In fact, even humans have some limited regenerative capacity. In addition to the ability of bone to heal after fraction and liver regenerative capabilities, children have the abilities to regenerate fingertips. For example, children who experience amputation of the proximal end of their distal phalanx are capable of regenerating a fully functional fingertip, even retaining the original fingerprint.[60,64] This regeneration of the phalanx has been shown to involve signaling through Msx and BMP signaling cascades.[64]

Non-regenerative Tissues

The fact that previously regenerative tissues can become nonregenerative at later stages of life indicates that some key requirement for regeneration is lost in the adult. The mechanism of this loss of regenerative capacity is not fully understood, however, several clues as to how this regenerative switch might be turned on and off have been uncovered. For example, methylation patterns and microRNAs both have been shown to regulate expression of genes important for regeneration. Enhanced methylation of the enhancer of the growth factor Shh has been shown to correlate with a loss of regenerative capacity in *Xenopus* and urodels.[62] Similarly, microRNA expression has been shown to hinder regeneration and in fact, the inhibition of miR-133 can help restore regeneration.[72]

Other key differences in the wound healing process exist when comparing tissues in which regeneration occurs and those in which only wound repair occurs. For example, although wound closure and inflammation are the first steps in wound healing in both regenerative and non-regenerative tissues, this process of healing has been shown to be much faster in regenerative tissues.[59,60,62,64] Development of the wound epithelium was shown to occur on the order of hours (typically less than 12 hours) in species which were able to regenerate, while wound closure takes days in non-regenerative species.[59,64] Interestingly, this regeneration occurs much faster than even original development and, after amputation, tadpoles are capable of "catching up" developmentally with non amputee peers.[62] The faster wound healing which occurs in regenerative tissues also results in a decrease in the duration of the inflammatory response in these tissues when compared to non-regenerative tissues.[60] Furthermore, the dedifferentiation and proliferation observed in regenerative tissues is not adequately recapitulated in non-regenerative tissues.[64] As the reduced proliferation is accompanied by a decrease in MMP expression, this limits the remodeling capacity in these tissues.[60] It is therefore not surprising that repair of non-regenerative tissues often results in fibrous tissue and scar formation.

Disease

In order to achieve functional recovery following damage to the hand, as may occur in traumatic injury, we rely on tissue engineering strategies to overcome the limited regenerative capacity inherent to humans. As such, the examination of natural developmental cascades, and differences between regenerative and non-regenerative tissues is an excellent starting point for the development of therapeutic strategies. Through examination of these pathways, tissue engineers can hope to develop a variety of "replacement parts" with which to repair damaged tissue. These investigations are also pertinent for situations where tissue must be replaced following tumor resection or for congenital birth defects such as syndactyly or Apert syndrome. In each of these cases, development of tissue replacement therapies makes sense as traumatic injury cannot be treated proactively. However, in the case of disease, proactive tissue engineering strategies may also be effective strategies for reducing the clinical disease phenotypes. As such, another important area of investigation for tissue engineering is the study of disease states in hopes of catching and correcting diseases before replacement therapies become necessary. Through enhanced understanding of joint diseases, a more proactive approach to disease therapy may be devised which treats the early indicators of disease, thereby preventing disease progression, circumventing the need for extensive repair.

Arthritis

Arthritis is prevalent in the elderly. For example, radiographic evidence of osteoarthritis (OA) in the carpometacarpal joint is exhibited by 40% of women and 25% of men over the age of 75.[73] In fact, recent estimates suggest that approximately 27 million Americans are affected by osteoarthritis[74] while the CDC reports that 1.293 million people suffer from rheumatoid arthritis (RA).[75] This is especially alarming since RA and OA both result in significant pain and loss of mobility.[73,76,77] Furthermore, OA is the leading cause of disability in the elderly,[78] while over 30% of patients with RA have ceased work citing health reasons within 10 years of disease onset.[77]

Osteoarthritis consists of a degradation of the articular cartilage at the joint surface.[78] As the disease progresses, the cartilage matrix degrades and its ability to function in smooth joint movement is compromised. Rheumatoid arthritis also involves destruction of the articular cartilage,[79] as it is invaded by proliferating synovial cells.[80] Both diseases result in changes in extracellular matrix composition, such as increased fibronectin concentration as a result of increased synthesis.[81]

Interestingly, although both OA and RA are generally thought of as diseases of the cartilage, significant disease manifestation is also observed in the subchondral bone. In OA, the subchondral bone is known to express increased levels of MMPs and other markers of tissue remodeling.[82] In addition, osteoarthritic subchondral bone layer seems to thicken, though the bone quality appears osteoporotic and has decreased mineralization.[83,84] In contrast, in rheumatoid arthritis, the bone and mineralized cartilage are resorbed by osteoclasts.[85]

Just as inflammation plays a role in a tissue's ability to repair or regenerate itself, inflammatory responses also play a large role in mediation of diseases such as osteoarthritis and rheumatoid arthritis. Though OA is considered to be more degenerative, while RA is considered to be more inflammatory, cartilage is lost in both. This is believed to be related to an imbalance between catabolism and anabolism, which is induced by inflammatory cytokines.[86]

In osteoarthritis, part of disease manifestation is believed to result from chondrocyte senescence which occurs in response to oxidative stress and proinflammatory cytokines.[87] In addition, there seems to be a relationship in OA between pro-inflammatory mediators such as tumor necrosis factor (TNF) alpha ($-\alpha$) and interleukin-1 (Il-1) and MMP expression.[79] Furthermore, proteomics analysis of OA chondrocytes has shown that they upregulate production of certain stress proteins while decreasing production of proteins involved in glycolysis.[88] Similarly, in RA certain disease attributes can be directly related to inflammation. For example, TNF-α and Il-1are believed to induce macrophages to differentiate into osteoclasts through induced synovial expression of receptor activator of nuclear factor (NF)kB ligand (RANKL).[85] In addition, TNF-α and Il-1 regulate MMP expression.[26]

In arthritis, increased production of MMP-1 (collagenase-1) and MMP-13 (collagenase-3), which both degrade Type II collagen, seems to overwhelm the production of tissue inhibitors of metalloproteinases (TIMPs).[26] Furthermore, a disintegrin and metalloproteinase with thrombospondin motif (ADAMTS) -4 and -5 mediate much of the aggrecan degradation in arthritis.[26] Although this proteinase activity was formerly believed to be related solely to the destruction of tissues in arthritis, recent evidence suggests another possible role as an attempt to provide remodeling for repair.[89] This makes sense given the vital role these proteinases have been shown to play in development and natural wound repair.

It is possible that much of the disease attributes of OA and RA are derived from inflammatory mediated production of reactive oxygen species. Adiponectin, a member of the TNF superfamily, is able to stimulate nitric oxide (NO) production through NO synthase Type II expression.[90] HSP90 is also believed to play a role in NO synthesis and was found to be increased in OA chondrocytes.[88] Other factors such as TNF-α, Il, and interferon gamma (IFN-γ) are also involved in the stimulation of NO production.[91] In contrast, certain agents such as transforming growth factor (TGF-B), and interleukins-4, 10, and -13 have been shown to inhibit NO production.[91] Furthermore, the effects of Il-1, which inhibits chondrocyte production of aggrecan and Type II collagen synthesis and thereby the mature cartilage phenotype are believed to be related to NO production.[92] This reactive oxygen species production is thought to play multiple roles in disease progression including decreased extracellular matrix synthesis, increased cell death, procollagen IIA, and increased MMP activity.[91,30] Several of these processes and the signaling pathways involved are reviewed by Henrotin, Y.E. *et al.*

Overall, OA chondrocytes produce more NO than RA chondrocytes, while both produced significantly higher levels than normal tissues.[86,93,94] Further, OA chondrocytes were shown to have increased oxidative DNA damage when compared to normal chondrocytes.[95]

Tissue Engineering of Cartilage and Bone

Drawing from the lessons of regenerative tissues and the disease manifestation of OA and RA, it makes sense that one aspect of effective repair in cartilage and bone will involve some sort of modulation of the immune response to allow for regenerative processes to occur rather than fibrous tissue formation. In addition, given the lack of a sufficient proliferative cell source, either through depleted stem cell population, senescence or death of fully differentiated cells, or some other process, delivery of a cell to serve as a cell source for tissue formation is also a promising area for tissue engineering research. Delivery of these cells may also aid in stimulation of regenerative processes by speeding tissue growth.

In addition to these primary lessons, other strategies originating from the natural development and wound repair processes are also being pursued. For example, temporal and spatial delivery of growth factors, cytokines, mitogens, and chemokines may be delivered through various drug delivery schemes to attempt to stimulate aspects of the repair process such as extracellular matrix production, cellular migration, and cell differentiation. Further, different scaffolds are being developed to study cells in a three-dimensional environment, building upon the traditional two-dimensional biological approach to *in vitro* cell growth. In addition, combination therapies which involve delivery of multiple factors such as cells and or growth factors delivered in a scaffold. Finally, natural development and wound repair illustrate the importance of the underlying relationships between different cell and tissue types. As a result, several tissue engineering strategies are focused on the examination of the interactions of two or more cell types in co-culture with one another.

Cell Choice

To repair diseased and damaged tissues, delivery of cells to help repopulate the defect with new tissue is a common line of investigation. Several key considerations are important when considering a cell choice for cellular therapy. Will the cell source be autologous or allogenic? Will the cells be fully differentiated before they are implanted (e.g. chondrocyte or osteoblast implantation) or will a stem or progenitor cell population be delivered and allowed to differentiate *in vivo*?

Current clinical procedures are using autologous delivery of fully differentiated cells in efforts to repair cartilage through ACI. However, given the limited number obtained through biopsy combined with limited proliferative capacity of these cells, a lot of research is focused on stem cell therapies for the development of future treatment modalities. The main cell responsible for building bone in native tissue, the osteoblast, also suffers from a low cell yield from biopsy and limited proliferative capacity.[96] As a result, two main classifications of stem cells are currently under investigation for cartilage and bone tissue engineering: mesenchymal stem cells and embryonic stem cells.

Mesenchymal Stem Cells

Mesenchymal stem cells (MSCs) are an adult stem cell that has been derived from a variety of sources including bone marrow, umbilical cord blood, and adipose tissue. MSCs have a proven capacity to differentiate into a variety of lineages including bone,[97–99] fat,[100] cartilage,[101,102] and other musculoskeletal tissues.[103,104] In addition, MSCs have been shown to possess a homing capacity allowing them to gather at sites of injury.[105,106] MSCs provide an enticing cell source for tissue engineering given their potential for mounting a responsive, multifaceted approach to repair.

As previously discussed mesenchymal cells are known to play a vital role as a cell source for development of limb structures such as bone and cartilage. Given the proven capacity for mesenchymal stem cells to differentiate into a wide variety of musculoskeletal lineages and home to sites of injury, many believe that native MSCs serve as a reservoir of reparative cells in the adult. This characteristic, in conjunction with their proliferative capabilities, makes them a popular choice for therapeutic development of stem cell therapies.

Mesenchymal stem cells possess several other potentially promising attributes for tissue engineering and regenerative strategies. Not only do MSCs possess the capacity to differentiate into a variety of cell lineages, they also are capable of secreting factors to stimulate other cells in their vicinity in a paracrine manner. This enables MSCs to serve as a support cell in engineered tissues,[107] which may be the primary mechanism of their observed therapeutic effect. Several studies have shown that MSCs are capable of secreting several classes of factors including chemokines, cytokines, growth factors, and protease inhibitors.[108] More specifically, MSCs have been shown to secrete factors which modulate matrix turnover in cartilage and bone, such as TGF-B, MMPs, aggrecanases, BMPs, and tissue inhibitors of MMPs (TIMPs).[108–110]

Another attractive function of MSCs is their role in the immune system and stress responses. MSCs have been shown in multiple instances to modulate immune response[111,112] and secrete immunosuppressive factors.[113] MSCs have been shown to help control graft-versus-host disease.[111,112] In addition, MSCs have been shown to secrete cytokines, such as IL-10 which inhibit inflammatory cytokines such as tumor necrosis factor alpha (TNF-α) and interferon gamma (IFN-γ).[111] In addition, MSCs have been shown to increase gene expression for inducible nitric oxide synthase (iNOS) in response to IFN-γ along with another of several proinflammatory cytokines. iNOS in turn catalyzes nitric oxide synthesis which consecutively modulates the immune responses through its effects on T-cells, macrophages, and cytokine receptors among others.[114,115] For example, NO production by MSCs has been found to inhibit T-cell proliferation.[116] Proteomic experiments have also illustrated that MSCs produce bioactive factors related to oxidative stress and other stress responses.[110] For example, MSCs have been shown to secrete numerous members of the heat shock protein family, such as Hsp60.[110]

Given that the population of MSCs has been shown to decrease significantly with age,[117] some have posed the possibility that the loss of regenerative capacity in adulthood may be in part due to a dwindling source of MSCs and other multipotent progenitor cells. This is further supported by evidence that the MSCs isolated from RA patients have a reduced proliferative capacity caused by decreased telomere length,[118] while MSCs from OA patients are shown to have reduced chondrogenic potential.[119]

The role of MSCs in the adult organism is likely multifaceted. MSCs are important developmentally as a cell source. Though their native role in the adult organism is as yet unproven, they possess a wide variety of potential roles which far surpass their use simply as a cell source for repair. The homing and immunomodulatory capabilities of MSCs are especially enticing given the evidence suggesting that a prolonged immune response is related to the lack of regenerative capabilities in adult mammalian tissues. Furthermore, MSCs role in immune regulation and oxidative stress is alluring for potential treatment of diseases such as arthritis in which the immune and stress responses play a seemingly direct role in disease manifestation. The production of factors stimulating proliferation and preventing apoptosis by MSCs[113] may also help restore regenerative capacity to cartilage and bone tissues. For example, MSCs might be able to help stimulate wound regeneration given their ability to increase the speed of wound healing,[120] a process which is slowed in non regenerative tissues compared to regenerative tissues.

Numerous studies have been conducted to examine the potential of MSCs to serve as a source for cartilage and bone tissue repair. In addition to the *in vitro* work which has illustrated their potential for differentiation, stimulation of other cells, and immunomodulation, abundant studies illustrating their *in vivo* potential have also been performed. For example, an intraarticular injection of MSCs was shown to diminish degeneration of the articular cartilage in a caprine model of osteoarthritis.[121] In addition, MSCs were shown to aid in repair of cartilage defects in a full-thickness defect model in the rabbit.[122] Finally, MSCs delivered to human patients through implantation in hydroxyapatite scaffolds or percutaneous injection, respectively, were shown to have potential for repair of large bone defects[123] and cartilage repair in degenerative joint disease.[124] In addition, Chondrogen, an Osiris product, is being examined as a potential injectable delivery of MSCs for treatment of knee arthritis.[125]

Embryonic Stem Cells

Embryonic stem cells provide an alternative cell source for tissue engineering applications. An advantage to the use of Embryonic stem (ES) cells over MSCs is their extensive proliferative and differentiation capacity. They are pluripotent and can divide into ectodermal and endodermal tissues in addition to mesodermal tissues. However, given the current ethical issues surrounding their use, the difficulty in adequately controlling their differentiation to obtain homogeneous cell populations, and fears about their potential to form tumors when implanted their clinical potential has not yet been fully realized.

Embryonic stem cells (ESCs) have been investigated for use in both bone and cartilage tissue engineering. For example, Hwang *et al.* investigated the chondrogenic differentiation of embryonic stem cells utilizing a combination of three dimensional culture in poly ethylene glycol hydrogels and growth factors[126] while *in vivo* bone tissue formation by ESCs has been investigated by Tremoleda, J.L. *et al.*[127]

Unfortunately, given that the study and clinical use of embryonic stem cells is currently limited in many countries,[128] the development of clinically relevant tissue engineering therapies utilizing these cell types may be delayed.

Scaffolds

Scaffold choice must take into account a wide variety of parameters including mechanical properties, degradation rates and products, and material type. Though an understanding of these material properties begins to guide scaffold choice, our attempts to mimic nature's complexity still remains a challenge.

Scaffolds may be designed out of natural materials such as collagen, demineralized bone matrix (DBM), or alginate, or they may be designed by synthetic polymers such as poly(l-lactic acid) (PLLA) or poly(glycolic acid) (PGA). For example, PEG based scaffolds have been used to encapsulate chondrocytes,[129] MSCs,[130] and embryonic stem cells[131] for cartilage tissue engineering. In addition, scaffolds such as fibrin and alginate have been used for engineering cartilage.[96] Composite scaffolds such as poly (lactic-co-glycolic acid) (PLGA) or poly (ethylene glycol) (PEG) and PLGA may also be used. For example, a polymer made of PLLA, PEG, and PLGA has been used for bone tissue engineering applications.[132]

In addition, scaffolds may be formed through a combination of natural and synthetic materials such as a DBM with polymer carries.[52] For example, products currently used as bone graft substitutes in the limb have been made of DBM along with hyaluronic acid (HA) and collagen, gelatin, or calcium sulfate.[52] Additionally, combinations of collagen, HA and tricalcium phosphate, and other additives have also been explored.[52]

Natural extracellular matrix materials, or mimics of extracellular matrix materials have also been used extensively in cartilage and bone tissue engineering. For example, collagen mimetic peptide modification of PEG gels was shown to augment the chondrogenesis of MSCs.[133] In addition, modification of PEG gels with an argine, glycine, aspartate (RGD) motif, which serves as an adhesion site, was shown to improve embryonic stem cell chondrogenesis[130] and osteogenesis of MSCs[134] and osteoblasts.[135] In addition, additives such as chondroitin sulfate have been used to enhance formation of cartilaginous tissue in scaffolds.[129] For tissue engineered bone, DBM has been used along with both polymer and mineral additives.[52,136] In addition to its osteoinductive capabilities, DBM has also been shown to stimulate chondrogenesis.[137]

It should be noted that the inherent bioactivity of a naturally occurring material can be exploited when included in a scaffold. For example, CS has anti-inflammatory properties, and its incorporation in a scaffold can result in an immunomodulating biomaterial.[138]

In addition to naturally occurring ECM components and whole extracellular matrix materials, such as DBM, histological sections may also be used as scaffolds for tissue engineering purposes. For example, histological sections have been shown to hold potential for directing differentiation of embryonic stem cells.[139] Though the use of these materials is still in its infancy, these materials hold significant promise for directing cellular differentiation and tissue formation in the future.

Scaffolds may be designed such that they are temporary structures which will degrade over time as they are replaced by tissue, or alternatively, as more permanent structures. The degradation times of polymers under consideration of bone tissue engineering can range from very short time scales to a period of several years depending on the composition, crystallinity, crosslinking density, molecular weight, and method of degradation.[140] These scaffolds include polymers such as polyanhydrides and PLGA, which undergo hydrolysis, Type 1 collagen and Hyaluronic acid, which undergo enzymatic degradation, and ceramics such as calcium phosphate or bioactive glass, which undergo dissolution and osteoclastic resorption.[140]

Concerns exist with the use of degradable scaffold materials. Although scaffolds may be designed such that their degradation products are metabolites, or other bioactive factors, their ultimate location and activity are still important variables. For example, the toxicity of local pH changes due to lactic acid release upon degradation of PLLA scaffolds is concerning.[47] This potential toxicity must be adequately addressed before use of PLLA matrices as delivery systems for transplanted cells. In addition, clearance through the liver and kidney are important considerations for any products intended to be implanted into the body.

In addition to the variability in derivative time scales, the polymers and ceramics used for bone tissue engineering matrices represent a wide variety of mechanical properties. As previously discussed, mechanical properties are immensely important to achieve the structural functionality of bone. In order to remain functionally sound during the repair process, any implant must possess appropriate mechanical properties. Several components currently under investigation for bone repair, such as chitosan and polyanhydrides, possess properties below that of trabecular bone.[140] Others, such as polypropylene fumarate and calcium sulfate, are similar to bone, while some possess mechanical properties that exceed trabecular bone, such as polycaprolactone and calcium phosphate.[140] A material with mechanical properties which do not match bone may be justified for situations in which the expectation is that the implant is temporary and thus will not have to withstand normal loading conditions. Mechanical property considerations are also important for cartilage tissue engineering.[96] For example, without the mechanical properties that facilitate stress resistance, the function of a tissue-engineered cartilage substitute is compromised.[96]

In addition to the static mechanical properties of a scaffold, dynamic mechanical stimulation has also been shown to affect tissue formation *in vitro*. For example, after initial differentiation towards a chondrocyte phenotype, mechanical stimulation can further augment the degree of chondrogenesis of human embryoid body derived progenitor cells.[141] In addition, mechanical compression enhances the chondrogenesis of MSCs *in vitro*.[141] Furthermore, mechanical stimulation was shown to regulate MSCs through modulation of MMP activity.[109] This is especially important since MMPs have been shown to play a role in regulation of proteins, enzymes, and receptors as part of their regulation of MSCs.[142]

Furthermore, processing and delivery of the scaffolds are important design considerations. For example, PEG derivatives, such as PEG diacrylate (PEODA) or a combination of PEODA and chondroitin sulfate methacrylate (CSMA), which can be polymerized directly in a defect can conform to a variety of shapes.[143,144] CS derived polymers have also been made which can form *in vivo* and bond directly to cartilage defects.[145] In contrast, fibrous or sponge like scaffolds may require prefabrication.[146]

Many reconstruction attempts of the cartilage and bone of the hand have involved cell seeded scaffolds. For example, bone marrow derived stromal cells have been seeded onto demineralized bone scaffolds for bone tissue engineering.[147] In addition, the osteogenesis of mesenchymal cells has been explored in calcium phosphate ceramics[148,149] and combinations of hydroxyapatite and tricalcium phosphate.[150]

Bioactive Factors

Much of the research on growth factors and other soluble factors has investigated their inherent role in development, repair, disease and induction of differentiation, proliferation, and migration of stem cells. However, the delivery of growth factors and other soluble mediators of tissue repair such as MMPs and their inhibitors is also important for engineered tissues. In addition, delivery of factors known to affect the immune and clotting responses may be investigated in hopes that modulation of these pathways could lead to better repair.

During *in vitro* experimentation, growth factor delivery is often achieved simply via media supplementation. However, delivery of *in vivo* therapeutics through simple supplementation such as injection or oral delivery is unlikely to be sufficient given the necessity for precise spatial and temporal delivery. As a result, one strategy of delivery currently under investigation is the use of scaffolds as a carrier for the growth factors. In these systems, growth factors may be delivered through linkage directly to the polymer carrier (e.g. covalent or ionic linkage to the polymer backbone).[47,96,128]

Alternatively, a bioactive factor may be non-covalently incorporated within the polymer and released through degradation of the polymer, polymer swelling, osmosis, or a variety of other specialized mechanisms.[47,96,128] For example, BMP has been delivered in PLLA, PEG, and PLGA,[132] a collagen Type IV carrier,[151] or even in demineralized bone matrix (DBM). In addition, the encapsulation and subsequent release of bioactive molecules in PLGA microspheres contained within calcium phosphate cements has been explored for tissue engineering applications.[152]

Co-culture Models for Cartilage and Bone Reconstruction

Successful reconstruction of cartilage and bone will ultimately require an enhanced understanding of the complex interplay between bone and cartilage. Given the complex interplay between cartilage and bone in development, it follows that the repair of cartilaginous and boney structures should include some examination of cell-cell interactions. For example, enhanced comprehension of cell-cell communications[128] and further understanding of the role of structural architecture on cellular processes[96] will enhance tissue engineering of limbs and joints such as the hand.

Initial studies to evaluate cell-cell interactions in the musculoskeletal system relied on conditioned media studies, directly mixed co-culture systems, and transwell co-culture studies. For example, Gerstenfeld *et al.* found that mesenchymal stem cells enhanced their osteogenesis when grown in transwell co-culture with chondrocytes.[153] In addition, chondrocyte conditioned media was shown to augment both osteogenic and chondrogenic differentiation of MSCs.[154] In addition, MSCs in transwell or direct mixed monolayer co-culture with osteoblasts were found to interact through Wnt and cadherin pathways depending on the co-culture method.[155] Co-culture of chondrocytes with osteoblasts in these two co-culture systems was also found to modulate the proliferation and differentiation of these cells.[156] A similar examination was performed in transwell co-culture which found that chondrogenesis of human embryonic stem cells increased with co-culture with chondrocytes, but not with fibroblasts.[157]

These co-culture systems were later enhanced with mechanisms to examine three-dimensional tissue formation. For example, Jiang, J. *et al.* examined the co-culture of osteoblasts and chondrocytes through use of a cell micromass of one cell population coated with the other cell type.[158] This study found that co-culture of chondrocytes and osteoblasts decreased the differentiation of both cell types when compared to controls.[158] The effect of chondrocytes on MSC differentiation was examined in a co-culture model of cartilage explants grown with MSCs encapsulated in alginate beads.[159]

These three-dimensional systems were also used to examine the interactions of diseased tissues. During co-culture of OA chondrocytes in alginate beads grown with monolayer OA subchondral osteoblasts, MMP production by the chondrocytes was found to increase while aggrecan production was decreased in co-culture as compared to controls.[160] Using this same co-culture scheme, the gene expression of the OA chondrocytes was also found to be influenced by the OA osteoblasts.[161]

Three dimensional scaffolds were also used to design osteochondral composites for tissue replacement. For example, osteochondral composites were made through combination of a chondrocyte pellet coated onto MSCs seeded in a PLA scaffold[162] as well as through suturing of chondrocytes encapsulated in PGA to periosteal cells seeded in a scaffold made of PLGA and PEG.[163] Furthermore, a small phalanx was successfully engineered with both bone and articular cartilage through use of a PLGA core polymer wrapped with bovine periosteum except at one end which was seeded with chondrocytes before *in vivo* subcutaneous implantation in a mouse.[164] Similar reconstructions were also previously performed with PLGA and periosteum *in vitro* that utilized chondrocytes delivered in polymer sheets and an additional population of tenocytes to include tendon in the reconstruction.[165]

Lingering Questions

Lingering questions still remain about the exact interplay between cells and tissues, signaling cascades, and the reasons adult mammals have limited regenerative capacity. Furthermore, the ideal tissue replacement strategy has yet to be fully defined. Although current repair strategies seek to replace the tissue with a functional equivalent, this may not be optimal for preventing disease recurrence and future tissue failure. Ultimately, accurate reconstruction of the hand will require enhanced understanding of natural growth and repair, as well as a better understanding of disease mechanisms. Finally a more refined definition of the optimal functional properties of replacement tissue would better guide the progress of tissue engineering.

References

1. Poole AR, Kojima T, Yasuda T, Mwale F, Kobayashi M, Laverty S. Composition and structure of articular cartilage: A template for tissue repair. *Clinical Orthopaedics and Related Research* **391** (Suppl.) (2001) S26–33.
2. Ulrich-Vinther M, Maloney MD, Schwarz EM, Rosier R, O'Keefe RJ. Articular cartilage biology. *The Journal of the American Academy of Orthopaedic Surgeons* **11**(6) (2003) 421–430.
3. Dowthwaite GP, Bishop JC, Redman SN, Khan IM, Rooney P, Evans DJ, *et al*. The surface of articular cartilage contains a progenitor cell population. *Journal of Cell Science* **117**(Pt 6) (2004) 889–897.
4. Wong M, Carter DR. Articular cartilage functional histomorphology and mechanobiology: a research perspective. *Bone* **33**(1) (2003) 1–13.
5. Seeman E. Bone quality: the material and structural basis of bone strength. *Journal of Bone and Mineral Metabolism* **26**(1) (2008) 1–8.
6. Sandberg MM, Aro HT, Vuorio EI. Gene expression during bone repair. *Clinical Orthopaedics and Related Research* **289** (1993) 292–312.
7. Giraud Guille MM, Mosser G, Helary C, Eglin D. Bone matrix like assemblies of collagen: from liquid crystals to gels and biomimetic materials. *Micron (Oxford, England: 1993)* **36**(7–8) (2005) 602–608.
8. Lamoureux F, Baud'huin M, Duplomb L, Heymann D, Redini F. Proteoglycans: Key partners in bone cell biology. *BioEssays: News and Reviews in Molecular, Cellular and Developmental Biology* **29**(8) (2007) 758–771.
9. Ham AW. A histological study of the early phases of bone repair. *The Journal of Bone and Joint Surgery American Volume* **12** (1930) 827–844.

10. Dalle Carbonare L, Giannini S. Bone microarchitecture as an important determinant of bone strength. *Journal of Endocrinological Investigation* **27**(1) (2004) 99–105.

11. Scott A, Khan KM, Duronio V, Hart DA. Mechanotransduction in human bone: *In vitro* cellular physiology that underpins bone changes with exercise. *Sports Medicine (Auckland, N.Z.)* **38**(2) (2008) 139–160.

12. Parfitt AM. Misconceptions (2): Turnover is always higher in cancellous than in cortical bone. *Bone* **30**(6) (2002) 807–809.

13. Tatari H. The structure, physiology, and biomechanics of articular cartilage: Injury and repair. [Eklem kikirdaginin yapisi, fizyolojisi ve biyomekanigi: Yaralanma ve onarim] *Acta Orthopaedica Et Traumatologica Turcica* **41**(Suppl. 2) (2007) 1–5.

14. Boyde A. The real response of bone to exercise. *Journal of Anatomy* **203**(2) (2003) 173–189.

15. Hollinger JO, Kleinschmidt JC. The critical size defect as an experimental model to test bone repair materials. *The Journal of Craniofacial Surgery* **1**(1) (1990) 60–68.

16 Goldring MB, Tsuchimochi K, Ijiri K. The control of chondrogenesis. *Journal of Cellular Biochemistry* **97**(1) (2006) 33–44.

17. Grassel S, Ahmed N. Influence of cellular microenvironment and paracrine signals on chondrogenic differentiation. *Frontiers in Bioscience: A Journal and Virtual Library* **12** (2007) 4946–4956.

18. Mackie EJ, Ahmed YA, Tatarczuch L, Chen KS, Mirams M. Endochondral ossification: how cartilage is converted into bone in the developing skeleton. *The International Journal of Biochemistry & Cell Biology* **40**(1) (2008) 46–62.

19. de Crombrugghe B, Lefebvre V, Nakashima K. Regulatory mechanisms in the pathways of cartilage and bone formation. *Current Opinion in Cell Biology* **13**(6) (2001) 721–727.

20. Davis LA, Zur Nieden NI. Mesodermal fate decisions of a stem cell: The wnt switch. *Cellular and Molecular Life Sciences: CMLS* **65**(17) (2008) 2658–2674.

21. Werb Z, Chin JR. Extracellular matrix remodeling during morphogenesis. *Annals of the New York Academy of Sciences* **857** (1998) 110–118.

22. Johansson N, Saarialho-Kere U, Airola K, Herva R, Nissinen L, Westermarck J, *et al.* Collagenase-3 (MMP-13) is expressed by hypertrophic chondrocytes, periosteal cells, and osteoblasts during human fetal bone development. *Developmental Dynamics: An Official Publication of the American Association of Anatomists* **208**(3) (1997) 387–397.

23. Nakamura M, Sone S, Takahashi I, Mizoguchi I, Echigo S, Sasano Y. Expression of versican and ADAMTS1, 4, and 5 during bone development in the rat mandible and hind limb. *The Journal of Histochemistry and Cytochemistry: Official Journal of the Histochemistry Society* **53**(12) (2005) 1553–1562.

24. Lind T, McKie N, Wendel M, Racey SN, Birch MA. The hyalectan degrading ADAMTS-1 enzyme is expressed by osteoblasts and up-regulated at regions of new bone formation. *Bone* **36**(3) (2005) 408–417.

25. Pogue R, Lyons K. BMP signaling in the cartilage growth plate. *Current Topics in Developmental Biology* **76** (2006) 1–48.

26. Malemud, CJ. Matrix metalloproteinases (MMPs) in health and disease: an overview. *Frontiers in Bioscience: A Journal and Virtual Library* **11** (2006) 1696–1701.

27. Storm EE, Kingsley DM. GDF5 coordinates bone and joint formation during digit development. *Developmental Biology* **209**(1) (1999) 11–27.

28. Hartmann C, Tabin CJ. Wnt-14 plays a pivotal role in inducing synovial joint formation in the developing appendicular skeleton. *Cell* **104**(3) (2001) 341–351.

29. Shum L, Nuckolls G. The life cycle of chondrocytes in the developing skeleton. *Arthritis Research* **4**(2) (2002) 94–106.

30. Khan IM, Gilbert SJ, Caterson B, Sandell LJ, Archer CW. Oxidative stress induces expression of osteoarthritis markers procollagen IIA and 3B3(-) in adult bovine articular cartilage. *Osteoarthritis and Cartilage/OARS, Osteoarthritis Research Society* **16**(6) (2008) 698–707.

31. Takahara M, Harada M, Guan D, Otsuji M, Naruse T, Takagi M, *et al.* Developmental failure of phalanges in the absence of growth/differentiation factor 5. *Bone* **35**(5) (2004) 1069–1076.

32. Tavella S, Biticchi R, Morello R, Castagnola P, Musante V, Costa D, *et al*. Forced chondrocyte expression of sonic hedgehog impairs joint formation affecting proliferation and apoptosis. *Matrix Biology: Journal of the International Society for Matrix Biology* **25**(7) (2006) 389–397.
33. Ehlen HW, Buelens LA, Vortkamp A. Hedgehog signaling in skeletal development. *Birth Defects Research. Part C, Embryo Today: Reviews* **78**(3) (2006) 267–279.
34. Zhou J, Meng J, Guo S, Gao B, Ma G, Zhu X, *et al*. IHH and FGF8 coregulate elongation of digit primordia. *Biochemical and Biophysical Research Communications* **363**(3) (2007) 513–518.
35. Montero JA, Lorda-Diez CI, Ganan Y, Macias D, Hurle JM. Activin/TGFbeta and BMP crosstalk determines digit chondrogenesis. *Developmental Biology* **321**(2) (2008) 343–356.
36. Bakrania P, Efthymiou M, Klein JC, Salt A, Bunyan DJ, Wyatt A, *et al*. Mutations in BMP4 cause eye, brain, and digit developmental anomalies: overlap between the BMP4 and hedgehog signaling pathways. *American Journal of Human Genetics* **82**(2) (2008) 304–319.
37. Khan IM, Redman SN, Williams R, Dowthwaite GP, Oldfield SF, Archer CW. The development of synovial joints. *Current Topics in Developmental Biology* **79** (2007) 1–36.
38. Zhao W, Sala-Newby GB, Dhoot GK. Sulf1 expression pattern and its role in cartilage and joint development. *Developmental Dynamics: An Official Publication of the American Association of Anatomists* **235**(12) (2006) 3327–3335.
39. Suzuki T, Hasso SM, Fallon JF. Unique SMAD1/5/8 activity at the phalanx-forming region determines digit identity. *Proceedings of the National Academy of Sciences of the United States of America* **105**(11) (2008) 4185–4190.
40. Redman SN, Oldfield SF, Archer CW. Current strategies for articular cartilage repair. *European Cells Materials* **9** (2005) 23–32; discussion 23–32.
41. Wakitani S, Kawaguchi A, Tokuhara Y, Takaoka K. Present status of and future direction for articular cartilage repair. *Journal of Bone and Mineral Metabolism* **26**(2) (2008) 115–122.
42. Beris AE, Lykissas MG, Papageorgiou CD, Georgoulis AD. Advances in articular cartilage repair. *Injury* **36**(Suppl 4) (2005) S14–23.
43. Goldring MB, Goldring SR. Osteoarthritis. *Journal of Cellular Physiology* **213**(3) (2007) 626–634.
44. Sato K, Nakamura T, Nakamichi N, Okuyama N, Toyama Y, Ikegami H. Finger joint reconstruction with costal osteochondral graft. *Techniques in Hand & Upper Extremity Surgery* **12**(3) (2008a) 150–155.
45. Pei GX, Ren GH, Ren YJ, Wei KH. Reconstruction of phalangeal articulations of the hand with vascularised phalangeal articulations of foot. *Injury* **39**(Suppl. 3) (2008) S109–115.
46. Kanczler JM, Oreffo RO. Osteogenesis and angiogenesis: The potential for engineering bone. *European Cells & Materials* **15** (2008) 100–114.
47. Yaszemski MJ, Payne RG, Hayes WC, Langer R, Mikos AG. Evolution of bone transplantation: Molecular, cellular and tissue strategies to engineer human bone. *Biomaterials* **17**(2) (1996) 175–185.
48. Keene DR, Sakai LY, Burgeson RE. Human bone contains type III collagen, type VI collagen, and fibrillin: Type III collagen is present on specific fibers that may mediate attachment of tendons, ligaments, and periosteum to calcified bone cortex. *The Journal of Histochemistry and Cytochemistry: Official Journal of the Histochemistry Society* **39**(1) (1991) 59–69.
49. Thompson Z, Miclau T, Hu D, Helms JA. A model for intramembranous ossification during fracture healing. *Journal of Orthopaedic Research: Official Publication of the Orthopaedic Research Society* **20**(5) (2002) 1091–1098.
50. Gerstenfeld LC, Cho TJ, Kon T, Aizawa T, Cruceta J, Graves BD, *et al*. Impaired intramembranous bone formation during bone repair in the absence of tumor necrosis factor-alpha signaling. *Cells, Tissues, Organs* **169**(3) (2001) 285–294.
51. Kakinoki R, Ikeguchi R, Matsumoto T, Nakamura T. Reconstruction of a phalangeal bone using a vascularised metacarpal bone graft nourished by a dorsal metacarpal artery. *Injury* **39**(Suppl. 4) (2008) 25–28.
52. Ladd Amy L, Pliam Nick B. Bone graft substitutes in the radius and upper limb. *Journal of the American Society for Surgery of the Hand* **3**(4) (2003) 227–245.

53. Innocenti M, Adani R, Boyer MI. Nonvascularized osteoarticular allograft replacement of the proximal interphalangeal joint after extensive loss of bone, joint, and extensor tendon. *Techniques in Hand & Upper Extremity Surgery* **11**(2) (2007) 149–155.

54. Exner GU, Dumont CE, Malinin TI, von Hochstetter AR. Recurrent aggressive chondrosarcoma of the middle phalanx of the index finger: excision and reconstruction with an osteocartilaginous allograft. *Archives of Orthopaedic and Trauma Surgery* **123**(8) (2003) 425–428.

55. Sato K, Sasaki T, Nakamura T, Toyama Y, Ikegami H. Clinical outcome and histologic findings of costal osteochondral grafts for cartilage defects in finger joints. *The Journal of Hand Surgery* **33**(4) (2008b) 511–515.

56. Vlaskalin T, Wong CJ, Tsilfidis C. Growth and apoptosis during larval forelimb development and adult forelimb regeneration in the newt (notophthalmus viridescens). *Development Genes and Evolution* **214**(9) (2004) 423–431.

57. Straube WL, Tanaka EM. Reversibility of the differentiated state: regeneration in amphibians. *Artificial Organs* **30**(10) (2006) 743–755.

58. Kumar A, Godwin JW, Gates PB, Garza-Garcia AA, Brockes JP. Molecular basis for the nerve dependence of limb regeneration in an adult vertebrate. *Science (New York, NY)* **318**(5851) (2007) 772–777.

59. Campbell LJ, Crews CM. Wound epidermis formation and function in urodele amphibian limb regeneration. *Cellular and Molecular Life Sciences: CMLS* **65**(1) (2008) 73–79.

60. Yokoyama H. Initiation of limb regeneration: the critical steps for regenerative capacity. *Development, Growth & Differentiation* **50**(1) (2008) 13–22.

61. Slack JM, Beck CW, Gargioli C, Christen B. Cellular and molecular mechanisms of regeneration in xenopus. *Philosophical Transactions of the Royal Society of London Series B, Biological Sciences* **359**(1445) (2004) 745–751.

62. Tseng AS, Levin M. Tail regeneration in xenopus laevis as a model for understanding tissue repair. *Journal of Dental Research* **87**(9) (2008) 806–816.

63. Pearl EJ, Barker D, Day RC, Beck CW. Identification of genes associated with regenerative success of xenopus laevis hindlimbs. *BMC Developmental Biology* **8** (2008) 66.

64. Han M, Yang X, Taylor G, Burdsal CA, Anderson RA, Muneoka K. Limb regeneration in higher vertebrates: developing a roadmap. *Anatomical Record. Part B, New Anatomist* **287**(1) (2005) 14–24.

65. Ho DM, Whitman M. TGF-beta signaling is required for multiple processes during xenopus tail regeneration. *Developmental Biology* **315**(1) (2008) 203–216.

66. Calderwood SK, Mambula SS, Gray PJ, Jr, Theriault JR. Extracellular heat shock proteins in cell signaling. *FEBS Letters* **581**(19) (2007a) 3689–3694.

67. Calderwood SK, Mambula SS, Gray PJ, Jr. Extracellular heat shock proteins in cell signaling and immunity. *Annals of the New York Academy of Sciences* **1113** (2007b) 28–39.

68. Stoick-Cooper CL, Moon RT, Weidinger G. Advances in signaling in vertebrate regeneration as a prelude to regenerative medicine. *Genes & Development* **21**(11) (2007). 1292–1315.

69. Niazi IA, Saxena S. Abnormal hind limb regeneration in tadpoles of the toad, bufo andersoni, exposed to excess vitamin A. *Folia Biologica* **26**(1) (1978) 3–8.

70. Maden M. Vitamin A and pattern formation in the regenerating limb. *Nature* **295**(5851) (1982) 672–675.

71. Stocum DL. Developmental biology. acceptable nAGging. *Science (New York, NY)* **318**(5851) (2007) 754–755.

72. Yin VP, Poss KD. New regulators of vertebrate appendage regeneration. *Current Opinion in Genetics & Development* **18**(4) (2008) 381–386.

73. Van Heest AE, Kallemeier P. Thumb carpal metacarpal arthritis. *The Journal of the American Academy of Orthopaedic Surgeons* **16**(3) (2008) 140–151.

74. Lawrence RC, Felson DT, Helmick CG, Arnold LM, Choi H, Deyo, RA, *et al.* Estimates of the prevalence of arthritis and other rheumatic conditions in the united states. part II. *Arthritis and Rheumatism* **58**(1) (2008) 26–35.

75. CDC. *Arthritis types-overview: Rheumatoid arthritis.* Retrieved November, Vol. 8 (2008), from http://www.cdc.gov/arthritis/arthritis/rheumatoid.htm.

76. Felson DT. Clinical practice. osteoarthritis of the knee. *The New England Journal of Medicine* **354**(8) (2006) 841–848.

77. Plasqui G. The role of physical activity in rheumatoid arthritis. *Physiology & Behavior* **94**(2) (2008) 270–275.

78. Tetlow LC, Adlam DJ, Woolley DE. Matrix metalloproteinase and proinflammatory cytokine production by chondrocytes of human osteoarthritic cartilage: associations with degenerative changes. *Arthritis and Rheumatism* **44**(3) (2001) 585–594.

79. Zwerina J, Redlich K, Polzer K, Joosten L, Kronke G, Distler J, *et al.* TNF-induced structural joint damage is mediated by IL-1. *Proceedings of the National Academy of Sciences of the United States of America* **104**(28) (2007) 11742–11747.

80. Chacko AT, Rozental TD. The rheumatoid thumb. *Hand Clinics* **24**(3) (2008). 307–314, vii.

81. Charni-Ben Tabassi N, Garnero P. Monitoring cartilage turnover. *Current Rheumatology Reports* **9**(1) (2007) 16–24.

82. Sakao K, Takahashi KA, Mazda O, Arai Y, Tonomura H, Inoue A, *et al.* Enhanced expression of interleukin-6, matrix metalloproteinase-13, and receptor activator of NF-kappaB ligand in cells derived from osteoarthritic subchondral bone. *Journal of Orthopaedic Science: Official Journal of the Japanese Orthopaedic Association* **13**(3) (2008) 202–210.

83. Grynpas MD, Alpert B, Katz I, Lieberman I, Pritzker KP. Subchondral bone in osteoarthritis. *Calcified Tissue International* **49**(1) (1991) 20–26.

84. Hopwood B, Tsykin A, Findlay DM, Fazzalari NL. Microarray gene expression profiling of osteoarthritic bone suggests altered bone remodelling, WNT and transforming growth factor-beta/bone morphogenic protein signalling. *Arthritis Research & Therapy* **9**(5) (2007) R100.

85. Schett G. Cells of the synovium in rheumatoid arthritis. osteoclasts. *Arthritis Research & Therapy* **9**(1) (2007) 203.

86. Mazzetti I, Grigolo B, Pulsatelli L, Dolzani P, Silvestri T, Roseti L, *et al.* Differential roles of nitric oxide and oxygen radicals in chondrocytes affected by osteoarthritis and rheumatoid arthritis. *Clinical Science (London, England: 1979)* **101**(6) (2001) 593–599.

87. Dai SM, Shan ZZ, Nakamura H, Masuko-Hongo K, Kato T, Nishioka K, *et al.* Catabolic stress induces features of chondrocyte senescence through overexpression of caveolin 1: Possible involvement of caveolin 1-induced down-regulation of articular chondrocytes in the pathogenesis of osteoarthritis. *Arthritis and Rheumatism* **54**(3) (2006) 818–831.

88. Ruiz-Romero C, Carreira V, Rego I, Remeseiro S, Lopez-Armada MJ, Blanco FJ. Proteomic analysis of human osteoarthritic chondrocytes reveals protein changes in stress and glycolysis. *Proteomics* **8**(3) (2008) 495–507.

89. Murphy G, Nagase H. Reappraising metalloproteinases in rheumatoid arthritis and osteoarthritis: destruction or repair? *Nature Clinical Practice Rheumatology* **4**(3) (2008) 128–135.

90. Lago R, Gomez R, Otero M, Lago F, Gallego R, Dieguez C, *et al.* A new player in cartilage homeostasis: adiponectin induces nitric oxide synthase type II and pro-inflammatory cytokines in chondrocytes. *Osteoarthritis and Cartilage/OARS, Osteoarthritis Research Society* (2008).

91. Henrotin YE, Bruckner P, Pujol JP. The role of reactive oxygen species in homeostasis and degradation of cartilage. *Osteoarthritis and Cartilage/OARS, Osteoarthritis Research Society* **11**(10) (2003) 747–755.

92. McCarty MF, Russell AL. Niacinamide therapy for osteoarthritis — does it inhibit nitric oxide synthase induction by interleukin 1 in chondrocytes? *Medical Hypotheses* **53**(4) (1999) 350–360.

93. Grimshaw MJ, Mason RM. Bovine articular chondrocyte function in vitro depends upon oxygen tension. *Osteoarthritis and Cartilage / OARS, Osteoarthritis Research Society* **8**(5) (2000) 386–392.

94. Coimbra IB, Jimenez SA, Hawkins DF, Piera-Velazquez S, Stokes DG. Hypoxia inducible factor-1 alpha expression in human normal and osteoarthritic chondrocytes. *Osteoarthritis and Cartilage/OARS, Osteoarthritis Research Society* **12**(4) (2004) 336–345.

95. Chen AF, Davies CM, De Lin M, Fermor B. Oxidative DNA damage in osteoarthritic porcine articular cartilage. *Journal of Cellular Physiology* **217**(3) (2008) 828–833.

96. Chong AK, Chang J. Tissue engineering for the hand surgeon: a clinical perspective. *The Journal of Hand Surgery* **31**(3) (2006) 349–358.

97. Friedenstein AJ, Petrakova KV, Kurolesova AI, Frolova GP. Heterotopic of bone marrow analysis of precursor cells for osteogenic and hematopoietic tissues. *Transplantation* **6**(2) (1968) 230–247.

98. Owen M, Friedenstein AJ. Stromal stem cells: marrow-derived osteogenic precursors. *Ciba Foundation Symposium* **136** (1988) 42–60.

99. Haynesworth SE, Goshima J, Goldberg VM, Caplan AI. Characterization of cells with osteogenic potential from human marrow. *Bone* **13**(1) (1992) 81–88.

100. Dennis JE, Merriam A, Awadallah A, Yoo JU, Johnstone B, Caplan AI. A quadripotential mesenchymal progenitor cell isolated from the marrow of an adult mouse. *Journal of Bone and Mineral Research: The Official Journal of the American Society for Bone and Mineral Research* **14**(5) (1999) 700–709.

101. Yoo JU, Johnstone B. The role of osteochondral progenitor cells in fracture repair. *Clinical Orthopaedics and Related Research* **355**(Suppl.) (1998) S73–81.

102. Pittenger MF, Mackay AM, Beck SC, Jaiswal RK, Douglas R, Mosca JD, et al. Multilineage potential of adult human mesenchymal stem cells. *Science (New York, NY)* **284**(5411) (1999) 143–147.

103. Wakitani S, Saito T, Caplan AI. Myogenic cells derived from rat bone marrow mesenchymal stem cells exposed to 5-azacytidine. *Muscle & Nerve* **18**(12) (1995) 1417–1426.

104. Young RG, Butler DL, Weber W, Caplan AI, Gordon SL, Fink DJ. Use of mesenchymal stem cells in a collagen matrix for achilles tendon repair. *Journal of Orthopaedic Research: Official Publication of the Orthopaedic Research Society* **16**(4) (1998) 406–413.

105. Granero-Molto F, Weis JA, Longobardi L, Spagnoli A. Role of mesenchymal stem cells in regenerative medicine: application to bone and cartilage repair. *Expert Opinion on Biological Therapy* **8**(3) (2008) 255–268.

106. Pittenger MF. Mesenchymal stem cells from adult bone marrow. *Methods in Molecular Biology (Clifton, NJ)* **449** (2008) 27–44.

107. Caplan AI, Dennis JE. Mesenchymal stem cells as trophic mediators. *Journal of Cellular Biochemistry* **98**(5) (2006) 1076–1084.

108. Liu CH, Hwang SM. Cytokine interactions in mesenchymal stem cells from cord blood. *Cytokine*, **32**(6) (2005) 270–279.

109. Kasper G, Glaeser JD, Geissler S, Ode A, Tuischer J, Matziolis G, et al. Matrix metalloprotease activity is an essential link between mechanical stimulus and mesenchymal stem cell behavior. *Stem Cells (Dayton, Ohio)* **25**(8) (2007) 1985–1994.

110. Park HW, Shin JS, Kim CW. Proteome of mesenchymal stem cells. *Proteomics* **7**(16) (2007) 2881–2894.

111. Aggarwal S, Pittenger MF. Human mesenchymal stem cells modulate allogeneic immune cell responses. *Blood* **105**(4) (2005) 1815–1822.

112. Fibbe WE, Nauta AJ, Roelofs H. Modulation of immune responses by mesenchymal stem cells. *Annals of the New York Academy of Sciences* **1106** (2007) 272–278.

113. Schinkothe T, Bloch W, Schmidt A. *In vitro* secreting profile of human mesenchymal stem cells. *Stem Cells and Development* **17**(1) (2008) 199–206.

114. Ren G, Zhang L, Zhao X, Xu G, Zhang Y, Roberts AI, et al. Mesenchymal stem cell-mediated immunosuppression occurs via concerted action of chemokines and nitric oxide. *Cell Stem Cell* **2**(2) (2008) 141–150.

115. Zheng ZH, Li XY, Ding J, Jia JF, Zhu P. Allogeneic mesenchymal stem cell and mesenchymal stem cell-differentiated chondrocyte suppress the responses of type II collagen-reactive T cells in rheumatoid arthritis. *Rheumatology (Oxford, England)* **47**(1) (2008) 22–30.

116. Sato K, Ozaki K, Oh, I, Meguro A, Hatanaka K, Nagai T, et al. Nitric oxide plays a critical role in suppression of T-cell proliferation by mesenchymal stem cells. *Blood* **109**(1) (2007) 228–234.

117. Caplan AI. Adult mesenchymal stem cells for tissue engineering versus regenerative medicine. *Journal of Cellular Physiology* **213**(2) (2007) 341–347.

118. Kastrinaki MC, Sidiropoulos P, Roche S, Ringe J, Lehmann S, Kritikos H, et al. Functional, molecular and proteomic characterisation of bone marrow mesenchymal stem cells in rheumatoid arthritis. *Annals of the Rheumatic Diseases* **67**(6) (2008) 741–749.

119. Murphy JM, Dixon K, Beck S, Fabian D, Feldman A, Barry F, Reduced chondrogenic and adipogenic activity of mesenchymal stem cells from patients with advanced osteoarthritis. *Arthritis and Rheumatism* **46**(3) (2002) 704–713.

120. McFarlin K, Gao X, Liu YB, Dulchavsky DS, Kwon D, Arbab AS, *et al.* Bone marrow-derived mesenchymal stromal cells accelerate wound healing in the rat. *Wound Repair and Regeneration: Official Publication of the Wound Healing Society [and] the European Tissue Repair Society* **14**(4) (2006) 471–478.

121. Murphy JM, Fink DJ, Hunziker EB, Barry FP. Stem cell therapy in a caprine model of osteoarthritis. *Arthritis and Rheumatism* **48**(12) (2003) 3464–3474.

122. Yan H, Yu C. Repair of full-thickness cartilage defects with cells of different origin in a rabbit model. *Arthroscopy: The Journal of Arthroscopic & Related Surgery: Official Publication of the Arthroscopy Association of North America and the International Arthroscopy Association* **23**(2) (2007) 178–187.

123. Quarto R, Mastrogiacomo M, Cancedda R, Kutepov SM, Mukhachev V, Lavroukov A, *et al.* Repair of large bone defects with the use of autologous bone marrow stromal cells. *The New England Journal of Medicine* **344**(5) (2001) 385–386.

124. Centeno CJ, Busse D, Kisiday J, Keohan C, Freeman M, Karli D. Increased knee cartilage volume in degenerative joint disease using percutaneously implanted, autologous mesenchymal stem cells. *Pain Physician* **11**(3) (2008) 343–353.

125. Osiris Therapeutics. (2008). *FDA fast-track clearance expedites stem cell therapy*. Retrieved November 8, 2008, from http://www.osiristx.com/clinical_trials.php.

126. Hwang NS, Kim MS, Sampattavanich S, Baek JH, Zhang Z, Elisseeff J Effects of three-dimensional culture and growth factors on the chondrogenic differentiation of murine embryonic stem cells. *Stem Cells (Dayton, Ohio)* **24**(2) (2006a) 284–291.

127. Tremoleda JL, Forsyth NR, Khan NS, Wojtacha D, Christodoulou I, Tye BJ, *et al.* Bone tissue formation from human embryonic stem cells *in vivo*. *Cloning and Stem Cells* **10**(1) (2008) 119–132.

128. Walgenbach KJ, Voigt M, Riabikhin AW, Andree C, Schaefer DJ, Galla TJ, *et al.* Tissue engineering in plastic reconstructive surgery. *The Anatomical Record* **263**(4) (2001) 372–378.

129. Hwang NS, Varghese S, Lee HJ, Theprungsirikul P, Canver A, Sharma B, *et al.* Response of zonal chondrocytes to extracellular matrix-hydrogels. *FEBS Letters* **581**(22) (2007a) 4172–4178.

130. Sharma B, Williams CG, Khan M, Manson P, Elisseeff JH. *In vivo* chondrogenesis of mesenchymal stem cells in a photopolymerized hydrogel. *Plastic and Reconstructive Surgery* **119**(1) (2007) 112–120.

131. Hwang NS, Varghese S, Zhang Z, Elisseeff J. Chondrogenic differentiation of human embryonic stem cell-derived cells in arginine-glycine-aspartate-modified hydrogels. *Tissue Engineering* **12**(9) (2006b) 2695–2706.

132. Miyamoto S, Takaoka K. Bone induction and bone repair by composites of bone morphogenetic protein and biodegradable synthetic polymers. *Annales Chirurgiae Et Gynaecologiae Supplementum* **207** (1993) 69–75.

133. Lee HJ, Yu C, Chansakul T, Hwang NS, Varghese S, Yu SM. *et al.* Enhanced chondrogenesis of mesenchymal stem cells in collagen mimetic peptide-mediated microenvironment. *Tissue Engineering Part A* **14**(11) (2008) 1843–1851.

134. Yang F, Williams CG, Wang DA, Lee H, Manson PN, Elisseeff J. The effect of incorporating RGD adhesive peptide in polyethylene glycol diacrylate hydrogel on osteogenesis of bone marrow stromal cells. *Biomaterials* **26**(30) (2005) 5991–5998.

135. Kim TI, Lee G, Jang JH, Chung CP, Ku Y. Influence of RGD-containing oligopeptide-coated surface on bone formation *in vitro* and *in vivo*. *Biotechnology Letters* **29**(3) (2007) 359–363.

136. Colnot C, Romero DM, Huang S, Helms JA. Mechanisms of action of demineralized bone matrix in the repair of cortical bone defects. *Clinical Orthopaedics and Related Research* **435** (2005) 69–78.

137. Gawande SR, Tuan RS. Characterization of bone-derived chondrogenesis-stimulating activity on embryonic limb mesenchymal cells *in vitro*. *Cell and Tissue Kinetics* **23**(5) (1990) 375–390.

138. Ronca F, Palmieri L, Panicucci P, Ronca G. Anti-inflammatory activity of chondroitin sulfate. *Osteoarthritis and Cartilage / OARS, Osteoarthritis Research Society* **6**(Suppl A) (1998) 14–21.

139. Takeuchi T, Ochiya T, Takezawa T. Tissue array substratum composed of histological sections: a new platform for orienting differentiation of embryonic stem cells towards hepatic lineage. *Tissue Engineering Part A* **14**(2) (2008) 267–274.

140. Khan Y, Yaszemski MJ, Mikos AG, Laurencin CT. Tissue engineering of bone: material and matrix considerations. *The Journal of Bone and Joint Surgery American Volume* **90**(Suppl 1) (2008) 36–42.

141. Terraciano V, Hwang N, Moroni L, Park HB, Zhang Z, Mizrahi J, *et al.* Differential response of adult and embryonic mesenchymal progenitor cells to mechanical compression in hydrogels. *Stem Cells (Dayton, Ohio)* **25**(11) (2007) 2730–2738.

142. Mannello F, Tonti GA, Bagnara GP, Papa S. Role and function of matrix metalloproteinases in the differentiation and biological characterization of mesenchymal stem cells. *Stem Cells (Dayton, Ohio)* **24**(3) (2006) 475–481.

143. Elisseeff J, Anseth K, Sims D, McIntosh W, Randolph M, Langer R. Transdermal photopolymerization for minimally invasive implantation. *Proceedings of the National Academy of Sciences of the United States of America* **96**(6) (1999) 3104–3107.

144. Li Q, Williams CG, Sun DD, Wang J, Leong K, Elisseeff JH. Photocrosslinkable polysaccharides based on chondroitin sulfate. *Journal of Biomedical Materials Research Part A* **68**(1) (2004) 28–33.

145. Wang DA, Varghese S, Sharma B, Strehin I, Fermanian S, Gorham J, *et al.* Multifunctional chondroitin sulphate for cartilage tissue-biomaterial integration. *Nature Materials* **6**(5) (2007) 385–392.

146. Ting V, Sims CD, Brecht LE, McCarthy, JG, Kasabian AK, Connelly PR, *et al. In vitro* prefabrication of human cartilage shapes using fibrin glue and human chondrocytes. *Annals of Plastic Surgery* **40**(4) (1998) 413–420; discussion 420–421.

147. Mauney JR, Jaquiery C, Volloch V, Heberer M, Martin I, Kaplan DL. *In vitro* and *in vivo* evaluation of differentially demineralized cancellous bone scaffolds combined with human bone marrow stromal cells for tissue engineering. *Biomaterials* **26**(16) (2005) 3173–3185.

148. Goshima J, Goldberg VM, Caplan AI. The osteogenic potential of culture-expanded rat marrow mesenchymal cells assayed *in vivo* in calcium phosphate ceramic blocks. *Clinical Orthopaedics and Related Research* **262** (1991a) 298–311.

149. Goshima J, Goldberg VM, Caplan AI. The origin of bone formed in composite grafts of porous calcium phosphate ceramic loaded with marrow cells. *Clinical Orthopaedics and Related Research* **269** (1991b) 274–283.

150. Ohgushi H, Okumura M, Tamai S, Shors EC, Caplan AI. Marrow cell induced osteogenesis in porous hydroxyapatite and tricalcium phosphate: a comparative histomorphometric study of ectopic bone formation. *Journal of Biomedical Materials Research* **24**(12) (1990) 1563–1570.

151. Gao TJ, Lindholm TS, Marttinen A, Puolakka T. Bone inductive potential and dose-dependent response of bovine bone morphogenetic protein combined with type IV collagen carrier. *Annales Chirurgiae Et Gynaecologiae Supplementum* **207** (1993) 77–84.

152. Habraken WJ, Wolke JG, Mikos AG, Jansen JA. PLGA microsphere/calcium phosphate cement composites for tissue engineering: *in vitro* release and degradation characteristics. *Journal of Biomaterials Science. Polymer Edition* **19**(9) (2008) 1171–1188.

153. Gerstenfeld LC, Barnes GL, Shea CM, Einhorn TA. Osteogenic differentiation is selectively promoted by morphogenetic signals from chondrocytes and synergized by a nutrient rich growth environment. *Connective Tissue Research* **44**(Suppl. 1) (2003) 85–91.

154. Hwang NS, Varghese S, Puleo C, Zhang Z, Elisseeff J. Morphogenetic signals from chondrocytes promote chondrogenic and osteogenic differentiation of mesenchymal stem cells. *Journal of Cellular Physiology* **212**(2) (2007b) 281–284.

155. Wang Y, Volloch V, Pindrus MA, Blasioli DJ, Chen J, Kaplan DL. Murine osteoblasts regulate mesenchymal stem cells via WNT and cadherin pathways: mechanism depends on cell-cell contact mode. *Journal of Tissue Engineering and Regenerative Medicine* **1**(1) (2007) 39–50.

156. Nakaoka R, Hsiong SX, Mooney DJ. Regulation of chondrocyte differentiation level via co-culture with osteoblasts. *Tissue Engineering* **12**(9) (2006) 2425–2433.

157. Vats A, Bielby RC, Tolley N, Dickinson SC, Boccaccini AR, Hollander AP, *et al.* Chondrogenic differentiation of human embryonic stem cells: the effect of the micro-environment. *Tissue Engineering* **12**(6) (2006) 1687–1697.

158. Jiang J, Nicoll SB, Lu HH. Co-culture of osteoblasts and chondrocytes modulates cellular differentiation *in vitro*. *Biochemical and Biophysical Research Communications* **338**(2) (2005) 762–770.

159. Ahmed N, Dreier R, Gopferich A, Grifka J, Grassel S. Soluble signalling factors derived from differentiated cartilage tissue affect chondrogenic differentiation of rat adult marrow stromal cells. *Cellular Physiology and Biochemistry: International Journal of Experimental Cellular Physiology, Biochemistry, and Pharmacology* **20**(5) (2007) 665–678.

160. Sanchez C, Deberg MA, Piccardi N, Msika P, Reginster JY, Henrotin YE. Osteoblasts from the sclerotic subchondral bone downregulate aggrecan but upregulate metalloproteinases expression by chondrocytes. this effect is mimicked by interleukin-6, -1beta and oncostatin M pre-treated non-sclerotic osteoblasts. *Osteoarthritis and Cartilage/OARS, Osteoarthritis Research Society* **13**(11) (2005a) 979–987.

161. Sanchez C, Deberg MA, Piccardi N, Msika P, Reginster JY, Henrotin YE. Subchondral bone osteoblasts induce phenotypic changes in human osteoarthritic chondrocytes. *Osteoarthritis and Cartilage/OARS, Osteoarthritis Research Society* **13**(11) (2005b) 988–997.

162. Tuli R, Nandi S, Li WJ, Tuli S, Huang X, Manner PA, *et al.* Human mesenchymal progenitor cell-based tissue engineering of a single-unit osteochondral construct. *Tissue Engineering* **10**(7–8) (2004) 1169–1179.

163. Schaefer D, Martin I, Shastri P, Padera RF, Langer R, Freed LE, *et al. In vitro* generation of osteochondral composites. *Biomaterials* **21**(24) (2000) 2599–2606.

164. Sedrakyan S, Zhou ZY, Perin L, Leach K, Mooney D, Kim TH. Tissue engineering of a small hand phalanx with a porously casted polylactic acid-polyglycolic acid copolymer. *Tissue Engineering* **12**(9) (2006) 2675–2683.

165. Isogai N, Landis W, Kim TH, Gerstenfeld LC, Upton J, Vacanti JP. Formation of phalanges and small joints by tissue-engineering. *The Journal of Bone and Joint Surgery American Volume*, **81**(3) (1999) 306–316.

CHAPTER 13

Blood Vessels

Masayuki Yamato* and Teruo Okano*

Artery and Vein

Each year in the United States, there are approximately 1.4 million procedures performed which require arterial prostheses. Most of these procedures are in small caliber (<6 mm) vessels, for which synthetic graft materials are not generally suitable. While autologous venous or arterial vessels are generally used, not all patients possess adequate conduits for revascularization. Therefore, the field of tissue engineering has paid much attention to fabricate functional and transplantable vascular conduits with autologous cells.[1] In this section, we will summarize several attempts to fabricate arteries *in vitro*.

The early tissue engineering of vascular grafts

In 1979, Professor E. Bell's group reported that fibroblasts seeded in hydrated collagen gel can condense a surrounding collagen lattice to a tissue-like form.[2] The final concentration of collagen after gel contraction was equivalent to that of dermis. The report pointed out the potential use of the system as an immunologically well-tolerated "tissue" for wound healing, heralding the beginning of tissue engineering. Following this, his group reported a blood vessel model in 1986. This model had a multilayered structure of endothelial cells and smooth muscle cells resembling that of an artery, and withstood physiologic pressures by utilizing multiple layers of collagen integrated with a Dacron mesh.[3] As far as we know, this type of device has never been applied to clinical use, but many efforts have been made to create functional tissue-engineered blood vessels after this report.[4,5]

Use of scaffolds for vascular grafts

In 1999, Dr. L.E. Niklason in Prof. R. Langer's group reported a tissue-engineered vascular graft utilizing biodegradable synthetic polymer scaffold seeded with bovine smooth muscle and endothelial cells which had been derived from a biopsy of vascular tissue.[6] Pulsatile culture conditions and intra-luminal flow in a newly designed bioreactor increased rupture strengths to greater than 2000 millimeters of mercury, suture retention strengths up to 90 grams, and collagen contents of up to 50 percent during eight-week culture. This tissue-engineered vascular graft also showed contractile responses to pharmacological agents. This tissue-engineered artery showed patency up to 24 days after the implantation in miniature swine.

---

*Corresponding authors.

More recently, her group reported that similar small-diameter vessel grafts can be fabricated from adult human bone marrow-derived mesenchymal stem cells (hMSCs).[7] They chose hMSCs as the cell source, because smooth muscle cells from elderly persons have limited proliferative capacity and reduced collagen production, which impair the mechanical strength of engineered vessels. hMSCs were differentiated to smooth muscle cells in culture plates, then subjected to a bioreactor system. The fabricated vessel walls were substantially similar to native vessels.

Prof. C.A. Vacanti's group utilized a unique hemodynamically-equivalent pulsatile bioreactor to fabricate a three-layered robust and elastic artery.[8] A polyglycolic acid (PGA)/polycaprolactone sheet seeded with porcine aortic smooth muscle cells, and a PGA sheet seeded with porcine aortic fibroblasts, were wrapped in turn on a 6-mm diameter silicone tube and incubated in culture medium for 30 days. After the supporting tube was removed, the lumen was seeded with porcine aortic endothelial cells and incubated for another 2 days. Finally, the pulsatile bioreactor culture was performed for an additional 2 weeks, and the engineered vessels were distinctly similar in appearance and elasticity to native arteries (Figure 1). Ample production of elastin and collagen in the engineered grafts was observed. Tensile tests demonstrated that engineered vessels acquired equivalent ultimate strength and similar elastic characteristics as native arteries.

Prof. J.E. Mayer's group utilized decellularized matrix as a scaffold and endothelial progenitor cells (EPCs) as a cell source to replace arterial endothelial cells for tissue-engineered small-diameter blood vessels.[9] The inner diameter of porcine iliac vessels was 4 mm. EPCs isolated from peripheral blood of sheep were expanded *ex vivo,* and then seeded on decellularized matrix. EPC-seeded grafts remained patent for 130 days as a carotid interposition graft in sheep, whereas non-seeded grafts occluded within 15 days. Interestingly, the EPC-explanted grafts exhibited contractile activity and nitric-oxide-mediated vascular relaxation that were similar to native carotid arteries.

Sheet-based tissue engineering

Dr. N. L'Heureux's group of Cytograft Tissue Engineering performed a clinical trial of tissue-engineered blood vessel for adult arterial revascularization.[10] Their technology is

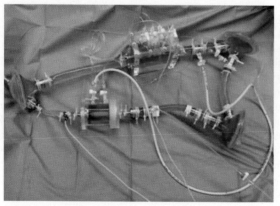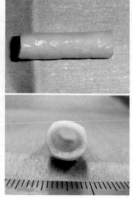

Figure 1. Tissue-engineered blood vessel fabricated in a hemodynamically-equivalent pulsatile bioreactor. (Provided by Prof. Iwasaki of Waseda Univ.)

termed sheet-based tissue engineering. Functional tissue-engineered blood vessels with physiologic mechanical properties were fabricated only from autologous cells.[11] No synthetic or exogenous materials were used; instead, the vessels were created with the use of autologous fibroblasts and endothelial cells harvested from a small biopsy specimen of skin and superficial vein. In brief, dermal fibroblasts are grown in conditions that promote the production of extracellular matrix proteins. After approximately 6 weeks in culture, sheets of dermal fibroblasts are detached from the culture substrate and rolled around a temporary support material. After a maturation period of approximately 10 weeks, the individual layers fuse into a homogenous tissue. This process can be repeated to produce multilayer vessels with burst pressures in excess of 3000 mmHg. A few days before implantation, the vessel is seeded with autologous endothelial cells. Among 54 vessels developed using this approach by the L'Heureux group, the average burst pressure was 3340 ± 849 mm Hg, which compares favorably with native veins.[12] One significant drawback of this approach is the time intensive nature of the protocol. Total production time is approximately 24 weeks.

In their clinical trials, patients (average 68 years old) receiving hemodialysis whose arteriovenous shunts were failing were enrolled. The results showed the tissue-engineered blood vessel produced *in vitro* could withstand arterial pressure produced by an arteriovenous fistula for at least 3 months. The tissue-engineered blood vessel in a patient (Figure 2) was used for more than 13 months, until the patient underwent successful kidney transplantation. In total, the 14-cm long graft was punctured more than 200 times. These results clearly highlight the advantages of using sheet-based tissue engineering.

Capillary Networks

While cornea and cartilage are avascular tissues, capillary networks branched from artery and vein are required to maintain metabolism and activities of cells in the other tissues. Without capillary networks distributed all over the tissue, supply of oxygen and nutrients as well as metabolite excretion would rely on simple diffusion. Relying on this process would limit the size and thickness of tissue-engineered constructs; a thickness of approximately 200 microns is the limit above which diffusion is insufficient to supply oxygen and nutrients. It is often observed that only on the surfaces of three-dimensional scaffolds do cells survive; however under static culture conditions, cells may be found at the inner space and core of tissue engineering constructs. In order to overcome this problem bioreactors are often employed. However, without sufficiently rapid migration of capillaries into the core of the constructs, necrosis inevitably occurs after transplantation.

Therefore, huge efforts have been made to introduce capillary networks into tissue-engineered constructs. Muscle tissue is energy-consuming so that dense capillary networks are extensively formed within it (Figure 3). In this section, we summarize several representative attempts to introduce capillary networks into tissues and fabricated constructs.

Drug delivery system (DDS) and controlled release

Several angiogenic growth factors, including vascular endothelial growth factor (VEGF) and basic fibroblast growth factor (bFGF), have been identified. Delivery or incorporation of these angiogenic factors into tissue-engineered constructs and/or implanted sites remains the simplest way to promote angiogenesis. However, direct injection of angiogenic proteins often requires delivery of supraphysiologic concentrations for a therapeutic effect owing to their short half-lives (on the order of minutes) after injection.[13] Injected

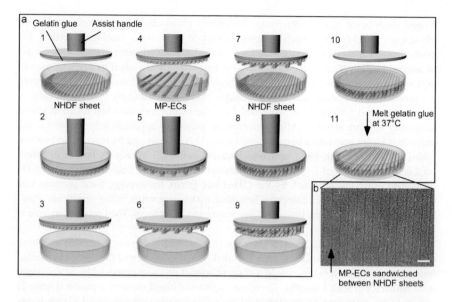

Figure 2. Schematic illustration of 3-D manipulation of fibroblast cell sheets (NHDF sheets) and micro-patterned endothelial cells (MP-ECs) with the gelatin-coated stacking manipulator. (a) NHDF sheet cultured on non-patterned thermo-responsive surfaces and MP-ECs cultured on TMP surfaces are prepared in advance. (1–3) The layering manipulator affixed with gelatin is placed onto a non-patterned NHDF sheet cultured on a thermo-responsive surface and incubated for 20 min at 20°C, then the NHDF sheet attached with gelatin glue is lifted. (4) The detached NHDF sheet is transferred and gently placed onto MP-ECs cultured on a TMP surface. (5–7) After incubating for 20 min at 20°C, the bilayer NHDF sheet and MP-ECs are placed onto another attached NHDF sheet in a similar manner. The process of (1)–(9) is repeated to create three-layer or five-layer cell constructs *in vitro*. (10) After stacking cell sheets, multi-layered tissue with the gelatin layer is removed from the manipulator and placed onto TCPS to allow adhesion onto the surface by incubating for 20 min at 20°C. (11) Finally, excess volumes of cell culture medium are added and dishes are warmed at 37°C to melt and remove the gelatin. (b) Phase contrast view of MP-ECs sandwiched between NHDF sheets. Scale bar: 100 μm.

proteins are often quickly degraded by proteolytic enzymes, and/or trapped by extracellular matrix and inactivated. Injection of excessive concentrations of angiogenic factors may cause dangerous side effects, such as leaky blood vessels or hemorrhage.[14] To minimize adverse effects, natural and synthetic polymeric scaffolds, including alginate, gelatin and poly(lactide-co-glycolide), have been developed for localized, controlled angiogenic growth factor release. Sustained VEGF delivery from a porous poly(lactide-co-glycolide) scaffold has been shown to enhance angiogenesis and survival of transplanted cells.[15]

Prof. D.J. Mooney's group has intensively pursued the development of tissue engineering techniques to promote capillary network formation.[16] Recently, they reported that biodegradable microspheres containing VEGF promoted formation of neointestinal cysts in tissue-engineered intestine by sustained release of biologically active VEGF.[17] Furthermore, intestinal constructs with VEGF microspheres were significantly larger than those containing empty microspheres. Capillary density was also significantly increased in the VEGF-containing neointestinal constructs. Interestingly, microcapillary formation resulted in higher proliferation of epithelial cells, although epithelial cells are not responsive

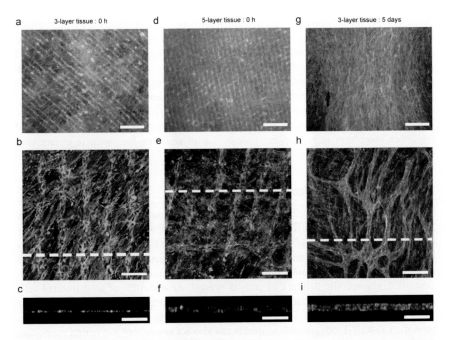

Figure 3. Fluorescently stained 3-D tissue constructs. Constructs contained both MP-ECs and NHDF sheets were stained with UEA-1 (*red*) for ECs and F-actin for both ECs and NHDFs (*green*). (a–c) Three-layer tissue after layering but before further culture. (d–f) Five-layer tissue after layering but before further culture. (g–i) Three-layer tissue after 5-day cultivation at 37°C after layering procedures: (a, d and g) fluorescent micrographs at lower magnification. Scale bars: 5 μm. (b, e and h) confocal laser micrographs, and (c, f and i) cross-sectional views of (b, e and h). Scale bars: 100 μm.

to VEGF. These findings support the importance of regeneration of capillary networks inside tissue-engineered constructs.

Although VEGF initiates the formation of new blood vessels, VEGF alone is unable to induce the formation of mature, functional blood vessels.[18] Other factors, such as angiopoietin-1 (Ang1) and platelet-derived growth factor (PDGF), are required to promote vessel maturation and stabilization.[19] When these additional factors are not provided, newly formed capillaries are highly leaky and ultimately disappear. In order to fabricate functional, mature vessels, delivering combinations of several angiogenic factors in the appropriate spatiotemporal sequence is important. Prof. Mooney's group also reported that dual delivery of VEGF for endothelial cells and PDGF for pericytes, each with distinct release kinetics, from a single poly(lactide-co-glycolide) scaffold, promoted formation of mature, stable vasculature in a mouse ischemic hind limb model.[20] The lumens of typical capillaries are lined with endothelial cells in contact with circulating blood and pericytes. Pericytes are a relatively undifferentiated cell residing outside the basement membrane. These cells serve to support capillary vessels, but can differentiate into fibroblasts, smooth muscle cells, or macrophages.

In place of VEGF protein, a plasmid DNA encoding the VEGF gene may also be delivered by hydrogel to provide local and sustained release of VEGF. Concurrent physical dissociation of the cross-linked structure and hydrolytic chain breakage of ionically cross-linked polysaccharide hydrogels allows for the titration of pDNA release by alteration of hydrogel properties. Local pDNA release significantly improved the recovery of blood perfusion in the ischemic hindlimbs of mice as compared with a bolus injection of VEGF encoding pDNA.[21] Localized gene-therapy approaches to the delivery of angiogenic growth factors may circumvent limitations of direct protein delivery, including the frequent requirement for supraphysiologic quantities. Delivery of genes encoding angiogenic growth factors allows sustained, localized protein production and can be used for either short-term (e.g. transient transfection of cells) or long-term (e.g. retroviral transduction) delivery.[22]

Endothelial cell transplantation

Transplantation of cells forming capillaries can also regenerate new vascular networks. Endothelial progenitor cells (EPCs) circulate in blood, and intravenous infusion of these cells enhances angiogenesis.[23] Similarly, injection of bone marrow cells also enhances angiogenesis.[24] However, since the possibility of tumorigenesis resulting from injection of these endothelial progenitors into the systemic circulation cannot be excluded, well-controlled, site-specific delivery of cells is needed. Around 10% of total injected cells remain in the transplanted tissue sites after simple cell suspension injection. Similar to the cases of delivery of protein and gene, biomaterial scaffolds can be useful in cell transplantation as well. Biodegradable scaffolds can also act as extracellular matrix supporting capillary tube formation of endothelial cells.[25] Co-delivery of endothelial cells and angiogenic factors can enhance vessel formation. A combined approach to promote neovascularization via delivery of human umbilical vein endothelial cells and VEGF from a poly(lactide-co-glycolide) scaffold has been demonstrated.[26]

Microfluidics

Prof. J.P. Vacanti's group proposed a novel microfluidics-based bilayer device with a discrete parenchymal chamber modeled on hepatic organ architecture. The microfluidics network was designed using computational models to provide appropriate flow behavior based on physiological data from human microvasculature. Patterned silicon wafer molds were used to generate films with the vascular-based microfluidics network designed by soft lithography. Inside the microfluidics network, endothelial cells were seeded, and the entire surface of the microfluidic device was covered by proliferated endothelial cells after a short time culture.[27] This device was designed for an extracorporeal use and drug screening, but it is likely that such microfabrication-based techniques will be utilized for regenerative medicine in the future.

Cell sheet engineering

We have pursued another type of sheet-based tissue engineering, named cell sheet engineering. To fabricate transplantable cell sheets, we developed temperature-responsive culture surfaces. A temperature-responsive culture dish was first reported by our group in 1990,[28] after which analyses of harvested cells and cell sheets were reported.[29] In brief, a temperature-responsive polymer, poly(N-isopropylacrylamide) is covalently immobilized to the surface of the dish, allowing cells to adhere and proliferate at 37°C, but also causing cells to detach without any need for protease by temperature reduction below

32°C. Therefore, by a reduction in temperature after cells reach confluency, all the cultured cells are harvested as a single contiguous cell sheet. This technique was first applied to clinical settings in plastic reconstructive surgery for skin,[30] followed by use in corneal reconstruction.[31] We reported that the transplantation of layered cardiomyocyte sheets can improve cardiac function after myocardial infarction.[32] Furthermore, we examined the effects of promoting neovascularization by controlling the densities of co-cultured endothelial cells within engineered myocardial tissues.[33] Neonatal rat cardiomyocytes were co-cultured with endothelial cells isolated from GFP-transgenic mice on temperature-responsive culture dishes. Co-cultured GFP-endothelial cells formed cell networks within the cardiomyocyte sheets, which were preserved during cell harvest from the dishes using simple temperature reduction. After layering of three cardiac cell sheets to create three-dimensional myocardial tissues, these patch-like tissue grafts were transplanted onto infarcted rat hearts. Four weeks after transplantation, recovery of cardiac function and the number of capillaries in myocardial tissue were significantly improved, due to increased endothelial cell density. Finally, blood vessels originating from GFP-endothelial cells in regenerated cardiac tissues bridged into the infarcted myocardium to connect with capillaries of the host heart. These results indicate that neovascularization induced co-cultured endothelial cells within cell sheets can contribute to improved tissue function after the transplantation.

We also reported a novel method to fabricate pre-vascularized tissue equivalents using multi-layered cultures which combine micro-patterned endothelial cells as vascular pre-cursors with fibroblast monolayer sheets as tissue matrix.[34] Stratified tissue equivalents are constructed by alternately layering fibroblast monolayer sheets with patterned endothelial cell sheets harvested from newly developed thermo-responsive micro-patterned surfaces, alternating 20 mm-wide cell-adhesive lanes with 60 mm non-adhesive zones (Figure 3). Endothelial cell patterning fidelity was maintained within the multi-layer tissue constructs after assembly, leading to self-organization into microvascular-like networks after 5-day tissue culture. This novel technique holds promise for the study of cell-cell communications and angiogenesis in reconstructed, three dimensional environments as well as for the fabrication of tissues with complex, multicellular architecture.

Application to Hand Surgery

The use of tissue engineered vascular prostheses would certainly have wide implications for hand surgery, not only for primary replacement of absent or damaged blood vessels but also for the revascularization, or microvascularization, of other tissue types. The need for engineered blood vessels in surgical correction of major trauma, congenital absence or acquired malformation of vessels, and hand replantation is clear. The ability to use a diverse array of technologies, from sheet based tissue engineering to hydrogel controlled release of VEGF/bFGF, may one day allow for microvascularization of many different types of engineered tissues, increasing their viability. Recent advances in promoting capillary invasion of prosthetic or transplanted tissues are particularly promising for use in the surgical setting. Given the relatively low flow and pressure waveforms found in distal arteries in the upper extremity and hand, the hand would seem to be the ideal clinical model for the testing of autologously grown blood vessels. As the financial and labor costs required to produce autologously engineering blood vessels falls, hand surgeons may find themselves at the forefront of adopting this exciting technology.

References

1. Niklason LE, Langer RS. Advances in tissue engineering of blood vessels and other tissues, *Transpl Immunol* **5** (1997) 303–306.
2. Bell E, Ivarsson B, Merrill C. Production of a tissue-like structure by contraction of collagen lattices by human fibroblasts of different proliferative potential *in vitro*, *Proc Natl Acad Sci USA* **76** (1979) 1274–1278.
3. Weinberg CB, Bell E. A blood vessel model constructed from collagen and cultured vascular cells, *Science* **231** (1986) 397–400.
4. Matsuda T. Recent progress of vascular graft engineering in Japan, *Artif Organs* **28** (2004) 64–71.
5. Zhang WJ, Liu W, Cui L, Cao Y. Tissue engineering of blood vessel. *J Cell Mol Med* **11** (2007) 945–957.
6. Niklason LE, Gao J, Abbott WM, Hirschi KK, Houser S, Marini R, Langer R. Functional arteries grown *in vitro*. *Science* **284** (1999) 489–493.
7. Gong Z, Niklason LE. Small-diameter human vessel wall engineered from bone marrow-derived mesenchymal stem cells (hMSCs), *FASEB J* **22** (2008) 1635–1648.
8. Iwasaki K, Kojima K, Kodama S, Paz AC, Chambers M, Umezu M, Vacanti CA. Bioengineered three-layered robust and elastic artery using hemodynamically-equivalent pulsatile bioreactor, *Circulation* **118**(14 Suppl.) (2008) S52–57.
9. Kaushal S, Amiel GE, Guleserian KJ, Shapira OM, Perry T, Sutherland FW, Rabkin E, Moran AM, Schoen FJ, Atala A, Soker S, Bischoff J, Mayer JE. Jr, Functional small-diameter neovessels created using endothelial progenitor cells expanded *ex vivo*, *Nat Med* **7** (2001) 1035–1040.
10. L'Heureux N, McAllister TN, de la Fuente LM. Tissue-engineered blood vessel for adult arterial revascularization, *N Engl J Med* **357** (2007) 1451–1453.
11. L'Heureux N, Dusserre N, Konig G, Victor B, Keire P, Wight TN, Chronos NA, Kyles, AE, Gregory CR, Hoyt G, Robbins RC, McAllister TN. Human tissue-engineered blood vessels for adult arterial revascularization, *Nat Med* **12** (2006) 361–365.
12. Schaner PJ, Martin ND, Tulenko TN, Shapiro IM, Tarola NA, Leichter RF, Carabasi RA, Dimuzio PJ. Decellularized vein as a potential scaffold for vascular tissue engineering, *J Vasc Surg* **40** (2004) 146–153.
13. Chen RR, Mooney DJ. Polymeric growth factor delivery strategies for tissue engineering, *Pharm Res* **20** (2003) 1103–1112.
14. Gu F, Amsden B, Neufeld R. Sustained delivery of vascular endothelial growth factor with alginate beads, *J Control Release* **96** (2004) 463–472.
15. Smith MK, Peters MC, Richardson TP, Garbern JC, Mooney DJ. Locally enhanced angiogenesis promotes transplanted cell survival, *Tissue Eng* **10** (2004) 63–71.
16. Kong HJ, Mooney DJ. Microenvironmental regulation of biomacromolecular therapies, *Nat Rev Drug Discov* **6** (2007) 455–463.
17. Rocha FG, Sundback CA, Krebs NJ, Leach JK, Mooney DJ, Ashley SW, Vacanti JP, Whang EE. The effect of sustained delivery of vascular endothelial growth factor on angiogenesis in tissue-engineered intestine, *Biomaterials* **29** (2008) 2884–2890.
18. Ash JD, Overbeek PA. Lens-specific VEGF-A expression induces angioblast migration and proliferation and stimulates angiogenic remodeling, *Dev Biol* **223** (2000) 383–398.
19. Yancopoulos GD, Davis S, Gale NW, Rudge JS, Wiegand SJ, Holash J. Vascular-specific growth factors and blood vessel formation, *Nature* **407** (2000) 242–248.
20. Richardson TP, Peters MC, Ennett AB, Mooney DJ. Polymeric system for dual growth factor delivery, *Nat Biotechnol* **19** (2001) 1029–1034.
21. Kong HJ, Kim ES, Huang YC, Mooney DJ. Design of biodegradable hydrogel for the local and sustained delivery of angiogenic plasmid DNA, *Pharm Res* **25** (2008) 1230–1238.
22. Partridge KA, Oreffo RO. Gene delivery in bone tissue engineering: progress and prospects using viral and nonviral strategies, *Tissue Eng* **10** (2004) 295–307.
23. Assmus B, Schachinger V, Teupe C, Britten M, Lehmann R, Dobert N, Grunwald F, Aicher A, Urbich C, Martin H, Hoelzer D, Dimmeler S, Zeiher AM. Transplantation of progenitor cells

and regeneration enhancement in acute myocardial infarction (TOPCARE-AMI), *Circulation*, **106** (2002) 3009–3017.

24. Tateishi-Yuyama E, Matsubara H, Murohara T, Ikeda U, Shintani S, Masaki H, Amano K, Kishimoto Y, Yoshimoto K, Akashi H, Shimada KIwasaka T, Imaizumi T. Therapeutic angiogenesis for patients with limb ischaemia by autologous transplantation of bone-marrow cells: a pilot study and a randomised controlled trial, *Lancet* **360** (2002) 427–435.

25. Nor JE, Peters MC, Christensen JB, Sutorik MM, Linn S, Khan MK, Addison CL, Mooney DJ, Polverini PJ. Engineering and characterization of functional human microvessels in immunodeficient mice, *Lab Invest* **81** (2001) 453–463.

26. Peters MC, Polverini PJ, Mooney DJ. Engineering vascular networks in porous polymer matrices, *J Biomed Mater Res* **60** (2002) 668–678.

27. Carraro A, Hsu WM, Kulig KM, Cheung WS, Miller ML, Weinberg EJ, Swart EF, Kaazempur-Mofrad M, Borenstein JT, Vacanti JP, Neville C. *In vitro* analysis of a hepatic device with intrinsic microvascular-based channels, *Biomed Microdevices* **10** (2008) 795–805.

28. Yamada N, Okano T, Sakai H, Karikusa F, Sawasaki Y, Sakurai Y. Thermo-responsive polymeric surfaces; control of attachment and detachment of cultured cells. *Makromol Chem Rapid Commun* **11** (1990) 571–576.

29. Yang J, Yamato M, Shimizu T, Sekine H, Ohashi K, Kanzaki M, Ohki T, Nishida K, Okano T. Reconstruction of functional tissues with cell sheet engineering, *Biomaterials,* **28** (2007) 5033–5043.

30. Yamato M, Utsumi M, Kushida A, Konno C, Kikuchi A, Okano T. Thermo-responsive culture dishes allow the intact harvest of multilayered keratinocyte sheets without dispase by reducing temperature, *Tissue Eng* **7** (2001) 473–480.

31. Nishida K, Yamato M, Hayashida Y, Watanabe K, Yamamoto K, Adachi E, Nagai S, Kikuchi A, Maeda N, Watanabe H, Okano T, Tano Y. Corneal reconstruction with tissue-engineered cell sheets composed of autologous oral mucosal epithelium, *N Eng J Med* **351** (2004) 1187–1196.

32. Miyagawa S, Sawa Y, Sakakida S, Taketani S, Kondoh H, Memon IA, Imanishi Y, Shimizu T, Okano T, Matsuda H. Tissue cardiomyoplasty using bioengineered contractile cardiomyocyte sheets to repair damaged myocardium: their integration with recipient myocardium, *Transplantation* **80** (2005) 1586–1595.

33. Sekine H, Shimizu T, Hobo K, Sekiya S, Yang J, Yamato M, Kurosawa H, Kobayashi E, Okano T. Endothelial cell coculture within tissue-engineered cardiomyocyte sheets enhances neovascularization and improves cardiac function of ischemic hearts *Circulation* **118**(14 Suppl.) (2008) S145–S152.

34. Tsuda Y, Shimizu T, Yamato M, Kikuchi A, Sasagawa T, Sekiya S, Kobayashi J, Chen G, Okano T. Cellular control of tissue architectures using a three-dimensional tissue fabrication technique. *Biomaterials* **28** (2007) 4939–4946.

CHAPTER 14

Bench to Bedside: Navigating Industry, the FDA and Venture Capital

Nicholas J. Panetta‡, Deepak M. Gupta‡, Michael T. Longaker‡ and Geoffrey C. Gurtner*,†,‡

Introduction

Steven P. Jobs, co-founder and CEO of Apple Inc., once said with regard to the process of innovation, "My experience has been that creating a compelling new technology is so much harder than you think it will be that you're almost dead when you get to the other shore."[1] Nowhere is this truer than in the world of surgical innovation. Certainly, surgeons are uniquely positioned to view patient care from a privileged vantage point, allowing for the identification of root causes of inadequacies in the treatment of surgical disease, and thus affording insight towards how improvements can be achieved. However, the process of successfully developing an idea conceptualized in the mind of a surgeon scientist is often burdened with a myriad of road blocks, including but not limited to the processes of securing funding to support product development, the protection of intellectual property, the conduct of clinical trials, and the attainment of federal approval for clinical implementation and distribution to the marketplace. Such burdens can prove to be cumbersome, resulting in only 10% of medical device start-ups coming to fruition.[2] Unfortunately, the level of commitment on the part of the surgeon required to surmount these obstacles has historically proved to be more than most are will to put forth.[3]

The reason for this frequently observed lack of enthusiasm in surgeons to initiate and/or participate in surgical innovation is multifaceted. The course of surgical device development can be protracted, requiring on average five to eight years to move from idea to application in the clinical arena.[4] The ever mounting demands of clinical practice, paired with precipitous declines in reimbursement, have placed time and financial constraints on physicians that make it exceedingly difficult to allocate sufficient resources to actively participate in the materialization of ideas conceptualized in the course of their own practice.[5] Navigating the convoluted processes innate to surgical innovation can be psychologically daunting, and often hinder the evolution of otherwise sound concepts. Moreover, surgeons as a group often prove to be their own worst enemy. Far too often

*Corresponding author.
†Stanford University School of Medicine, 257 Campus Drive, Stanford, CA 94305-5148, Email: ggurtner@stanford.edu
‡Hagey Laboratory for Pediatric Regenerative Medicine, Division of Plastic and Reconstructive Surgery, Department of Surgery, Stanford University School of Medicine.

clinical decision making is based in authoritative ideas and opinions known as "surgical dogma." Though frequently a paucity of evidence exists to support these principals, they are firmly entrenched in surgical teachings and serve to oppose the process of surgical innovation. Thwarted by the fear of personal failure and professional ridicule, surgeons often unknowingly become averse to pursuing new approaches to old problems, thus stifling the advancement of patient care. However, all is not lost. As will be illustrated in the following discussion, overcoming all of these obstacles to surgical innovation can be accomplished through the acquisition of learned skills, attainable by any surgeon motivated to do so.

Clinical Need Driven Innovation

History provides us with many examples of surgeons who, in the course of caring for their patients, have recognized a void in the tools and techniques at their disposal to utilize in the intervention upon surgical disease. Committed to investing the necessary effort, these individuals achieved success in revolutionizing the treatment of surgical patients through the development of novel techniques and devices. Cardiac and solid organ transplantation — procedures performed routinely in centers across the nation today — were inconceivable prior to the efforts of Norman E. Shumway, Joseph Murray, and Thomas E. Starzl. Regarding device innovation, no individual better exemplifies the surgeon as an inventor than Thomas J. Fogarty. In addition to developing the balloon catheter, dramatically improving the outcomes of surgical intervention for vascular disease, Dr. Fogarty's name appears on more than 100 patents for surgical devices.[6,7] When the accomplishments of such individuals are considered, one can truly appreciate how realizing the transformation of an idea inspired by a "real world" problem, into a codified surgical innovation holds significant potential to be personally, professionally, and economically rewarding for those willing to make the necessary sacrifices.

Today, the fields of tissue engineering and regenerative medicine are the focus of significant surgical innovation efforts. Current reconstructive modalities, intended to repair or replace diseased or damaged tissue, remain somewhat inadequate. Too frequently their implementation results in outcomes with both poor form and inferior functionality. Sometimes one of these factors can be optimized at the expense of the other. However, as hand surgeons, rarely are we afforded this luxury. Further compounding the difficulties of hand surgery is the fact that tissues requiring reconstruction are often composite in nature, consisting of skin, bone, muscle, tendon, cartilage, nerve, and vasculature.

As such, the potential for tissue engineering to address these challenges, meeting a significant need for the development of novel reconstructive modalities, is substantial. Indeed, the potential to recapitulate the native form and function of deficient structures with *de novo* engineered tissues represents a true paradigm shift in current reconstructive techniques. It is exciting to conceptualize a single, biologically active, progenitor cell-based reconstructive modality capable of undergoing regionally mediated tissue specific differentiation, resulting in an outcome with excellent form and function. Early studies probing either autogenous, allogeneic, xenogeneic, or synthetic scaffold based repair; novel growth factor mediated augmentation of endogenous tissue regeneration, or cell-based reconstructive modalities involving all of the above tissues have been initiated, with tissue engineered skin having appreciated the largest successful translation to the marketplace.[8-18] However, these innovations in tissue engineering will not be appreciated through the efforts of a single or small group of surgeon inventors. True multidisciplinary collaboration will be imperative for the development of translational interventions to be realized. To date, studies resulting from collaboration by investigators in the disciplines of developmental biology, organogenesis, stem cell biology, and material sciences have yielded a solid foundation of knowledge upon which the field of tissue engineering may develop, and continued collaboration will be imperative to refine novel tissue engineering strategies and achieve optimal patient outcomes. In this chapter, we will explore the process of how novel surgical innovations move from concept to reality, and how understanding this process will be essential to the forward progress of tissue engineering and regenerative medicine — fields still in their infancy.

Innovation in the Care of the Surgical Patient

All surgical innovation is based on the desire of a physician to improve the care of their patients using a new technical or technologic approach. However, despite the necessity of this act of innovation, it is far from sufficient to drive the innovative process to completion. Innovation in surgery can largely be attributed to many of the same guiding principles which differentiate successful from failed innovative endeavors in other fields. Critical aspects of the innovative process include a well defined goal, alignment of actions to goals, successful interaction of multiple groups, close monitoring of progress, and open lines of communication and access to information between involved parties.[19] Failure to afford due diligence to any of these factors may significantly undermine progress, producing unfruitful efforts to realize developmental milestones.

However, the process of surgical innovation is adorned with an additional layer of complexity not appreciated by the business world and elsewhere. Unique to surgical innovation is the fact that the implementation of novel techniques or devices is directly and inseparably connected to the health and well being of a human life. This, paired with the necessary integration of the process with corporations driven by financial motivations, as well as the potential for financial gain by the inventors themselves, creates a situation where conflicts of interest arise from the need to maintain a balance between stringent safety and efficacy standards with the desire for a novel innovation to be profitable. Such situations require an increased in the degree of scrutiny and regulation with which the process is overseen; yet, these assurances are often viewed as excessively encumbering the innovative process.

Surgical innovations can be broadly categorized as sustaining or disruptive. While sustaining innovations aim to make improvements on existing techniques and technologies, disruptive innovations put forth totally novel means to approach old problems.[3] Gains to be appreciated from sustaining innovations are evidenced by the evolution of central venous catheter technology. Causative in the majority of blood stream infections in critically ill patients, a clear potential for catheter improvement was identified.[20] In an attempt to meet this need, the concept of antibiotic impregnated catheters came to light. Central venous catheters impregnated with antibiotics such as minocycline and rifampin were introduced to the clinical arena, and since their implementation significant reductions in catheter-related blood stream infections, including drug resistant bacteremia, compared with standard catheters have been observed.[20–22] This seemingly simplistic modification in an existing, widely used device has since resulted in decreased hospital and intensive care unit stays, improved patient morbidity and mortality, and a significant reduction in costs to the healthcare system arising secondary to catheter related blood stream infections.[20,23–25]

From this example alone, one is able to appreciate the significant strides in the care of surgical patients that can arise from sustaining technological advancements. In reality, most would agree that to achieve excellence as a surgeon, he or she must involve themselves in the process of incremental innovation on a daily basis; continually submerging themselves in the process of critically analyzing their own patient care practices, identifying specific areas where improvement can potentially be appreciated, and devising and implementing modifications in techniques and decision making algorithms to achieve this end.

Beyond the sustaining innovations that move patient care ahead one small step at a time, truly revolutionary ideas are the source of disruptive innovations with the potential to change the landscape of current medical practice. Disruptive surgical innovations that have accomplished this end can been seen in the examples of inhaled anesthetics, cardiopulmonary bypass, and video laparoscopy.[3] It is hard for the modern surgeon to imagine practicing without the availability of such tools being commonplace. However, such ideas are often not readily accepted by the surgical community when initially introduced. The propagation of disruptive innovations is inherently associated with substantial risk that extends to all parties instrumental to the innovative process.

Initially, the observations of a surgeon innovator must yield sufficient evidence that a particular novel technique or technology may realistically convey significant improvements in patient care and outcomes. The necessity of strong supporting evidence is critical due to the fact that disruptive innovations by nature frequently deviate significantly from what is considered safe and accepted at the time. Furthermore, proof of profitability is essential to mitigate the substantial risk that accompanies disruptive innovations if successful acquisition of corporate investment necessary to conduct the costly research and development of such concepts is to occur.[4] Finally, information regarding the benefits to be gained from a new technique or technology must be conveyed clearly and empathetically to patients whom are involved early in its implementation, as individuals are often reluctant to forgo the tried-and-true for a relative "unknown." Certainly, the protracted periods of time that consistently accompany the movement of disruptive innovations from bench to bedside are based in the hesitation of all parties concerned to accept these risks.

Intellectual Property and the Patent Process

Long before significant funding is pursued, and clinical trials are still a fragile thought in the back of a surgeon scientist's mind, the time period immediately following the concretion of an original surgical innovation may be the most critical towards ensuring that the idea successfully comes to fruition. Yet, far too often failure to place sufficient focus on the completion of multiple necessary steps that must be taken to properly protect the intellectual property of the inventor results from ignorance of the consequences resulting from failure to do so, paired with an overzealous enthusiasm to see a concept realized.

First and foremost, a thorough investigation should be conducted, in the form of a patent and literature search, to bear out that an idea is indeed novel in nature prior to initiating the patent application process. For someone who is well versed in the process of patent submission, the potential to complete an adequate patent literature search may be realistic. To assist in the completion of this review process, multiple online resources are available to search current patents on existing art, including www.freepatentsonline.com, www.google.com/patents, and http://www.uspto.gov/patft/index.html. However, most parties will be served well by retaining the services of an intellectual property attorney to not only assist in the performance of patent reviews, but also to provide sound consultation regarding the defensibility of a patent should an award be made.[26] In the setting of an academic institution, such services are acquired through the technology licensing office. It behooves surgeon inventors to seek out such council early, whether privately or in the university setting, in order to avoid the myriad of pitfalls that can potentially result from infringement upon existing art, a misunderstanding of tenets surrounding public disclosure, and poor documentation of intellectual property.

Having a well defined understanding of what activities fall under the umbrella of public disclosure, in addition to what forms of information transfer constitute the disclosure of an innovative idea, is critical to the avoidance of unforeseen involvement in the transfer of information that may potentially thwart the successful award of a patent. In the U.S. patent system, any exchange of information in the form of publication, public use, sale, or offer for sale constitutes a public disclosure. Participation in any of the above is known as a *barring event*, and applies regardless of where the event took place in the world. U.S. patent law states that an inventor has one year from the barring event to file a patent application, known as the bar date. After this window of opportunity has passed, the inventor is disallowed from submitting an application to patent the invention that was publicly disclosed. Of note, a similar one year grace period to file a patent application following public disclosure of an idea does not exist for foreign patents. Once public disclosure of an idea has been made, an inventors right to file for a foreign patent is permanently compromised. As a point of clarification, it is important to realize that the term "publication" is not confined to printed scientific material in the traditional sense. Today, in a world where information in conveyed between professionals and to the public in a vast array of different media, the breadth of the term publication in the patent world has been expanded to be inclusive of printed material in the form of formal publications, photocopies, typed manuscripts, microfilms, abstracts, and master's theses, as well as verbal communications including Ph.D. dissertations, overhead presentations, poster sessions, and tape recordings of speeches.[27]

In the U.S., "first-to-invent" remains the principle guiding the award of patents, whereas the remainder of the world operates under a "first-to-file" system. Initiatives have previously been proposed in Congress to convert to a "first-to-file" U.S. patent system, and significant debate remains over the pros and cons of such a change.[28] Nonetheless, the "first-to-invent" system remains intact, and as such the importance of comprehensive documentation of all work involved in the development of any innovation is critical to the surgeon. Inclusive in this should be the maintenance of a clear and detailed record of the works progress. The record should be signed, dated, and witnessed regularly to contribute to the validity of the studies should a dispute over the originality of the innovation develop.

An additional mechanism that exists which can contribute significantly to documenting the progress of an inventive process is the filing of a provisional patent. Provisional patents explicitly state whom the inventors are involved in the project of interest, and consist of a written communication describing an invention. Though it serves to establish a filing date, information contained within a provisional patent is not reviewed by the USPTO, and a formal utility patent must be filed within one year from the receipt of the provisional patent.[29]

It has been demonstrated to this point that traveling the lengthy road from conceptualization to implementation of a novel surgical invention can be arduous, complex, and will require the devotion of significant time and effort from involved parties. Thus, taking proper steps to ensure that rights to not only the concept of interest, but additional ideas derived directly from the original model, are properly assigned to the inventor is critical. In the United States, this protection is afforded in the form of utility patents, issued by the

United States Patent and Trademark Office (USPTO). The protection put forth by utility patents is clearly defined as a property right granted by the Government of the United States of America to an inventor to exclude others from making, using, offering for sale, or selling the invention throughout the United States, or importing the invention into the United States, for a limited time in exchange for public disclosure of the invention when the patent is granted. This umbrella provides shelter to the means by which something is produced, machines, articles of manufacture, composition of matter, or a new and useful process.[30] Furthermore, patents pertaining to significant improvements on any of the afore-mentioned categories can be procured. When a patent is issued for a particular medical device, the protection extends for a period of 20 years. Upon patent submission, three main criteria have been established to guide the approval process conducted by the USPTO: 1) the novelty requirement, 2) the utility requirement, and 3) the nonobviousness require-ment.[26] The first of these requirements expresses the stipulation that the invention for which the patent is being applied for cannot have previously existed and/or been utilized, nor can the concept encompassing the invention have been put forth in printed literature. The second criterion simply outlines that the invention must serve a defined function and successfully works towards this end. Finally, implied in the nonobviousness requirement is the proviso that the invention cannot be obvious to someone skilled in the art to which the invention is directed. When a novel invention is in its infancy, frequently the emotion is to achieve progress as expeditiously as possible. However, due diligence is warranted on the part of the surgeon scientist to ensure that such haste does not come at the expense of not properly ensuring that all conditions for successful patent approval are met.

Proof of Concept

From the point when an idea is first conceived, it is important to outline the steps neces-sary to develop a product which will be received well when presented to industry and have the highest potential to achieve FDA approval for delivery to the marketplace. Inherent to the successful innovation of a medical product is the clear demonstration of its potential to improve patient care and outcomes. However, this alone is insufficient for even the most innovative ideas to be realized. The processes involved in the development of a medical innovation require significant financial support. Research and development, prototyping, patent acquisition, and the conduct of clinical trials can consume considerable financial resources, without which great ideas remain just that.

The importance of risk analysis cannot be overstated. The level of risk that will be assumed through participation in the development of a novel technology, as well as the potential to develop a viable risk mitigation strategy, hold significant weight in an investor's decision to support a project. Unfortunately, the risk posed to those investing in the development of medical products founded in the biologic sciences, and especially tissue engineering strategies, exceeds that appreciated by other industries. As opposed to research and development investments in other high-tech fields, the risk-to-return ratio is significantly higher in the biotechnology industry. This is founded in the fact that, whereas other industries largely focus on the evolution of concepts known to work in principal, investment in biotech is cloaked in ". . . profound and persistent uncertainty, rooted in the limited knowledge of human biological systems and processes. . ."[31] By considering the private venture capital support of U.S. based biotech companies investigating embryonic and adult stem cells over the time period spanning from 1994 to 2004, one can begin to appreciate the effect this risk has on the willingness to make significant financial commit-ments to such product development endeavors. During that time period, only 1% of the total venture capital investment activity was parsed for these projects.[32] Undoubtedly, reservation to invest in embryonic stem cell research is fueled by the political and ethical issues encumbering the efforts of investigators, and dampening their potential for use in the development of translational therapies.[33–35] This is despite the fact many feel that therapeutic modalities utilizing human embryonic stem cells, due to their pluripotency and regenerative capacity, have the potential to positively impact vast patient populations afflicted with a multitude of human disease.[36,37] Thus, in order to secure funding for a project, inventors must be able to do more than validate that an innovation will enhance patient care. Additionally, it must be clearly demonstrated that it can be developed and produced in a cost-effective means, and that its introduction to the marketplace will be profitable and readily accepted by society at large. If these factors can be demonstrated to a reasonable extent, investors will put forth significant commitments to see a revolutionary concept realized, as with any high risk investment comes the potential for significant financial gains.

FDA and the Regulation of Surgical Inventions

Oversight of the development of novel surgical techniques and medial products in the U.S. is largely accomplished by the Food and Drug Administration (FDA). Additionally, AdvaMed (Advanced Medical Technology Association), a medical technologies industry trade association, has put forth a code of ethics which clearly defines a set of ethical standards by which members abide in their interactions with medical professionals in the development of novel medical technologies.[32] The rapid expansion of new medical technologies being introduced to the market, accompanied by their frequent use in "off label" applications, has dramatically increased focus on the importance of adequate post-market surveillance following FDA approval. Thus, a collaborative interaction between both federal and industry regulatory bodies is needed to develop comprehensive post-market surveillance systems that will allow for the safest, most efficient progress of surgical innovation.[38] Not only understanding, but having an appreciation of the need for such collaborative regulatory efforts in what may seem to be bureaucratic oversight, is of utmost importance to the novice surgeon innovator, and prevents one from becoming unduly frustrated and fatigued by the process itself.

Medical products are defined as drugs (FD&C Act (at 201(h))(l))), devices (FD&C Act (at 201(h))), or biologic products (PHS Act (at 351(a))). The overwhelming majority of surgeon driven innovations to date have been classified as devices, and the branch of the FDA responsible for overseeing the development and production of medical devices is The Center for Devices and Radiological Health (CDRH). Upon giving consideration to a novel medical device for approval, the CDRH, under the guidance of the Medical Devices Amendment of the Federal Food, Drug, and Cosmetic Act of 1976, first classify a device into one of three categories: Class I, Class II, or Class III; subject to general controls, special controls, or premarket approval, respectively. These classifications are broadly based upon the level of regulation warranted to ensure the safety and effectiveness of the device. Currently, the FDA has classified roughly 1,700 generic categories of devices, which are further partitioned into 16 panels of medical specialties.[39] Subsequently, based upon the classification assigned to a particular device, it is assigned to one of three approval pathways which guide its transition from prototype to market approval, namely exemption, 510(k) premarket notification, and premarket approval (PMA). Unless a device exemption applies, all Class I/II devices are subject to 510(k) premarket notification. Class III devices almost universally undergo the process of PMA.

Devices categorized as Class I, held to the least stringent regulatory standards, are those in which compliance with the FDA regulations outlined by general controls can affirm that the necessary level of safety and effectiveness exists. General controls necessitate establishment registration, medical device listing, device production in accordance with good manufacturing practices, and proper compliance with described labeling regulations.[40] In addition to being held to the standards outlined by general controls, regulations outlined by special controls are applicable to Class II devices. In these situations, it is determined that general controls are inadequate to guarantee an adequate level of safety and effectiveness. Therefore, special controls afford the opportunity to apply more stringent regulations including special labeling requirements, performance standards to which a device can be held accountable, as well as postmarket surveillance.[40] Categorization of a device as Class III is dictated by whether the device fulfills the criteria that it is intended to support/sustain human life, prevent the impairment of human health, or impart a risk of illness or injury to those involved in their use.[40] In general, a Class III device is one whose approval would pose significantly greater risk than that posed by Class I/II devices secondary to the paucity of information available to support its safety and efficacy. For these reasons, the level of scrutiny with which Class III devices are reviewed far exceeds that applied to Class I/II devices.

Class I/II devices that are not exempt, in addition to Class III devices exempt from PMA, and are intended for application in humans are subject to the process of 510(k) premarket notification.[41] The function of a 510(k) premarket notification application is to establish that a particular device of interest has substantial equivalence (SE) to one or more legally marketed devices, otherwise known as predicates. The FDA has clearly defined four categories of individuals or groups who are required to submit a 510(k) premarket notification application prior to any device marketing in the United States. Additionally, under current law, many Class I/II devices are exempt from the need for 510(k) premarket notification. Guidelines are published by the FDA which specifically outline both who is required to submit a 510(k) application, as well as which generic Class I/II devices are exempt from the process.[42,43] Succeeding the submission of a 510(k) application for Class I/II devices, the FDA will return an initial response within 90 days to the applicant. Upon receipt of notification declaring the device to be SE, marketing of the device may commence immediately. However, if the device in question is found to not demonstrate SE, Class III classification will be assigned to the device, at which time the applicant may avail themselves to several available courses of action.

Prior to being approved for delivery to the marketplace, Class III devices are subjected to the detailed and rigorous scientific review process of PMA.[44] Such stringent regulatory processes are implemented in the review of Class III devices given the undetermined level of risk that they pose to the health and well being of humans involved with their use. From the date of PMA submission, the FDA has 180 days to perform its initial review of the application. In addition to review by the FDA advisory committee, the committee may elect to additionally review the application in a public forum. The cumulative information derived from the advisory committee's, as well as the public's review of the submission, may be communicated to the FDA to guide their decision with regard to acceptance or rejection of the submission. Should a decision to reject a PMA be reached, submitters are allotted a period of 30 days to petition the FDA for reconsideration.

Not infrequently, situations arise when conditions surrounding device development require clinical evaluation to gather data in order to validate its safety and efficacy prior to submitting a 510(k) premarket notification or PMA application. When clinical trials to examine an investigational device are deemed necessary, an investigational device exemption (IDE) can be sought out.[45] Application approval is overseen by appropriate institutional review boards (IRB) as well as the FDA. Once IDE approval is acquired, several additional criteria must be met for device evaluation to commence. Informed consent must be obtained from all involved patients, the device must be labeled for investigational use only, a method of monitoring the progress of the study must be outlined and adhered to, and required records and reports must be submitted upon request. When all of these IDE criteria are satisfactorily met and maintained, a device with IDE approval may be distributed to investigators free from the binds of FDA compliance regulations discussed heretofore.

As the field of tissue engineering continues to mature, light is being shed upon the important role that non-device interventions will play in the field's future. Knowledge regarding the molecular signaling pathways critical to the lineage specific differentiation of pluripotent cells for the purpose of tissue engineering has evolved, and ever increasing attention is being turned to focus upon how involved cytokines, growth factors, and other essential proteins can be applied in the development of new tissue engineering technologies to enhance the regenerative potential of both resident and implanted progenitor cells. Such proteins have been classified by the FDA as Therapeutic Biological Products, and the Center for Drug Evaluation and Research (CDER) is responsible for their oversight.[46] Because these Therapeutic Biological Products are classified as drugs, their clinical evaluation requires the submission and approval of an Investigational New Drug (IND) application that demonstrates adherence to stringent standards.[47] To assist with the application process, the FDA has made available the Pre-IND Consultation Program, the goal of which is to provide consultation regarding whether appropriate data has been obtained for each section of the application to warrant IND submission.[47] Subsequent to submission of an IND, the FDA has 30 days to review the IND to ensure that due diligence has been taken on the part of investigators to demonstrate that the level of risk arising from exposure of subjects to the drug during the course of clinical investigation will not be excessive.[47]

Initially, we described the three categories of medial products regulated by the FDA as drugs, devices, and biologics. Our discussion has centered on the FDA approval process of novel technologies falling into one of these discrete categories, as such products have historically been the significant focus of surgeon inventors. However, as the fields of tissue engineering and regenerative medicine continue to evolve, an exponentially increasing number of medical products are being introduced which are derived from a combination of biomaterials and cell-based technologies. These products have been labeled by the FDA as Tissue Engineered Medical Products (TEMPs), and significant regulatory considerations associated with their development include what manipulations of cells to be implemented are necessary, what the cell source is and where they are to be applied, as well as the biologic characteristics of a particular scaffold if one is to be utilized.[48] TEMPs cannot easily be partitioned into any of the aforementioned three categories due to the nature of their composition. As such, regulatory bodies came to realize that these products could not be defined as medical devices in the traditional sense, and that necessary oversight of the development and implementation of these products would require restructuring of current regulatory bodies to some extent, in addition to codifying new oversight committees with the knowledge and capacity to address ethical, safety, and efficacy issues specific to TEMPs.

Efforts to this end were initiated in 1997 at a joint meeting between the InterCenter FDA Tissue Engineering Working Group and the American Society for Testing and Materials (ASTM). From this meeting, Division IV of the Committee F04 on Medical and

Surgical Material and Devices was formed, and the mission of the division was outlined in the following:

"The development of standards and promotion of related materials for tissue engineered medical products focusing on components of combination medical products intended to repair, replace or regenerate human tissue. These comprise the biological components such as the cells, tissue, cellular products, and/or biomolecules and biomaterials used in combination, including biologic, biomimetic and/or synthetic materials. This division will work with other committees within ASTM and other organizations having mutual interests."[49]

To accomplish this goal, ASTM formed multiple task groups to oversee specific aspects TEMP development. With these task groups in place, a comprehensive system to guide the evolution of TEMPs was established.

Within the FDA, the evolution of medial products combining multiple components previously defined separately (TEMPs; human cell, tissue, and cellular and tissue-based products (HCT/IP); device-biologic medical products; etc.) has lead to the materialization multiple novel regulatory bodies. Examples include the Office of Combination Products (OCP), and the Tissue Reference Group (TRG).[50] Additionally, the CDRH has developed a standards program (Tissue Engineering Specialty Task Group) ". . . to provide guidance to CDRH reviewers and industry on the recognition and use of national and international consensus standards, including declarations of conformity to these standards, during the evaluation of premarket submissions for medical devices."[51,52] The overarching goal of establishing consensus standards is to ensure that premarket review transpires as efficiently as possible, while making certain that any and all requirements of the process are met.

Multiple cell-based TEMPs have successfully traversed the FDA regulatory process and realized delivery to the marketplace. Two examples which have lead to significant improvements in patient outcomes, and are currently FDA approved for the treatment of chronic non-healing venous and diabetic ulcers, include Dermagraft® and Apligraf®. Dermagraft®, a fibroblast-derived temporary skin coverage for the treatment of burns approved in 1997, is a dermal substitute consisting of neonatal fibroblasts grown on an absorbable polyglactin scaffold.[53] Extending this technology, Apligraf® is a bilayered skin substitute, consisting of neonatal fibroblasts suspended in bovine collagen type I, on top of which cultured neonatal keratinocytes are seeded.[53] Multiple studies have found that both applications result in significant acceleration of healing compared to traditional wound dressings.[54–56] Cell-based tissue engineering strategies have also been successfully employed to recapitulate damaged articular cartilage. The technique, known as autologous chondrocyte implantation (ACI), involves the harvest, *in vitro* expansion, and reimplantation of autologous chondrocytes where deficient in diseased knee joints.[57] Multiple investigations have demonstrated the successful regeneration of cartilage utilizing this technique, but debate remains over whether its implementation is cost-effective and produces superior outcomes relative to traditionally surgical interventions, and room for significant refinement of the technique exits.[58–61] Examples put forth in the foregoing represent encouraging advances in the application of TEMPs to the clinical arena, as they provide proof of principal that tissue engineering can be successfully translated to the bedside.

Clearly, the oversight of TEMPs remains a work in progress. The complex evolution of multiple regulatory bodies intended to address the specific needs of tissue engineering and regenerative medicine acknowledges both the importance of these fields to the future of modern medicine, as well as the deficit of knowledge that remains with regard to how the translation of technologies arising from the fields to the care of patients can be accomplished by the most safe and efficient means. As our understanding of the biology of TEMPs continues to evolve, so will the ability of regulatory bodies to differentiate between real and unsubstantiated risks associated with their use, resulting in the streamlined approval of novel therapeutics.

Funding Acquisition

The level of investment required to develop a new medical technology can vary depending upon the novelty of the concept, and the amount of research and development that will be required for that concept to be realized. Broadly, investments can be divided into two categories depending upon the size of the financial commitment, usually delineated by being less than or greater than $500,000. Innovations that are relatively simple in nature may be realized with relatively small amounts of financial backing. However, truly disruptive innovations almost universally require the commitment of substantial capital for successful delivery to the marketplace.

Initial investments to support product research and development are commonly known as "seed" financing. Investors can look to a variety of both private and federal sources to seek out such funding. Private sources of seed money can come from Angel Investors, or Angel Groups/Networks. The latter are groups of Angel Investors who choose to increase their investing power by pooling capital. These are individuals, or a group of individuals, who provide capital for product development, usually in exchange for an equity stake in the company in which the investment is made. Other private groups from which investment capital is commonly sought out include family, friends, or grants from medical products companies. On a federal level, grant monies are made available by the U.S. Small Business Association (USSBA), through the Small Business Innovation Research Program (SBIR) and the Small Business Technology Transfer Program (STTR). The stated mission of the Office of Technology of the USSBA is to ". . . strengthen and expand the competitiveness of U.S. small high technology research and development businesses in the federal marketplace."[32] These grants are authorized and distributed under the guidance of the Small Business Innovation Development Act of 1982, the Small Business Research and Enhancement Act of 1992, and the Small Business Innovation Research Program Reauthorization Act of 2000.[32] The nature of the investment typically obtained by means of the above private and federal funding avenues ranges from $50,000 to $500,000, and is intended to provide development support for up to one and a half years.[4]

For a sustaining innovation, this may prove to be sufficient. However, the development of groundbreaking concepts typically spans a significantly greater period of time, and resource consumption vastly exceeds that of smaller projects.[4] Although the funding sources listed above may be adequate to establish some basic scientific principal and/or put forth a highly preliminary device prototype, the realization of such revolutionary ideas requires investment on a much larger scale. Large device companies may have the ability to provide the necessary amounts of capital. However, start-up companies pursuing the development of a novel medical technology are frequently backed by venture capital. Again, this capital may come from an individual, but is commonly derived from pooled resources known as a venture capital fund. Venture capitalists usually invest in young, small companies with significant potential for growth, but too immature to obtain funding by traditional means. Similar to seed money, these investments are usually made in exchange for a negotiated amount of preferred equity in the company.

Prior to making any investment, venture capitalists will want to establish answers to several key questions in advance of formalizing any commitments. These include identifying what problem will be addressed by the new technology, whether or not adequate research and development will lead to the creation of one or more techniques or tools capable of addressing the problem, and if current or anticipated future market demand will make implementation or production of the new technology profitable?[32] Once these questions are answered and the risk associated with investment is determined to be acceptable, acquisition of investment funds should be made as expeditiously a possible in order to "stay ahead" of the market, and avoid a missed opportunity arising from failing to remain actively engaged in the process.

Investment from venture capital is usually broken up into phases, and input of additional funding beyond the initial investment is usually pending the realization of sub-goals occurring throughout the development process. Initial venture capital investments are known as "Series A" investments, while secondary, more substantial investments are known as "Series B" investments. Series A investments, typically ranging from $500,000 to $10 million, are usually directed towards early goals including the establishment of necessary basic science principals, prototyping, patent acquisition, clinical trials, and traversing the FDA approval process. Series B investments, ranging from $2 to $50 million dollars, are distributed when the early development goals are met, and are allocated towards large scale implementation and delivery to the marketplace.[4]

Commercialization

The innate way in which the medical products industry operates and interacts with universities and government institutions, historically responsible for the vast majority of basic science research, has evolved dramatically over the span of the past 30 years. Despite the necessary level of interaction between the business world and scientists in order for medical product development to progress, work conducted in the two fields largely transpired independently until the mid 1970's. It was at this time that the landscape of the medical products industry changed. In 1976, Genentech was founded by Robert Swanson and Herbert Boyer to investigate and develop recombinant DNA technology.[31] This joining of forces between a venture capitalist and scientist in the quest to advance towards a common goal was the first endeavor of its kind, and represented the birth of the biotechnology industry.

Evolution of the biotech industry, melded from the university and business worlds, generated significant excitement. The belief was widely held that such an arrangement would allow for the direct investment of capital into the evolution of a novel scientific concept, and deliver on the promise of rapid and efficient transition of technology from the laboratory to the marketplace. The early successes of biotech pioneers such as Genentech and Amgen further instilled faith that the industry could successfully address these goals. The years to follow bore witness to the founding of a staggering number of start-up biotech companies, approximately 4,000 over the past 30 years, many of which have been industry offspring of university based investigators.[31] However, all factors considered, many argue that the industry has fallen short of delivering upon expectations. Despite biotech companies experiencing significant financial growth, profit has been scarce.[31] Furthermore, the anticipated increases in production of novel medical technologies have not been appreciated in many medical technology sectors.[31]

As no surprise, research and development efforts in the fields of tissue engineering and regenerative medicine have not been immune to the challenges plaguing the biotech industry as a whole. Increasing focus in the field centering on the development of cell-based technologies has been met in stride with mounting hesitation of risk adverse venture capitalists and industry players to invest in what largely remains a "black box." This hesitation is at least partially based in the questionable potential for profitability of cell-based therapeutics. Whereas successful drugs are mass produced and distributed widely, cell-based engineered tissues may be patient specific, have a shortened shelf life, and require the intervention of a physician to administer.[32] All of these are factors which could limit the potential for production on a large scale. Furthermore, as briefly mentioned earlier, the inherent risk associated with financing the development of biologic innovations not known to be technologically feasible leads potential investors to think twice. That being said, the future of tissue engineering and regenerative medicine remains promising. The essence of this promise resides in the profoundly positive impact that translational tissue engineering strategies could impart on patients, afflicted with a host of different disease process, through the replacement of diseased or deficient tissue. To this end, as of 2007 there were 110 development-stage companies with focus on tissue engineering and regenerative medicine operating, responsible for 55 products in FDA clinical trials, the employment of 2500 investigators and staff, and the commitment of $850 million towards product development; a dramatic increase relative to trends observed at the start of the decade.[62]

Summary

As history has shown, revolutionary surgical innovations have appreciated setbacks and been met with skepticism over the course of their evolution. Tissue engineering, while still in its infancy, has been subject to similar barriers. Yet, lessons learned from early failures have proved to be valuable to ongoing investigations, and the future is arguably brighter than ever. The past five years have seen a dramatic economic growth in the fields of tissue engineering regenerative medicine, and the number of patients which have been treated with tissue-regenerative products has well surpassed the million mark.[62]

Recent studies have provided a wealth of information regarding the ideal composition of biosynthetic scaffolds to maximize endogenous and *de novo* tissue regeneration. Furthermore, *in vitro* and *in vivo* studies continue to expand our understanding of the genetic and molecular underpinnings guiding the differentiation from multipotent progenitor cells down differentiated lineage specific pathways. This information will be essential for identifying environmental stimuli as well as targeted molecular manipulations that can be utilized to optimize the regenerative process.

These advances in our understanding of such complex processes are of utmost importance to the translation of tissue engineering and regenerative medicine to the clinical arena. Not only does this knowledge bring us closer to the realization of medical products that can be used in patient care, but it also additionally reduces the level of risk appreciated by investors and industry that these products will come to fruition at all. In turn, enthusiasm is fostered to provide the financial support needed for progress to be achieved expeditiously.

With increasing investment, scientific endeavors focusing on the fields of tissue engineering and regenerative medicine have expanded rapidly and are now transpiring globally. As such, the importance of open communication and the efficient dissemination of information regarding new discoveries continues to expand. In 2005, awareness of this need lead to the organization of the Tissue Engineering and Regenerative Medicine International Society (TERMIS).[63] To achieve the aforementioned goals, this charter has outlined a plan to convene for a World Congress triennially, with each contributing chapter holding annual meetings on interim years. Such "global" collaboration will help to eliminate the redundancy of efforts on the part of investigators, stimulate original thinking towards existing ideas, and be vital to efficiently acquiring a comprehensive understanding of the biologic principals in which tissue engineering is based.

Finally, early growing pains felt by the biotechnology industry have led to the realization that current approaches to product development will need to be modified to achieve large-scale delivery to the market-place. As such the Multi-Agency Tissue Engineering Science Interagency Working Group (MATES) was formed, and in 2007 published the results of their studies, which outline the necessary steps that basic science investigators and industry must take to ensure widespread translation of tissue engineering and regenerative medicine to the bedside.[63] With a blueprint derived from such a diverse set of perspectives in hand, both investigators and the medical products industry are better armed than ever with the tools necessary to ensure that surgical patients of the future will benefit from all of the potential advances in tissue engineering and regenerative medicine yet to be appreciated.

References

1. Steven P. Jobs on creating new technology. http://www.1000ventures.com/business_guide/innovation_technology.html. 3rd November (2008).
2. Kermit EL. From concept to exit strategies — medical device innovation. *Conf Proc IEEE Eng Med Biol Soc* **7** (2004) 5130.
3. Riskin DJ, Longaker MT, Gertner M, Krummel TM. Innovation in surgery: a historical perspective. *Ann Surg* **244** (2006) 686–693.
4. Wall J, Longaker MT, Gurtner, G. (eds.). *From Idea to Bedside: The Process of Surgical Invention and Innovation* (2008).
5. Taylor M. Experiments in payment. *Hosp Health Netw* **82** (21) (2008) 28–33.
6. Fogarty TJ. *Patent Genius — search for Thomas J. Fogarty*. http://www.patentgenius.com/inventedby/FogartyThomasJPortolaValleyCA.html. 30th October (2008).
7. Fogarty TJ. Patent Genius — search for Thomas J. Fogarty. http://www.patentgenius.com/inventedby/FogartyThomasJPaloAltoCA.html. 30th October (2008).
8. Horch RE, Kopp J, Kneser U, Beier J, Bach AD. Tissue engineering of cultured skin substitutes. *J Cell Mol Med* **9** (2005) 592–608.
9. Calve S, *et al.* Engineering of functional tendon. *Tissue Eng* **10** (2004) 755–761.
10. Song L, Baksh D, Tuan RS. Mesenchymal stem cell-based cartilage tissue engineering: cells, scaffold and biology. *Cytotherapy* **6** (2004) 596–601.
11. Meek MF, Coert JH. US Food and Drug Administration/Conformit Europe-approved absorbable nerve conduits for clinical repair of peripheral and cranial nerves. *Ann Plast Surg* **60** (2008) 466–472.
12. Iwasaki K, *et al.* Bioengineered three-layered robust and elastic artery using hemodynamically-equivalent pulsatile bioreactor. *Circulation* **118** (2008) S52–57.
13. Bessa PC, Casal M, Reis RL. Bone morphogenetic proteins in tissue engineering: the road from laboratory to clinic, part II (BMP delivery). *J Tissue Eng Regen Med* **2** (2008) 81–96.
14. Bessa PC, Casal M, Reis RL. Bone morphogenetic proteins in tissue engineering: the road from the laboratory to the clinic, part I (basic concepts). *J Tissue Eng Regen Med* **2** (2008) 1–13.
15. Slater BJ, Kwan MD, Gupta DM, Panetta NJ, Longaker MT. Mesenchymal cells for skeletal tissue engineering. *Expert Opin Biol Ther* **8** (2008) 885–893.
16. Bian W, Bursac N. Tissue engineering of functional skeletal muscle: challenges and recent advances. *IEEE Eng Med Biol Mag* **27** (2008) 109–113.
17. Sekarski N, *et al.* Right ventricular outflow tract reconstruction with the bovine jugular vein graft: 5 years' experience with 133 patients. *Ann Thorac Surg* **84** (2007) 599–605.

18. Abdullah B, Shibghatullah AH, Hamid SS, Omar NS, Samsuddin AR. The microscopic biological response of human chondrocytes to bovine bone scaffold. *Cell Tissue Bank* (2008).
19. O'Sullivan D. Framework for Managing development in the Networked Organizations. *Journal of Computers in Industry* **1** (2002) 77–88.
20. Hanna HA *et al.* Antibiotic-impregnated catheters associated with significant decrease in nosocomial and multidrug-resistant bacteremias in critically ill patients. *Chest* **124** (2003) 1030–1038.
21. Hanna H *et al.* Long-term silicone central venous catheters impregnated with minocycline and rifampin decrease rates of catheter-related bloodstream infection in cancer patients: a prospective randomized clinical trial. *J Clin Oncol* **22** (2004) 3163–3171.
22. Chatzinikolaou I, *et al.* Antibiotic-coated hemodialysis catheters for the prevention of vascular catheter-related infections: a prospective, randomized study. *Am J Med* **115** (2003) 352–357.
23. Gilbert RE, Harden M. Effectiveness of impregnated central venous catheters for catheter related blood stream infection: a systematic review. *Curr Opin Infect Dis* **21** (2008) 235–245.
24. Raad I, *et al.* Central venous catheters coated with minocycline and rifampin for the prevention of catheter-related colonization and bloodstream infections. A randomized, double-blind trial. The Texas Medical Center Catheter Study Group. *Ann Intern Med* **127** (1997) 267–274.
25. Darouiche RO, *et al.* Efficacy of antimicrobial-impregnated bladder catheters in reducing catheter-associated bacteriuria: a prospective, randomized, multicenter clinical trial. *Urology* **54** (1999) 976–981.
26. Baker S, Weinzweig J. Bioentrepreneurialism for the plastic surgeon. *Plast Reconstr Surg* **122** (2008) 295–301.
27. The University of Wisconsin-Madison — Public Disclosure. http://www.grad.wisc.edu/research/ip/publicdisclosure.html. 1st November (2008).
28. Ambrogi RJ. Guide to Current Patent Reform Legislation. *Bullseye: Experts and the Law* (2007).
29. USPTO Provisional Patent. http://www.uspto.gov/web/offices/pac/provapp.htm. 1st November (2008).
30. USPTO — general information concerning patents. http://www.uspto.gov/web/offices/pac/doc/general/index.html#whatpat. 3rd November (2008).
31. Pisano GP. Can science be a business? Lessons from biotech. *Harv Bus Rev* **84** (2006) 114–124,150.
32. Prescott C. The Promise of Stem Cells: a venture capital perspective. in *Advances in Tissue Engineering* (ed. Polak, J.) 491–500 (Imperial College Press, London, 2008).
33. Weissman IL. Stem cells — scientific, medical, and political issues. *N Engl J Med* **346** (2002) 1576–1579.
34. Bahadur G. The moral status of the embryo: the human embryo in the UK Human Fertilisation and Embryology (Research Purposes) Regulation 2001 debate. *Reprod Biomed Online* **7** (2003) 12–16.
35. Chin JJ. Ethical issues in stem cell research. *Med J Malaysia* **58** Suppl A, (2003) 111–118.
36. Marshall E. The business of stem cells. *Science* **287** (2000) 1419–1421.
37. Metallo CM, Azarin SM, Ji L, de Pablo JJ, Palecek SP. Engineering tissue from human embryonic stem cells. *J Cell Mol Med* **12** (2008) 709–729.
38. Mehran R, *et al.* Post-market approval surveillance: a call for a more integrated and comprehensive approach. *Circulation* **109** (2004) 3073–3077.
39. Device classification. http://www.fda.gov/cdrh/devadvice/313.html. 3rd November (2008).
40. Device classification — general controls, special controls, premarket approval. http://www.fda.gov/cdrh/devadvice/3132.html. 3rd November (2008).
41. Premarket Notification 510(k). http://www.fda.gov/cdrh/devadvice/314.html. 3rd November (2008).
42. Class I devices — limitations on exemption. http://www.fda.gov/cdrh/modact/fr0202af.html. 3rd November (2008).
43. Class II devices — limitations on exemption. http://www.fda.gov/cdrh/modact/frclass2.html. 3rd November (2008).

44. Premarket approval. http://www.fda.gov/cdrh/devadvice/pma/. 3rd November (2008).

45. Investigational device exemption. http://www.fda.gov/cdrh/devadvice/ide/index.shtml. 3rd November (2008).

46. Therapeutic Biological Products. http://www.fda.gov/cder/biologics/default.htm. 26th November (2008).

47. Investigational New Drug (IND) Application Process. http://www.fda.gov/cder/Regulatory/applications/ind_page_1.htm. 26th November (2008).

48. Russell AJ, Bertram T. Moving into the Clinic. in *Principles of Tissue Engineering* (eds. Lanza RP, Langer RS, Vacanti J.) 15–31 (Elsevier, New York, 2008).

49. Picciolo GL, Hellman KB, Johnson PC. Meeting Report: tissue engineered medical products standards: The Time is Ripe. *Tissue Engineering* **4** (1998) 5–7.

50. Tissue Reference Group. http://www.fda.gov/CBER/tissue/tisrefgrp.htm. 3rd November (2008).

51. Standards Program — Recognition and Use of Consensus Standards. http://www.fda.gov/cdrh/osel/guidance/321.html. 3rd November (2008).

52. Standards Program — Specialty Task Groups. http://www.fda.gov/cdrh/standards/stg.html. 3rd November (2008).

53. Wong T, McGrath JA, Navsaria H. The role of fibroblasts in tissue engineering and regeneration. *Br J Dermatol* **156** (2007) 1149–1155.

54. Falanga V, *et al.* Rapid healing of venous ulcers and lack of clinical rejection with an allogeneic cultured human skin equivalent. Human Skin Equivalent Investigators Group. *Arch Dermatol* **134** (1998) 293–300.

55. Jones JE, Nelson EA. Skin grafting for venous leg ulcers. *Cochrane Database Syst Rev* CD001737 (2007).

56. Marston WA, Hanft J, Norwood P, Pollak R. The efficacy and safety of Dermagraft in improving the healing of chronic diabetic foot ulcers: results of a prospective randomized trial. *Diabetes Care* **26** (2003) 1701–1705.

57. Parenteau NL, Sullivan SJ, Brockbank KGM, Young JH. Lessons Learnt. in *Advances in Tissue Engineering* (ed. Polak, J.) 469–490 (Imperial College Press, London, 2008).

58. Bentley G, *et al.* A prospective, randomised comparison of autologous chondrocyte implantation versus mosaicplasty for osteochondral defects in the knee. *J Bone Joint Surg Br* **85** (2003) 223–230.

59. Dozin B, *et al.* Comparative evaluation of autologous chondrocyte implantation and mosaicplasty: a multicentered randomized clinical trial. *Clin J Sport Med* **15** (2005) 220–226.

60. Knutsen G, *et al.* A randomized trial comparing autologous chondrocyte implantation with microfracture. Findings at five years. *J Bone Joint Surg Am* **89** (2007) 2105–2112.

61. Clar C, *et al.* Clinical and cost-effectiveness of autologous chondrocyte implantation for cartilage defects in knee joints: systematic review and economic evaluation. *Health Technol Assess* **9**, iii–iv, ix–x, (2005) 1–82.

62. Lysaght MJ, Jaklenec A, Deweerd, E. Great expectations: private sector activity in tissue engineering, regenerative medicine, and stem cell therapeutics. *Tissue Eng Part A* **14** (2008) 305–315.

63. Nerem RM. An Introduction. in *Advances in Tissue Engineering* (ed. Polak, J.) 3–10 (Imperial College Press, London, 2008).

Index